T0291647

CAMBRIDGE LIBRARY COLLECTION

Books of enduring scholarly value

Mathematical Sciences

From its pre-historic roots in simple counting to the algorithms powering modern desktop computers, from the genius of Archimedes to the genius of Einstein, advances in mathematical understanding and numerical techniques have been directly responsible for creating the modern world as we know it. This series will provide a library of the most influential publications and writers on mathematics in its broadest sense. As such, it will show not only the deep roots from which modern science and technology have grown, but also the astonishing breadth of application of mathematical techniques in the humanities and social sciences, and in everyday life.

Mathematical and Physical Papers

Sir George Stokes (1819-1903) established the science of hydrodynamics with his law of viscosity describing the velocity of a small sphere through a viscous fluid. He published no books, but was a prolific lecturer and writer of papers for the Royal Society, the British Association for the Advancement of Science, the Victoria Institute and other mathematical and scientific institutions. These collected papers (issued between 1880 and 1905) are therefore the only readily available record of the work of an outstanding and influential mathematician, who was Lucasian Professor of Mathematics in Cambridge for over fifty years, Master of Pembroke College, President of the Royal Society (1885-90), Associate Secretary of the Royal Commission on the University of Cambridge and a Member of Parliament for the University.

Cambridge University Press has long been a pioneer in the reissuing of out-of-print titles from its own backlist, producing digital reprints of books that are still sought after by scholars and students but could not be reprinted economically using traditional technology. The Cambridge Library Collection extends this activity to a wider range of books which are still of importance to researchers and professionals, either for the source material they contain, or as landmarks in the history of their academic discipline.

Drawing from the world-renowned collections in the Cambridge University Library, and guided by the advice of experts in each subject area, Cambridge University Press is using state-of-the-art scanning machines in its own Printing House to capture the content of each book selected for inclusion. The files are processed to give a consistently clear, crisp image, and the books finished to the high quality standard for which the Press is recognised around the world. The latest print-on-demand technology ensures that the books will remain available indefinitely, and that orders for single or multiple copies can quickly be supplied.

The Cambridge Library Collection will bring back to life books of enduring scholarly value across a wide range of disciplines in the humanities and social sciences and in science and technology.

Mathematical and Physical Papers

VOLUME 4

GEORGE GABRIEL STOKES

CAMBRIDGE UNIVERSITY PRESS

Cambridge New York Melbourne Madrid Cape Town Singapore São Paolo Delhi

Published in the United States of America by Cambridge University Press, New York

www.cambridge.org
Information on this title: www.cambridge.org/9781108002653

This edition first published 1904
This digitally printed version 2009

ISBN 978-1-108-00265-3

MATHEMATICAL

AND

PHYSICAL PAPERS.

London: C. J. CLAY AND SONS,
CAMBRIDGE UNIVERSITY PRESS WAREHOUSE,
AVE MARIA LANE.
Glasgow: 50, WELLINGTON STREET.

Leipzig: F. A. BROCKHAUS.
New York: THE MACMILLAN COMPANY.
Bombay and Calcutta: MACMILLAN AND CO., Ltd.

G. G. Stokes

MATHEMATICAL

AND

PHYSICAL PAPERS

BY THE LATE

SIR GEORGE GABRIEL STOKES, BART.,

SC.D., LL.D., D.C.L., PAST PRES. R.S.,

KT PRUSSIAN ORDER *POUR LE MÉRITE*, FOR. ASSOC. INSTITUTE OF FRANCE, *ETC.*
MASTER OF PEMBROKE COLLEGE AND LUCASIAN PROFESSOR OF MATHEMATICS
IN THE UNIVERSITY OF CAMBRIDGE.

Reprinted from the Original Journals and Transactions,
with brief Historical Notes and References.

VOL. IV.

CAMBRIDGE:
AT THE UNIVERSITY PRESS.
1904

Cambridge:
PRINTED BY J. AND C. F. CLAY,
AT THE UNIVERSITY PRESS.

PREFACE.

IN the preface to the third volume Sir George Stokes expressed his intention, should life and health last, to complete the republication of his Scientific Papers without delay. These conditions were not destined to be fulfilled; and the task of completing this memorial of his genius has fallen to other hands.

This fourth volume includes the Papers that were published between 1853 and 1876. In accordance with the plan followed in the previous volumes, only memoirs and notes involving distinct additions to scientific knowledge have been included; this restriction has however had little effect except in the omission of some addresses, extracts from which will be collected together in another place. The texts of the papers have been reproduced as they originally appeared, with the exception that obvious misprints have been corrected. A few historical and elucidatory footnotes, those within square brackets, have been added by the Editor.

Advice has been received, in the selection and treatment of the material, from Lord Kelvin and Lord Rayleigh, who have seen most of the proof sheets. Acknowledgment is also due to Prof. Liveing who has very kindly examined the papers dealing with technical chemical and spectroscopic subjects, and to Mr F. G. Hopkins, who has gone through the work on the reactions of haemoglobin.

The remaining papers, together with a biographical notice which is being prepared for the Royal Society by Lord Rayleigh, will appear in a fifth volume. It is hoped that it will be possible to arrange a selection from Sir George Stokes' scientific correspondence, for publication either there or in a separate form: some extracts from it, relating to the contents of the present volume, are here printed in an Appendix.

The portrait prefixed to this volume is taken from an oil painting made in 1874 by Mr Lowes Dickinson, which belongs to Pembroke College.

J. LARMOR.

ST JOHN'S COLLEGE, CAMBRIDGE,
January, 1904.

CONTENTS.

MATHEMATICAL AND PHYSICAL PAPERS.

ON THE CHANGE OF REFRANGIBILITY OF LIGHT.—No. II.

[From the *Philosophical Transactions* for 1853, pp. 385—396.
Received *June* 16, read *June* 26, 1853.]

THE chief object of the present communication is to describe a mode of observation, which occurred to me after the publication of my former paper, which is so convenient, and at the same time so delicate, as to supersede for many purposes methods requiring the use of sun-light. On account of the easiness of the new method, the cheapness of the small quantity of apparatus required, and above all, on account of its rendering the observer independent of the state of the weather, it might be immediately employed by chemists in discriminating between different substances.

I have taken the present opportunity of mentioning some other matters connected with the subject of these researches. The articles are numbered in continuation of those of the former paper *.

Method of observing by the use of Absorbing Media.

241. Conceive that we had the power of producing at will media which should be perfectly opaque with regard to rays belonging to any desired regions of the spectrum, from the extreme red to the most refrangible invisible rays, and perfectly transparent with regard to the remainder. Imagine two such media prepared, of which the second was opaque with regard to those rays of the *visible* spectrum with regard to which the first was transparent, and *vice versâ*. It is clear that if both media were held in front of the eye no light would be perceived. The same would still be the case if the first medium were removed

* [*Ante*, Vol. III. p. 267.]

from the eye, and placed so as to intercept all the rays which fell on certain objects, which were then viewed through the second, provided the objects did nothing more than reflect, refract, scatter, or absorb the incident rays. But if any of the objects had the property of emitting rays of one refrangibility under the influence of rays of another, it might happen that some of the rays so emitted were capable of passing through the second medium, in which case the object would appear luminous in a dark field.

242. Let us consider now how the media must be arranged so as to bring out to the utmost the sensibility of a given substance. To take a particular instance, suppose the substance to be glass coloured by uranium. In this case the sensibility of the medium begins, with almost absolute abruptness, near the fixed line *b* of Fraunhofer, and continues from thence onwards. The dispersed light has the same, or at least almost rigorously the same, composition throughout, and consists exclusively of rays less refrangible than *b*. Consequently, we should have to prepare a first medium which was opaque with regard to the visible rays less refrangible than *b*, and transparent with respect to the rays, whether visible or invisible, more refrangible, and a second medium complementary to the former in the manner described in the preceding article. If the pair of media were still strictly complementary in this manner, but the point of the spectrum at which the transparency of the first medium began and that of the second ended were situated at some distance from *b*, the sensibility of the glass would be exhibited as before, only the maximum effect would not be produced, on account of the absorption of a portion either of the active or of the dispersed rays, according as the point in question was situated above or below *b*.

Now, although the commencement of the sensibility of canary glass is unusually abrupt, it generally happens that the sensibility of a medium, or at least the main part of it, comes on with great rapidity, and lasts throughout the rest of the spectrum, though frequently it is most considerable in a region extending not very greatly beyond the point where it commenced. In those cases in which the dispersion of different tints commenced at two or three different places in the spectrum, I have almost always had evidence of the independent presence of different sensitive principles, to which the observed effects were respectively due.

Hence, if we could prepare absorbing media at pleasure, we should get ready for general use in these observations a few pairs of media complementary in the particular manner already described, but having the points of the spectrum at which the transparency of the first medium commenced and that of the second ended different in different pairs, situated say in the yellow for one pair, in the blue for another, in the extreme violet for a third.

243. It is not of course possible to prepare media in this manner at pleasure, and all we can do is to select from among those which occur in nature. Nevertheless it is useful, as a guide in the selection, to consider what constitutes the ideal perfection of absorbing media for this particular purpose. But before proceeding to mention the media which I have found convenient, I will describe the arrangement which I have adopted for admitting the light.

A hole was cut in the window-shutter of a darkened room, and through this the light of the clouds and external objects entered in all directions. The diameter of the hole was four inches, and it might perhaps have been still larger with advantage. A small shelf, blackened on the top, which could be screwed on to the shutter immediately underneath the hole, served to support the objects to be examined, as well as the first absorbing medium. This, with a few coloured glasses, forms all the apparatus which it is absolutely necessary to employ, though for the sake of some experiments it is well to be provided also with a small tablet of white porcelain, and an ordinary prism, and likewise with one or two vessels for holding fluids.

244. In the observation, the first medium is placed resting on the shelf so as to cover the hole; the object is placed on the shelf immediately in front of the hole; the second medium is held anywhere between the eyes and the object. As it is not possible to obtain media which are strictly complementary, it will happen that a certain quantity of light is capable of passing through both media. This might no doubt be greatly reduced by increasing the absorbing power of the media, but it is by no means advisable to do so to any great extent, because it is important that the second medium should transmit as many as possible of the rays which are of such refrangibilities as to be stopped by the first.

Accordingly, it might sometimes be doubtful whether the illumination perceived on the object were due merely to scattered light which had passed through both media, or to really "degraded" light *. To remove all doubt, it is generally sufficient to transfer the second medium from before the eyes to the front of the hole. The light merely scattered by the object will necessarily be the same as before, if the room be free from stray light; and even if there be a little stray light, the illumination, so far as it is concerned, will be increased instead of diminished; whereas if the illumination previously observed were due to fluorescence, and the media were properly chosen, the object which before was luminous will now be comparatively dark.

Sometimes, in the case of substances which have only a low degree of sensibility, it is better to leave the second medium in front of the eyes, and use a third medium, which is held alternately in the path of the rays incident on the object and between the

* This term, which was suggested to me by my friend Prof. Thomson, appears to me highly significant. The expression *degradation of light* might be substituted with advantage for *true internal dispersion* to designate the general phenomenon; but it is perhaps a little too wide in its signification, and might be taken to include phosphorescence (if indeed in this case the refrangibility be really always lowered), as well as the emission of non-luminous radiant heat by a body which had been exposed to the red rays of the spectrum. As to the term *internal dispersion,* though I employed it, following Sir David Brewster, I confess I never liked it. It seems especially awkward when applied to a washed paper or dyed cloth; it was adopted at a time when the phenomenon was confounded with opalescence; and, so far as it implies theoretical notions at all, it seems rather to point to a theory now no longer tenable: I allude to the theory of suspended particles. Indeed, this theory, as it seems to me, ceased to be tenable as soon as Sir John Herschel had discovered the peculiar analysis of light connected with epipolic dispersion, and Sir David Brewster had connected the phenomenon with internal dispersion, so far at least as the common appearance of a continuous and coloured dispersed beam formed a connexion. The expression *dependent emission* is awkward, but would be significant, because the light is emitted in the manner of self-luminous bodies, but only in dependence upon the active rays, and so long as the body is under their influence. In this respect the phenomenon differs notably from phosphorescence. It is quite conceivable that a continuous transition may hereafter be traced by experiment from the one phenomenon to the other, but no such transition has yet been traced, nor is it by any means certain that the phenomena are not radically distinct. On this account it would, I conceive, be highly objectionable to call true internal dispersion *phosphorescence.* In my former paper I suggested the term *fluorescence*, to denote the general appearance of a solution of sulphate of quinine and similar media. I have been encouraged to give this expression a wider signification, and henceforth, instead of true internal dispersion, I intend to use the term fluorescence, which is a single word not implying the adoption of any theory.

object and the eyes. Such a medium, though not at all necessary, may be used also in the case of highly sensitive substances, for the sake of varying the experiment and rendering the result more striking.

As it will be convenient to have names for the media fulfilling these different offices, I will call the first medium, or that with which the whole is covered, the *principal absorbent*, the second medium the *complementary absorbent*, and the third medium, when such is employed, the *transfer medium*. For the transfer medium we may choose a medium of the same nature as the complementary absorbent, but paler. This is perhaps the best kind to employ in the methodical examination of various substances; but if the object of the observer be merely to illustrate the phenomena of the change of refrangibility of light, he may vary the experiments by using other media.

245. I have hitherto spoken only of the increase of illumination due to the sensibility of the substance under examination. But independently of illumination, the colour of the emitted light affords in most cases a ready means of detecting fluorescence. Thus, suppose the principal absorbent to transmit no visible rays but deep blue and violet, and the substance examined to appear, when viewed through the complementary absorbent, of a bright orange colour. Since no combination of the rays transmitted by the principal absorbent can make an orange, we may instantly conclude that the substance is sensitive. However, I do not consider it safe, at least for a beginner, to trust very much to *absolute* colour, for few who have not been used to optical experiments can be aware to what an extent the eye under certain circumstances is liable to be deceived. The *relative* colour of two objects seen at the same time under similar circumstances may usually be judged of safely enough; that is, of two such colours it may be possible to say with certainty that one inclines more to blue or to red than the other. Of course in many cases the change of colour is so great that there can be no mistake; still I think it a safe rule for a person employing these modes o observation without having been previously used to optical ex periments, to require some other proof of a change of refrangibility than merely a change of colour. Experience will soon show what appearances may safely be relied on.

246. If it be desired to view the object isolated as much as possible, it may be placed directly on the shelf, or better still, on black velvet. But it is generally preferable to have for comparison a standard object which reflects freely the visible rays, of whatever kind, incident upon it, and does not possess any sensible degree of fluorescence. It is in this way that the white porcelain tablet is useful; and in observing, I generally place the tablet on the shelf and the object on it. A white plate would answer, but a tablet is better, on account both of its shape and of the comparative dullness of its surface. It is true that the tablet used exhibited a very sensible amount of fluorescence when examined in a linear spectrum formed by a quartz train; still the effect was so small, and so much of it was due to those highly refrangible rays which are stopped by glass, that for practical purposes the tablet may be regarded as insensible. However, an observer is not obliged either to assume that all tablets are alike, or to apply to the particular tablet which he proposes to use, methods of observation requiring the use of apparatus which he is not supposed to possess. The methods of observation described in the present paper are complete in themselves; the observer has it in his power to test for himself the tablet he proposes to employ; and he is bound to do so before taking it for a standard of comparison. It may easily be tested by means of a prism, as will be explained presently.

247. The following are the combinations of media which I have chiefly employed:—

First Combination.—In this combination the principal absorbent is a glass coloured deep violet by manganese with a little cobalt, or else a glass coloured deep purple by manganese alone, combined with a rather pale blue glass coloured by cobalt, or with a deeper blue glass in case the day be bright. It is very easy to tell by means of a prism, with candle-light, whether a purple glass contains any sensible quantity of cobalt, on account of the very peculiar mode of absorption which is characteristic of this metal. In the examination of substances by this combination no complementary absorbent is usually required; but if it be wished to employ one, a pale yellow glass, of the kind mentioned in connexion with the next combination, may be used.

SECOND COMBINATION.—In this case the principal absorbent is a solution of the ammoniaco-sulphate of copper, employed in such thickness as to give a deep blue. In my experiments the fluid was contained in a cell with parallel sides of glass, which was closed at the top for greater convenience; but a very broad flat bottle would answer as well, because in the case of the principal absorbent the regularity of refraction of rays across it is of no consequence. Such a bottle, however, would have to be ordered expressly. The complementary absorbent in this combination is a yellow glass coloured by silver, and slightly overburnt. These glasses, as commonly prepared, are opaque with regard to most of the violet, but become transparent again with regard to the invisible rays beyond; and, in the case of a pale glass, the commencing transparency in the extreme violet may even be perceived by means of light received directly into the eye. I have got a glass of a pretty deep orange-yellow colour, which is more transparent than common window-glass with regard to rays of such high refrangibility as to be situated near the end of the region of the solar spectrum which it requires a quartz train to show. But when too much heat is used in the preparation, the glass acquires, on the interior of the coloured face, a delicate blue appearance, having a good deal of the aspect of a solution of sulphate of quinine, though it has in reality nothing to do with fluorescence; and in this state the glass is nearly opaque with regard to the invisible rays of the solar spectrum beyond the violet, though it still transmits a few among those which are nearly the most refrangible. Of course, if the complementary absorbent were always left in its position between the eyes and the object, its transparency or opacity with regard to invisible rays would be a matter of indifference; but as it is desirable that its transference from that position to the front of the hole should produce as much difference as possible, it is important that it should be opaque, or nearly so, with regard to the ultra-violet rays transmitted by the principal absorbent. Hence one of these slightly over-burnt glasses should be selected for the present purpose, and such are very commonly met with.

THIRD COMBINATION.—In this case everything is the same as in the preceding combination, except that the fluid is replaced by a glass of a pretty deep blue, coloured by cobalt.

FOURTH COMBINATION.—In this the principal absorbent is a solution of nitrate of copper, and the complementary absorbent a light red or deep orange glass.

248. In the first combination the darkness is tolerably complete without the use of any complementary absorbent, since no visible rays are transmitted except violet and some extreme red. The latter are no inconvenience, but rather help to set off the tint of the light due to fluorescence. This is, I think, the best combination to employ when the fluorescent light is blue, or at least deep blue; because in that case much of the light is lost by absorption in the yellow glass employed in the second combination. It has the advantage, too, of allowing the fluorescent light to enter the eye without being modified by absorption. Nevertheless no correct estimate can be formed of the absolute colour of the fluorescent light without making very great allowance for the effects of contrast, especially when the body, instead of being isolated as far as possible, is placed on the porcelain tablet.

The second combination is on the whole the most powerful. The media in this case make a very fair approach towards the ideal perfection explained in Art. 242. The darkness is so far complete, or else may easily be made so by increasing a little the strength of either absorbent, that if the tablet be written on with ink, and placed on the shelf between the media, the writing cannot be read. It forms a striking experiment, after having treated the tablet in this manner, to introduce between the media a piece of canary glass or a similar medium. The glass is not only luminous itself, but it emits so much light as to illuminate the whole tablet, so that the writing is instantly visible. In those cases in which the fluorescent light is yellow, orange, or red, it is shown a good deal more strongly by this combination than by the preceding.

The third combination is applicable to the same cases as the second. The blue glass answers extremely well, but is not quite so good as the blue fluid. The darkness is less complete, on account of the red and yellow transmitted by the glass. Nevertheless this combination is sometimes useful in observing with a prism, and at any rate it may very well be employed by a person who does not happen to have a vessel of the proper shape for holding the fluid.

In the second and third combinations the point of the spectrum at which the transparency of the principal absorbent begins, and that of the complementary absorbent ends, or rather the point which most nearly possesses this character, is situated in the blue. Thinking that the fluorescence of those substances which emit light of low refrangibility might be better brought out if this point was situated lower down in the spectrum, I tried the fourth combination. In this case the media have very fairly the required complementary character; the darkness is pretty complete, and the fluorescence of scarlet cloth and similar substances is very well exhibited. However, the effect in these cases is shown so well by the second combination, that, except it be for the sake of varying the experiment, I do not think it worth while to employ the fourth combination, more especially as it has the disadvantage of leaving the observer in doubt whether the red or orange light perceived constitutes the whole of the fluorescent light, or only that part of it which alone has been able to get through the complementary absorbent.

249. The mode of observation may be altered in various ways which afford pleasing illustrations of the theory, though in the regular examination of a set of substances it is best to proceed in a more methodical manner. Thus, if nothing but a violet or blue glass or a blue fluid be used as a principal absorbent, and the substances under examination be highly sensitive, their appearance will be remarkably changed if the coloured medium be transferred from before the hole to before the eyes. Again, if the complementary absorbent be made to exchange places with the principal absorbent, the result will be similar, although the very same media are merely interposed in different parts of the compound path of the light from the clouds to the eye. If a transfer medium be employed, and it be, as has hitherto been supposed, of the same general nature as the complementary absorbent, it will not produce much effect when it is interposed between the object and the eyes, but when it is placed in the path of the rays incident on the object, the fluorescent light will be nearly if not entirely cut off. If, however, we take for a transfer medium a glass or fluid having the same general character as the principal absorbent, the effect

will be just the reverse. This is strikingly shown in the case of a substance, which, like scarlet cloth, emits a red fluorescent light, by taking for a transfer medium a solution of nitrate of copper, and in the case of turmeric paper or yellow uranite, by taking the same solution, or else a blue glass. In the case of the two substances last mentioned, if we take for a transfer medium a red solution of mineral chameleon, diluted so as to be merely pink, the intensity of the light emitted will, under certain conditions, be not much different in the two positions of the medium, because a portion of the active rays in one position and a portion of the degraded rays in the other will be absorbed; but the colour of the portion of the emitted light which reaches the eye will be altogether different in the two positions of the transfer medium.

Mode of observing by means of a Prism.

250. In this method no absorbing medium is required except the principal absorbent. The white tablet being laid on the shelf, a slit is first held in such a position as to be seen projected against the sky, and the light thus coming directly into the eye, after having passed through the principal absorbent, is analysed by a prism held in the other hand. The slit is now held so that the tablet is seen through it, and the light coming from the tablet is analysed. It will be found that the spectrum seen in the first instance is faithfully reproduced, being merely less luminous, as must necessarily happen. At least, this was the case in those tablets which I have examined; and in this way each observer ought to test for himself the tablet he proposes to employ. After having been thus tested, the tablet may be used as a standard of comparison.

Suppose now that it is wished to examine a slip of turmeric paper, or a riband, or other similar object. The object is laid on the tablet, and the slit held immediately in front of it, in such a manner that one part, suppose the central portion, of the slit is seen projected on the object, and the remainder on the tablet. The light coming through the slit is then analysed by the prism, and the fluorescence, if any, of the object is indicated by light appearing in those regions of the spectrum in which, in the case of the light scattered by the tablet, there is nothing but darkness.

Occasionally in these observations a blue glass is preferable to a solution of the ammoniaco-sulphate of copper, because the extreme red and the greenish yellow bands transmitted by the glass, while too faint to interfere with the fluorescent light, are useful as points of reference.

251. The general appearance of the spectrum in this mode of observation may be gathered from the accompanying figures, of which the first represents turmeric paper seen under the blue glass, and the second represents a mass of crystals of nitrate of uranium seen under the copper solution. In fig. 1, RR', YY'

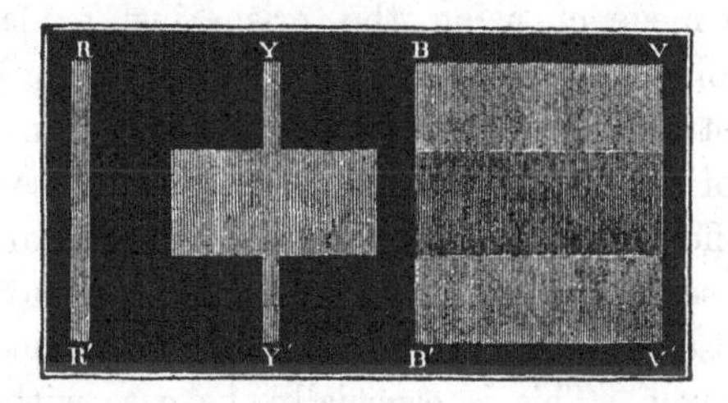

Fig. 1.

are the red and yellow bands transmitted by the glass, which are seen equally in the light scattered by the tablet and that scattered by the paper. $BVB'V'$ is the blue and violet light transmitted by the glass. Of this a considerable portion, especially in the more refrangible part, is absorbed by the turmeric paper, which on the other hand emits a quantity of red, yellow, and green light, not found among the incident rays. Fig. 2 sufficiently explains itself. In this case the fluorescent

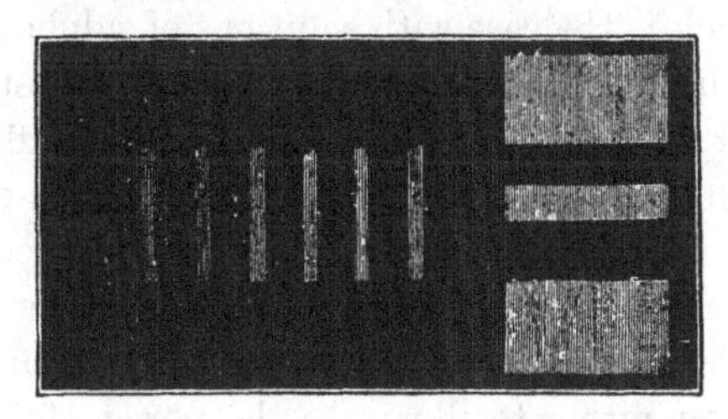

Fig. 2.

light is decomposed by the prism into bright bands, of which six may be readily made out. No blue or violet light enters the eye from the part of the slit which is seen projected on the mass of crystals, except where a crystalline face happens to be situated in such a position as to reflect the light of the sky into the eye, as represented in the figure. In the case of a substance so highly sensitive as nitrate of uranium, and which does not, like a slip of paper, lie flat on the tablet, the spectrum of the fluorescent light in reality extends, at least on the side next the window, though with less intensity, to some distance beyond the part of the slit which corresponds to the object, because the tablet is lighted up by the rays emitted by the object; but this is not represented in the figure.

252. The mode of using the prism just explained is that by which the phenomenon of the change of refrangibility is most strikingly illustrated; but in the actual examination of substances the chief use of the prism is to determine, in the case of substances which are sufficiently sensitive to admit with advantage of such a mode of observation, the composition of the fluorescent light. For this purpose it is often better to isolate the object by placing it on black velvet. This is especially the case with very minute crystals, or other objects, which are best placed on black velvet, and viewed through the prism as a whole, no slit being required.

Examples of the application of the preceding methods of observation.

253. The peculiar properties of paper washed with tincture of turmeric or stramonium seeds, of yellow uranite, and other highly sensitive substances, come out in a remarkable manner under the modes of examination described in this paper. I need not say that such is the case with solutions of sulphate of quinine, or horse-chestnut bark, or other clear and highly sensitive media, seeing that in this case the appearance due to fluorescence is obvious to common observation. If a piece of horse-chestnut bark be put to float in a glass of water close to the hole covered by the principal absorbent, the appearance of the descending streams of solution of esculine is very singular and beautiful. My present object is however rather to illustrate the power of these methods by their application to substances which stand much lower in the scale.

By the use of absorbing media alone, as well by a principal absorbent and a prism, I have been able to detect without difficulty the sensibility of white paper on a day of continuous clouds and rain. Even cotton-wool, which stands very much lower in the scale, is shown by the use of absorbing media with ordinary daylight to be sensitive. In the case of such substances as bone, ivory, white leather, the white part of a quill, which stand much higher in the scale, the most inexperienced observer could hardly fail instantly to detect the fluorescence. All plates of colourless glass which I have examined, and other pieces which were of such a shape as to admit of being looked into edgeways to a considerable depth, were found by the second combination to be sensitive. Crystals of sulphate of quinine, which may be readily prepared from the disulphate of commerce, show their fluorescence extremely well by the first combination. These crystals are much less sensitive than their solution, and the light which they emit is of a much deeper blue. It must in reality be of a very deep blue colour; for it nearly matches the fluorescent light of fluor-spar, although when the crystals are viewed under the violet glass the tint in both cases appears comparatively pale, from contrast with the violet. A solution of nitrate of uranium on the other hand has only a low degree of sensibility compared with the crystals of that salt. If a drop of the solution be placed on the porcelain tablet when the hole is covered with the deep blue copper solution, it appears comparatively dark, because much more illumination is lost by the absorption of the indigo and violet than is gained by the fluorescence of the solution. But when the tablet is viewed through the complementary absorbent, the solution is seen to be more luminous than the tablet, and to emit yellow rays, which are not found in the incident light.

The reactions of quinine mentioned in my former paper (Arts. 205–208), may very conveniently be observed by means of drops of the solution placed on the tablet; and in this way it is possible to work in a perfectly satisfactory manner with excessively minute quantities of quinine. The statement there made, that the blue colour was *destroyed* by hydrochloric acid, etc., must be understood only with reference to the mode of observation there supposed to be adopted, which was sufficient for the object in view. When the solutions are examined in a pure spectrum formed by sunlight, or even by the method described in the present paper, it is seen that the blue colour is not *absolutely* destroyed by hydrochloric acid, and is even developed to a *slight* extent on the addition of hydrochloric acid to a previously alkaline solution. Still there is a broad distinction between the two classes of solutions, which is all that is required. I have since extended a good deal these results, and mean to pursue the subject further. Meanwhile I may be permitted to correct an error in Art. 205, relative to the effect of hydrocyanic acid, which was there stated to develope the blue colour. The experiment was made with the acid of commerce, containing a foreign acid, to which the effect was probably due.

Comparison of the relative advantages of different modes of observation.

254. At first sight it might have been supposed that daylight could never be more than a poor substitute for sunlight in any observations relating to fluorescence. Such, however, I consider to be by no means the case. In the first place, when sunlight is used it is made to enter a room in a definite direction; whereas in using absorbing media all the rays are employed whose directions lie within a solid angle having the object examined for vertex, and the hole for base. If we leave out the part of this solid angle which corresponds to trees or houses, the part which corresponds to sky will still be so large as to make up in a good measure for the superior brilliancy of the light of the sun. In the second place, stray light is much more perfectly excluded than when a beam of sunlight, containing rays of all kinds, is admitted into a room. When indeed the use of sunlight is combined with that of absorbing media, it is possible to detect very minute degrees of sensibility. Still for general purposes I consider the methods depending upon the use of absorbing media with ordinary daylight quite comparable with, if not equal to, those methods involving the use of sunlight which are applicable to opaque bodies; I allude especially to the method of a linear spectrum. The peculiarities in the composition of the fluorescent light, when such exist, can be made out about equally well by both methods.

But when the substance to be investigated is a solution, or a clear solid of sufficient size to be examined as such, methods of observation can be put in practice with sunlight which surpass anything that can be done merely by the use of absorbing media. In consequence of the absence of stray light, which would otherwise dazzle the eye, an amount of concentration of the rays can be brought to bear on the object, which enables the observer to detect excessively minute degrees of sensibility. Thus, when the sun's light is condensed by a rather large lens, and made to pass through a strong solution of the ammoniaco-sulphate of copper, the condensed beam of violet and invisible rays serves to detect fluorescence in almost all fluids. This, however, is no great advantage; the method is in fact too powerful; and the observer is left in doubt whether the effect perceived be due to the fluid deemed to be examined, or to some impurity which it contains in an amount otherwise perhaps inappreciable. The great advantage which sunlight observations possess in the examination of substances, which, however, is only applicable to clear media, is that they enable the observer to make out the distribution of activity in the incident spectrum. In some cases this constitutes the chief peculiarity in the mode of fluorescence of a particular substance; in other cases it enables the observer to see, as it were, independently of each other, different sensitive substances existing together in solution. Another advantage of sunlight, which applies equally to clear and to opaque media, is that it enables the observer, with the assistance of a quartz train, to make out fluorescence which does not commence till that region of the spectrum which it requires a quartz train to show. But such cases are too rare to render this a point of much consequence. Of course there are observations such as those which relate to the fixed lines of the invisible rays, or to the determination of the absorbing action of a medium with regard to invisible rays of each degree of refrangibility in particular, which imperatively require sunlight: I am speaking at present only with reference to those observations of which the object is to investigate the mode of fluorescence of a particular substance.

As to the method of observation in which a prism is combined with a principal absorbent, its chief use is to determine, in the case of the more sensitive substances, the composition of the fluorescent light. It is not generally so convenient as the method which involves the use of absorbing media alone for determining which among a group of objects are sensitive, and which not, especially when the objects are minute.

255. Although the description of the mode of observing by means of absorbing media has run to some little length, the reader must not suppose that the observations are at all difficult. Of course observations of all kinds become more or less difficult when they are pushed to the extreme limit of refinement of which they are susceptible. But in the case of substances which are at all highly sensitive, and this comprises almost all the more interesting instances, the observations are extremely easy. I have spoken of a darkened room, which is certainly the most convenient when it can be had. But I have no doubt that an observer who could not procure such might easily arrange for himself a darkened box, which would answer the purpose. Indeed the fluorescence of highly sensitive substances, though they be opaque, may be exhibited by means of absorbing media in broad daylight.

Platinocyanides.

256. In the Report of the Twentieth Meeting of the British Association (Edinburgh, 1850, Transactions of the Sections, p. 5), is a notice by Sir David Brewster of "The Optical Properties of the Cyanurets of Platinum and Magnesia, and of Barytes and Platinum," salts which he had received from M. Haidinger of Vienna. The notice is chiefly devoted to the properties of the reflected light; but with respect to the latter of the salts, Sir David remarks that "it possesses the property of internal dispersion, the dispersed light being a *brilliant green*, while the transmitted light is *yellow*." Although the distinction between true internal dispersion and opalescence was not at the time understood, there could be little doubt from the nature of the case that the internal dispersion mentioned by Sir David Brewster was, in fact, an instance of the former of these phenomena; but I could not try for want of a specimen of the salt. Some months ago I received from M. Haidinger a specimen of the first of the salts mentioned at the beginning of this paragraph, namely M. Quadrat's cyanide of platinum and magnesium, a salt of great optical interest on account of the remarkable metallic reflexion which it exhibits. On examining the salt, I was greatly interested by finding that it was highly sensitive, the fluorescent light being red. This induced me to form some of Gmelin's platinocyanide of potassium, and I found that the blue light which this salt exhibits in certain

aspects is, in fact, due to fluorescence, a property which the salt possesses in an eminent degree. Having afterwards received some of the same salt pure from Professor Gregory, I applied it to the formation, on a small scale, of the platinocyanides of calcium, barium, strontium, and two or three others. The three salts last named, of which the second is that mentioned by Sir David Brewster, are all eminently sensitive, the fluorescent light being of different shades of green. It is only in the solid state that the platinocyanides are sensitive; their solutions look like mere water. The precipitates which a solution of platinocyanide of potassium gives with salts of the heavy metals are, in most cases that I have yet observed, insensible. With a solution of pernitrate of mercury, however, a bright yellow precipitate is produced which is exceedingly sensitive, so as to look brighter than even yellow uranite. The light emitted is yellow. It forms a very striking experiment to place side by side on the tablet with the second combination a drop of a solution of platinocyanide of potassium, and another of pernitrate of mercury. The drops look like water on the dark field of view; but when they are mixed, a precipitate is produced which glows like a self-luminous body with a yellow light. The precipitate with nitrate of silver is also sensitive, though not in so high a degree. The platinocyanides are extremely interesting; first, as forming a third case, or rather class of cases, in which the property of fluorescence is attached to substances chemically isolated in a satisfactory manner (though I believe chemists are acquainted with a few other organic compounds to which the property belongs), the other two cases being salts of quinine and of peroxide of uranium; and secondly, as exhibiting a new and remarkable feature, which consists in the polarization of the fluorescent light. I content myself at present with this notice; the salts require a more extended study.

On the Optical Properties of a recently discovered Salt of Quinine.

[From the *Report of the British Association*, Belfast, 1852, pp. 15–16.]

This salt is described by Dr Herapath in the *Philosophical Magazine* for March 1852, and is easily formed in the way there recommended, namely, by dissolving disulphate of quinine in warm acetic acid, adding a few drops of a solution of iodine in alcohol, and allowing the liquid to cool, when the salt crystallizes in thin scales reflecting (while immersed in the fluid) a green light with a metallic lustre. When taken out of the fluid the crystals are yellowish-green by reflected light, with a metallic aspect. The following observations were made with small crystals formed in this manner; and an oral account of them was given at a meeting of the Cambridge Philosophical Society, shortly after the appearance of Dr Herapath's paper.

The crystals possess in an eminent degree the property of polarizing light, so that Dr Herapath proposed to employ them instead of tourmalines, for which they would form an admirable substitute, could they be obtained in sufficient size. They appear to belong to the prismatic system; at any rate they are symmetrical (so far as relates to their optical properties and to the directions of their lateral faces) with respect to two rectangular planes perpendicular to the scales. These planes will here be called respectively the *principal plane of the length* and the *principal plane of the breadth*, the crystals being usually longest in the direction of the former plane.

When the crystals are viewed by light directly transmitted, which is either polarized before incidence or analysed after transmission, so as to retain only light polarized in one of the principal planes, it is found that with respect to light polarized in the principal plane of the length the crystals are transparent, and nearly colourless, at least when they are as thin as those which are usually formed by the method above mentioned. But with respect to light polarized in the principal plane of the breadth, the thicker crystals are perfectly black, the thinner ones only transmitting light, which is of a deep red colour.

When the crystals are examined by light reflected at the smallest angle with which the observation is practicable, and the reflected light is analysed, so as to retain, first, light polarized in the principal plane of the length, and secondly, light polarized in the other principal plane, it is found that in the first case the crystals have a vitreous lustre, and the reflected light is colourless; while in the second case the light is yellowish-green, and the crystals have a metallic lustre. When the plane of incidence is the principal plane of the length, and the angle of incidence is increased from 0° to 90°, the part of the reflected pencil which is polarized in the plane of incidence undergoes no remarkable change, except perhaps that the lustre becomes somewhat metallic. When the part which is polarized in a plane perpendicular to the former is examined, it is found that the crystals have no angle of polarization, the reflected light never vanishing, but only changing its colour, passing from yellowish-green, which it was at first, to a deep steel-blue, which colour it assumes at a considerable angle of incidence. When the light reflected in the principal plane of the breadth is examined in a similar manner, the pencil which is polarized in the plane of incidence undergoes no remarkable change, continuing to have the appearance of being reflected from a metal, while the other or colourless pencil vanishes at a certain angle, and afterwards reappears, so that in this plane the crystals have a polarizing angle.

If, then, for distinction's sake, we call the two pencils which the crystals, as belonging to a doubly refracting medium, transmit independently of each other, *ordinary* and *extraordinary*, the former being that which is transmitted with little loss, we may say, speaking approximately, that the medium is transparent with respect to the ordinary ray and opaque with respect to the extraordinary, while, as regards reflexion, the crystals have the properties of a transparent medium or of a metal, according as the refracted ray is the ordinary or the extraordinary. If common light merely be used, both refracted pencils are produced, and the corresponding reflected pencils are viewed together; but by analysing the reflected light by means of a Nicol's prism, the reflected pencils may be viewed separately, at least when the observations are confined to the principal planes. The crystals are no doubt biaxal, and the pencils here called ordinary and extraordinary are those which in the language of theory correspond to different sheets of the wave surface. The reflecting properties of the crystals may be embraced in one view by regarding the medium as not only doubly refracting and doubly absorbing, but *doubly metallic*. The *metallicity*, so to speak, of the medium of course alters continuously with the point of the wave surface to which the pencil considered belongs, and doubtless is not mathematically null even for the ordinary ray.

If the reflexion be really of a metallic nature, it ought to produce a relative change in the phases of vibration of light polarized in and perpendicularly to the plane of incidence. This conclusion the author has verified by means of the effect produced on the rings of calcareous spar. Since the crystals were too small for individual examination in this experiment, the observation was made with a mass of scales deposited on a flat black surface, and arranged at random as regards the azimuth of their principal planes. The direction of the change is the same as in the case of a metal, and accordingly the reverse of that which is observed in total internal reflexion.

In the case of the extraordinary pencil the crystals are least opaque with respect to red light, and accordingly they are less metallic with respect to red light than to light of higher refrangibility. This is shown by the green colour of the reflected light when the crystals are immersed in fluid, so that the reflexion which they exhibit as a transparent medium is in a good measure destroyed.

The author has examined the crystals for a change of refrangibility, and found that they do not exhibit it. Safflower-red, which possesses metallic optical properties, does change the refrangibility of a portion of the incident light; but the yellowish-green light which this substance reflects is really due to its metallicity and not to the change of refrangibility, for the light emitted from the latter cause is red, besides which it is totally different in other respects from regularly reflected light.

In conclusion, the author observed that the general fact of the reflexion of coloured polarized pencils had been discovered by Sir David Brewster in the case of chrysammate of potash*, and in a subsequent communication he had noticed, in the case of other crystals, the difference of effect depending upon the azimuth of the plane of incidence†. Accordingly, the object of the present communication was merely to point out the intimate connexion which exists (at least in the case of the salt of quinine) between the coloured reflexion, the double absorption, and the metallic properties of the medium.

Note added during printing.—When the above communication was made to the Association, the author was not aware of M. Haidinger's papers on the subject of the coloured reflexion exhibited by certain crystals. The general phenomenon of the reflexion of oppositely polarized coloured pencils had in fact been discovered independently by M. Haidinger and by Sir David Brewster, in the instances, respectively, of the cyanide of platinum and magnesium, and of the chrysammate of potash. A brief notice of the optical properties of the former crystal will be found in *Poggendorff's Annalen*, Bd. LXVIII. (1846), S. 302, and further communications from M. Haidinger on the subject are contained in several of the subsequent volumes of that periodical. The relation of the coloured reflexion to the azimuth of the plane of incidence was noticed by M. Haidinger from the first.

* Report of the Meeting of the British Association at Southampton, 1846, Part II. p. 7.

† *Ibid.* Edinburgh, 1847, p. 5.

On the Change of Refrangibility of Light and the exhibition thereby of the Chemical Rays.

[From the *Proceedings of the Royal Institution of Great Britain*, I, pp. 259–264; Friday evening lecture, *Feb.* 18, 1853. Also *Pogg. Ann.* LXXXIX, 1853, pp. 627–8.]

Before proceeding to the more immediate subject of the Lecture, it was necessary to refer to certain discoveries of Sir John Herschel and Sir David Brewster, more especially as it was the discovery by the former of these philosophers of the epipolic dispersion of light, and of the peculiar analysis of light which accompanies the phenomenon, that led to the researches respecting the change of refrangibility.

When a weak acid solution of quinine is prepared, by dissolving, suppose, one part of the commercial disulphate in 200 parts of water acidulated with sulphuric acid, a fluid is obtained which appears colourless and transparent when viewed by transmitted light, but which exhibits nevertheless in certain aspects a peculiar sky-blue colour. This colour of course had frequently been noticed; but it is to Sir John Herschel that we owe the first analysis of the phenomenon*. He found that the blue light emanates in all directions from a very thin stratum of fluid adjacent to the surface (whether it be the free surface or the surface of contact of the fluid with the containing glass vessel), by which the incident rays enter the fluid. His experiments clearly show that what here takes place is not a mere *subdivision* of light into a portion which is dispersed and a portion which passes on, but an actual *analysis*. For after the rays have once passed through the stratum from which the blue dispersed light comes, they are deprived of the power of producing the same effect; that is, they do not exhibit any blue stratum when they are incident a second time on a solution of quinine. To express the modification which

* *Philosophical Transactions* for 1845, p. 143.

the transmitted light had undergone, the further nature of which did not at the time appear, Sir John Herschel made use of the term "epipolized."

Sir David Brewster had several years before discovered a remarkable phenomenon in an alcoholic solution of the green colouring matter of leaves, or, as it is called by chemists, chlorophyll. This fluid when of moderate strength and viewed across a moderate thickness is of a fine emerald green colour; but Sir David Brewster found that when a bright pencil of rays, formed by condensing the sun's light by a lens, was admitted into the fluid, the path of the rays was marked by a *bright beam* of a *blood red colour**. This singular phenomenon he has designated *internal dispersion.* He supposed it to be due to suspended particles which reflected a red light, and conceived that it might be imitated by a fluid holding in suspension an excessively fine coloured precipitate. A similar phenomenon was observed by him in a great many other solutions, and in some solids; and in a paper read before the Royal Society of Edinburgh in 1846 he has entered fully into the subject†. In consequence of Sir John Herschel's papers, which had just appeared, he was led to examine a solution of sulphate of quinine; and he concluded from his observations that the "epipolic" dispersion of light exhibited by this fluid was only a particular instance of internal dispersion, distinguished by the extraordinary rapidity with which the rays capable of dispersion were dispersed.

The Lecturer stated, that, having had his attention called some time ago to Sir John Herschel's papers, he had no sooner repeated some of the experiments than he felt an extreme interest in the phenomenon. The reality of the epipolic analysis of light was at once evident from the experiments; and he felt confident that certain theoretical views respecting the nature of light had only to be followed fearlessly into their legitimate consequences, in order to explain the real nature of epipolized light.

The exhibition of a richly coloured beam of light in a perfectly clear fluid, when the observation is conducted in the manner of Sir David Brewster, seemed to point to the dispersions exhibited

* *Edinburgh Phil. Trans.* Vol. XII. p. 542.

† Vol. XVI. Part II., and *Phil. Mag.* June, 1848.

by the solutions of quinine and chlorophyll as one and the same phenomenon. The latter fluid, as has been already stated, disperses light of a blood red colour. When the transmitted light is subjected to prismatic analysis, there is found a remarkably intense band of absorption in the red, besides certain other absorption bands, of less intensity, in other parts of the spectrum. Nothing at first seemed more likely than that, in consequence of some action of the ultimate molecules of the medium, the incident rays belonging to the absorption band in the red, withdrawn, as they certainly were, from the incident beam, were given out in all directions, instead of being absorbed in the manner usual in coloured media. It might be supposed that the incident vibrations of the luminiferous ether generated synchronous vibrations in the ultimate molecules, and were thereby exhausted, and that the molecules in turn became centres of disturbance to the ether. The general analogy between the phenomena exhibited by the solutions of chlorophyll and of quinine would lead to the expectation of absorption bands in the light transmitted by the latter. If these bands were but narrow, the light belonging to them might not be missed in the transmitted beam, unless it were specially looked for; and the beam might be thus "epipolized," without, to ordinary inspection, being changed in its properties in any other respect. But on subjecting the light to prismatic analysis, first with the naked eye, and then with a magnifying power, no absorption bands were perceived.

A little further reflection showed that even the supposition of the existence of these bands would not alone account for the phenomenon. For the rays producing the dispersed light (if we confine our attention to the thin stratum in which the main part of the dispersion takes place) are exhausted by the time the incident light has traversed a stratum the fiftieth of an inch thick, or thereabouts, whereas the dispersed rays traverse the fluid with perfect freedom. This indicates a *difference of nature* between the blue-producing rays and the blue rays produced. Now, as the Lecturer stated, he felt very great confidence in the principle that the nature of light is completely defined by specifying its refrangibility and its state as to polarization. The difference of nature, then, indicated by the phenomenon, must be referred to a difference in one or other of these two respects.

At first he took for granted that there could be no change of refrangibility. The refrangibility of light had hitherto been regarded as an attribute absolutely invariable*. To suppose that it had changed would, on the undulatory theory, be equivalent to supposing that periodic vibrations of one period could give rise to periodic vibrations of a different period, a supposition presenting no small mechanical difficulty. But the hypotheses which he was *obliged* to form on adopting the other alternative, namely, that the difference of nature had to do with the state of polarization, were so artificial as to constitute a theory which appeared utterly extravagant. He was thus led to contemplate the possibility of a change of refrangibility. No sooner had he dwelt in his mind on this supposition, than the mystery respecting the nature of epipolized light vanished; all the parts of the phenomenon fell naturally into their places. So simple did the whole explanation become, when once the fundamental hypothesis was admitted, that he could not help feeling strongly impressed that it would turn out to be true. Its truth or fallacy was a question easily to be decided by experiment; the experiments were performed, and resulted in its complete establishment.

The Lecturer then described what may be regarded as the fundamental experiment. A beam of sunlight was reflected horizontally through a vertical slit into a darkened room, and a pure spectrum was formed in the usual manner, namely, by transmitting the light through a prism at the distance of several feet from the slit, and then through a lens close to the prism. In the actual experiment, two or three prisms were used, to produce a greater angular separation of the colours. Instead of a screen, there was placed at the focus of the lens a vessel containing a solution of sulphate of quinine. It was found that the red, orange, etc., in fact, nearly the whole of the visible rays, passed through the fluid as if it had been mere water. But on arriving about the middle of the violet, the path of the rays within the fluid was marked

* It is true that the phenomenon of phosphorescence is in a certain sense an exception; but the effect is in this case a work of time, which seems at once to remove it from all the ordinary phenomena of light, which, as far as sense can judge, take place instantaneously. It is true that there now appears a close analogy in many respects between true internal dispersion and phosphorescence. But while the nature of epipolized light remained yet unexplained, there was nothing in the former phenomenon to point to the latter.

by a sky-blue light, which emanated in all directions from the fluid, as if the medium had been self-luminous. This blue light continued throughout the region of the violet, and far beyond, in the region of the invisible rays. The posterior surface of the luminous portion of the fluid marked the distance to which the incident rays were able to penetrate into the medium before they were exhausted. This distance, which at first exceeded the diameter of the vessel, decreased with great rapidity, so that in the greater part of the invisible region it amounted to only a very small fraction of an inch. The fixed lines of the extreme violet, and of the more refrangible invisible rays, were exhibited by dark planes interrupting the dispersed light. When a small portion of the incident spectrum was isolated, by stopping the rest by a screen, and the corresponding beam of blue dispersed light was refracted sideways by a prism held to the eye, it was found to consist of light having various degrees of refrangibility, with colour corresponding, the more refrangible rays being more abundant than the less refrangible. The nature of epipolized light is now evident; it is nothing but light from which the highly refrangible invisible rays have been withdrawn by transmitting it through a solution of quinine, and does not differ from light from which those rays have been withdrawn by any other means.

The fundamental experiment, excepting that part of it which relates to the analysis of the dispersed light, was then exhibited by means of the powerful voltaic battery belonging to the Institution, which was applied to the combustion of metals. The rays emanating from the voltaic arc were applied to form a pure spectrum, which was received on a slab of glass coloured by peroxide of uranium, a medium which possesses properties similar to those of a solution of sulphate of quinine in a still more eminent degree.

The difference of nature of the illumination produced by a change of refrangibility, or "true internal dispersion," from that due to the mere scattering of light, may be shown in a very instructive form by placing paper washed with sulphate of quinine, or a screen of similar properties, so as to receive a long narrow horizontal spectrum, and refracting this upwards by a prism held to the eye. Were the luminous band formed on the paper due merely to the scattering of the incident rays, it ought of course to be thrown obliquely upwards; whereas it is actually decomposed by the prism into two bands, one ascending obliquely, and consisting of the usual colours of the spectrum in their natural order, the other running horizontally, and extending far beyond the more refrangible end of the former. Whatever be the screen, the horizontal band is always situated below the oblique, since there appears to be no exception to the law, that when the refrangibility of light is changed in this manner it is *always lowered.*

The general appearance of some highly "sensitive" media in the invisible rays was then exhibited by means of the flame of sulphur burning in oxygen, a source of these rays which Dr Faraday, to whose valuable assistance the Lecturer was much indebted, had in some preliminary trials found very efficacious. The chief media used were articles made of glass coloured by uranium, and solutions of quinine, of horse-chestnut bark, and of the seeds of the datura stramonium. A tall cylindrical jar filled with water showed nothing remarkable; but when a solution of horse-chestnut bark was poured in, the descending fluid was strongly luminous. The experiment was varied by means of white paper on which words had been written with a pretty strong solution of sulphate of quinine, an alcoholic solution of the seeds of the datura stramonium, and a purified aqueous solution of horse-chestnut bark. By gas-light the letters were invisible; but by the sulphur light, especially when it had been transmitted through a blue glass, which transmits a much larger proportion of the invisible than of the visible rays, the letters appeared luminous, on a comparatively dark ground. A glass vessel containing a thin sheet of a very weak solution of chromate of potash allowed the letters to be seen as well, or very nearly as well as before, when it was interposed between the eye and the paper; but when it was interposed between the flame and the paper the letters wholly disappeared,—the medium being opaque with respect to the rays which caused the letters to be luminous, but transparent with respect to the rays which they emitted.

It was then remarked what facilities are thus afforded for the study of the invisible rays. When a pure spectrum is once formed, it is as easy to determine the mode of absorption of an absorbing medium with respect to the invisible, as with respect to the visible rays. It is sufficient to interpose the medium in the path of the incident rays, and to notice the effect. Again, the effect of various flames and other sources of light on solutions of quinine, and on similar media, indicates the richness or poverty of those sources with respect to the highly refrangible invisible rays. Thus, the flames of alcohol, of hydrogen, etc., of which the illuminating power is so feeble, were found to be very rich in invisible rays. This was still more the case with a small electric spark, while the spark from a Leyden jar was found to abound in rays of excessively high refrangibility. These highly refrangible rays were stopped by glass, but passed freely through quartz. These results, and others leading to the same conclusion, had induced the Lecturer to order a complete train of quartz. A considerable portion of this was finished before the end of last August, and was applied to the examination of the solar spectrum. A spectrum was then obtained extending beyond the visible spectrum, that is, beyond the extreme violet, to a distance at least double that of the formerly known chemical spectrum. This new region was filled with fixed lines like the regions previously known.

But a spectrum far surpassing this was obtained with the powerful electrical apparatus belonging to the Institution. The voltaic arc from metallic points furnished a spectrum no less than *six or eight times* as long as the visible spectrum. This was in fact the spectrum which had already been exhibited in connexion with the fundamental experiment. The prisms and lens which the Lecturer had been employing in forming the spectrum were actually made of quartz. The spectrum thus obtained was filled from end to end with bright bands. When a piece of glass was interposed in the path of the incident rays, the length of the spectrum was reduced to a small fraction of what it had been, all the more refracted part being cut away. A strong discharge of a Leyden jar had been found to give a spectrum at least as long as the former, but not, like it, consisting of nothing but isolated bright bands.

The Lecturer then explained the grounds on which he concluded that the end of the solar spectrum on the more refrangible side had actually been reached, no obstacle existing to the exhibition of rays still more refrangible if such were present.

He stated also that during the winter, even when the sun shone clearly, it was not possible to see so far as before. As spring advanced he found the light continually improving, but still he was not able to see so far as he had seen at the end of August. It was plain that the earth's atmosphere was by no means transparent with respect to the most refrangible of the rays belonging to the solar spectrum.

In conclusion, there was exhibited the effect of the invisible rays coming from a succession of sparks from the prime conductor of a large electrifying machine, in illuminating a slab of glass coloured by uranium.

On the Cause of the Occurrence of Abnormal Figures in Photographic Impressions of Polarized Rings.

[From the *Philosophical Magazine*, VI, *Aug.* 1853, pp. 107—113. Also *Pogg. Ann.* XC, 1853, pp. 488—497.]

THE object of the following paper is to consider the theory of some remarkable results obtained by Mr Crookes in applying photography to the study of certain phenomena of polarization. An account of these results, taken from the *Journal of the Photographic Society*, is published in the last number of the *Philosophical Magazine**.

In the ordinary applications of photography, certain objects and parts of objects are to be represented which differ from one another in colour, or in brightness, or in both, according to the nature of the substances, and the way in which the lights and shadows fall. In the photograph the objects are represented as simply light or dark. Inasmuch as the photographic power, in relation to a given sensitive substance, of a heterogeneous pencil of rays is not proportional to its illuminating power, the darkness of the objects in a negative photograph is not proportional to, nor even always in the same order of sequence as, their brightness as they appear to the eye. Still, the outlines of the objects and of their parts are faithfully preserved. For although it is conceivable that two adjacent parts of an object, which the eye instantly distinguishes by their colour, should reflect rays of almost exactly equal photographic power in relation to the particular sensitive substance employed, so as to be absolutely undistinguishable on the photograph, or on the other hand, that two parts of an object between which the eye can see absolutely no distinction should yet come out distinct on the photograph, the conditions which would have to be satisfied in order that

* Vol. VI. p. 73.

the forms of a set of objects, suppose coloured patterns, or a painting of the rings of crystals, should be changed in this way by the substitution of one set of outlines for another are so very peculiar, that the chances may be regarded as infinity to one that no such changes of form will be produced to any material extent.

But when photography is applied to the representation of phenomena of interference, such as the rings of crystals seen by means of polarized light, the case is in some respects materially different. To take a particular instance, let us suppose the rings of calcareous spar to be viewed by white light, the planes of polarization of the polarizer and analyser being crossed, so as to give the black cross; and consider the alternations which take place in going outwards from the centre of the field, suppose in a direction inclined at angles of 45° to the arms of the cross. At first there are evident alternations of intensity; but very soon the eye, which under such circumstances is but a bad judge of differences of intensity even when the lights to be compared have the same colour, can no longer perceive the differences of illumination, but judges entirely by the difference of tint. The same takes place with nitre, sugar, and other colourless biaxal crystals. Except in the immediate neighbourhood of the optic axis or axes, the rings, which owe their existence and their forms in the first instance to the laws of double refraction and of the interference of polarized light, are in other respects created and their forms determined by the condition of maximum contrast of tint.

Now consider what takes place when an image of such a system is thrown on a sensitive plate, prepared suppose by means of bromide of silver. The rays of any one refrangibility would together form a regular system of rings, which, if these rays were alone present, and if the refrangibility were comprised within the limits between which the substance is acted on, would impress on the plate a system of rings exactly like those seen by means of the same homogeneous rays, provided they belong to the visible spectrum. The same would take place for rays of each refrangibility in particular, and the several elementary systems of rings thus formed are actually superposed when heterogeneous light is used. When the photograph is finished, it exhibits certain alternations of light and shade corresponding to alternations in the *total photographic intensity* of the rays which had acted on the plate, without any distinction being preserved between the action of rays of one refrangibility and that of rays of another; whereas, when the rings are viewed directly, the eye catches the differences of tint without noticing the difference of intensity, except in the neighbourhood of the optic axis or axes. Of course I am now speaking only of the alternations perceived in following a line drawn across the rings, not of the dark brushes, or of the variation of intensity perceived in passing along a given ring. Hence, when heterogeneous light is used, the circumstances which determine the rings are so different in the two cases that it is no wonder that the character of the rings seen on a photograph should differ in some respects from that of the rings seen directly.

But not only is a difference of character indicated as likely to take place; a more detailed consideration of the actual mode of superposition will serve to explain some of the leading features of the abnormal rings as observed by Mr Crookes. Let us take for example calcareous spar, and suppose the transmitted rays to be all of the same refrangibility. In this case the intensity along a given radius vector, drawn from the centre of the cross, varies as the square of the sine of half the retardation of phase of the ordinary relatively to the extraordinary pencil (see Airy's Tract). If i be the angle of incidence, the retardation varies nearly as $\sin^2 i$; and if $\sin^2 i = r$, we may take, as representing the variations of intensity I,

$$I = \sin^2 (mr^2) = \tfrac{1}{2}(1 - \cos 2mr^2) \text{ ....(1).}$$

In this expression m is a constant depending upon the refrangibility of the rays. In the case of calcareous spar the tints of the rings follow Newton's scale, and m is very nearly proportional to the reciprocal of the wave-length.

Suppose now that rays of two different degrees of refrangibility pass through the crystal together, and that the photographic intensities of the two kinds are equal. Suppose also that the aggregate effect which the two systems produce together on the plate is the sum of the effects which they are capable of producing separately. The latter supposition, if not strictly true, will no doubt be approximately true if the plate be not too long exposed. Then, if m' be what the parameter m becomes for the second system, we may represent the variation of intensity along a given radius vector by

$$I = \sin^2(mr^2) + \sin^2(m'r^2) = 1 - \cos(\overline{m - m'}\,r^2)\cos(\overline{m + m'}\,r^2) \ldots (2).$$

Suppose the refrangibilities of the two systems to be moderately different; then the difference between the two parameters m, m' will be small, but not extremely small, compared with either of them. Hence of the two factors in the expression for I the second will fluctuate a good deal more rapidly than the first, and will be that which mainly determines the radii, etc. of the rings. If the first factor were constant and equal to 1, its value when $r = 0$, the expression (2) would be of precisely the same form as (1), the parameter being the mean of the two, m, m'. However, the first factor is not constant, but decreases as r increases, and presently vanishes, and then changes sign. Hence the rings become less distinct than with homogeneous rays, and presently there takes place a sort of dislocation amounting to half an order, that is, the bright rings beyond a certain point, or in other words, outside the circle determined by a certain value of r, correspond, in regard to the series formed by their radii, with the dark rings inside this circle, and *vice versâ*. At some distance beyond that at which the dislocation takes place the rings become very distinct again; but it is useless to trace further the variations of the expression for I, because the circumstances supposed to exist in forming that expression are too remote from those of actual experiment to allow the interpretation of the formula to be carried far.

According to the numerical values of m and m', a dark ring might be converted, by the change of sign of the factor $\cos(m - m')\,r^2$, into a bright ring, or a bright ring into a dark ring, or a ring of either kind might be rendered broader or narrower than it would regularly have been. The coalescence of the fourth and fifth bright rings in Mr Crookes's photographs when bromide of silver was used seems to be merely an effect of this nature.

But in order that a dislocation of the kind above explained should take place, it is not essential that two kinds only of rays should act, nor even that the curve of photographic intensity should admit of two distinct maxima within the spectrum. Suppose that rays of all refrangibilities lying within certain limits pass through the crystal and fall on the plate. For the sake of obtaining an expression which admits of being worked out numerically without too much trouble, and yet results from a hypothesis not very remote from the circumstances of actual experiment, I will suppose the total photographic power of the rays whose parameters lie between m and $m + dm$ to be proportional to $\sin m dm$ between the limits $m = 2\pi$ and $m = 3\pi$, and to vanish beyond those limits. Since m is very nearly proportional to the reciprocal of the wave-length, and the ratio 3π to 2π or 3 to 2 is nearly that of the wave-lengths of the fixed lines D, H, this assumption corresponds to the supposition that the rays less refrangible than D are inefficient, that the action there commences, then increases according to a certain law, attains a maximum, decreases, and finally vanishes at H. The action would really terminate at H if a bath of a solution of sulphate of quinine of a certain strength were used. On this assumption, and supposing, as before, that the rays of different refrangibilities act independently of each other, we have

$$I = \int_{2\pi}^{3\pi} \sin^2 (mr^2) \sin m dm.$$

On working out this expression, and writing x for $2r^2$, we find

$$I = 1 + \frac{\cos(\frac{1}{2}\pi x)\cos(\frac{5}{2}\pi x)}{x^2 - 1} \quad \text{..............(3).}$$

As the full discussion of this formula presents no difficulty it may be left to the reader. The last factor in the numerator of the fraction is that where fluctuations correspond to the rings. Whenever x passes through an odd integer greater than 1 the first factor changes sign, and there is a dislocation or displacement of half an order, but when x passes through the value 1 the denominator changes sign along with both factors of the numerator, and there is no dislocation. When x becomes considerable the denominator $x^2 - 1$ becomes very large, and the fluctuations of intensity become insensible.

The following table contains the values of I calculated from the formula (3) for 16 values of x in each of the first 7 orders of rings. In passing from one ring to its consecutive the angle $\frac{5}{2}\pi x$ increases by 2π, and therefore x by 0·8. The sixteenth part of this, or 0·05, is the increment of x in the table. Each vertical column corresponds to one order. The value of x corresponding to any number in the table will be found by adding together the numbers in the top and left-hand columns.

x	0·00	0·80	1·60	2·40	3·20	4·00	4·80
·00	0·000	0·142	0·481	0·830	1·033	1·067	1·014
·05	0·080	0·223	0·543	0·860	1·037	1·060	1·010
·10	0·295	0·418	0·667	0·905	1·032	1·044	1·005
·15	0·619	0·692	0·829	0·955	1·020	1·021	1·001
·20	1·000	1·000	1·000	1·000	1·000	1·000	1·000
·25	1·377	1·293	1·154	1·033	0·977	0·979	1·001
·30	1·692	1·527	1·268	1·051	0·956	0·964	1·004
·35	1·898	1·669	1·327	1·054	0·939	0·956	1·008
·40	1·963	1·702	1·333	1·045	0·932	0·956	1·012
·45	1·881	1·791	1·286	1·030	0·936	0·963	1·013
·50	1·669	1·465	1·205	1·014	0·950	0·974	1·012
·55	1·356	1·243	1·103	1·004	0·973	0·987	1·007
·60	1·000	1·000	1·000	1·000	1·000	1·000	1·000
·65	0·654	0·775	0·913	1·004	1·027	1·010	0·991
·70	0·373	0·600	0·853	1·013	1·049	1·015	0·983
·75	0·192	0·499	0·826	1·024	1·063	1·016	0·977

A curve of intensity might easily be constructed from this table by taking ordinates proportional to the numbers in the table, and abscissæ proportional to the values of r, and therefore to the square roots of the numbers 0, 1, 2, 3, 4, etc. But the form of the curve will be understood well enough either from the formula (3), or from an inspection of the numbers in the table.

It will be seen that in the first three columns the numbers lying between the horizontal lines beginning ·20 and ·60 correspond to bright rings, and the remainder of each column, together with the beginning of the next, corresponds to a dark ring. But the dark ring, which would regularly follow the fourth bright ring, is converted, by the change of sign of the factor cos $(\frac{1}{2}\pi x)$, into a bright ring, forming with the former one broad bright ring having a minimum corresponding to $x=3$, where however the intensity falls only down to its mean value unity. A similar displacement occurs in the seventh column, but here the whole variation of intensity is comparatively small.

In the case of calcareous spar the character of the rings is the same all the way round, but in the photographs of the rings of nitre a new feature presents itself. Mr Crookes's figure of the abnormal rings of nitre is rather too small to be clear, but with the assistance of his description there is no difficulty in imagining what takes place. With reference to these photographs he observes, "But here a remarkable dislocation presented itself; each quadrant of the interior rings, instead of retaining its usual regular figure, appeared as if broken in half, the halves being alternately raised and depressed towards the neighbouring rings." This effect admits of easy explanation as a result of the superposition of systems of rings which separately are perfectly regular, when we consider that the poles of the lemniscates of the several elementary systems do not coincide, since in nitre the angle between the optic axes increases from the red to the blue. Now the change of character which may be described as a displacement of half an order is due to the circumstance that the smaller rings corresponding to the more refrangible rays are, as it were, overtaken by the larger rings corresponding to the less refrangible. It is plain that the variation of position of the poles of the lemniscates would tend to retard this effect in directions lying outside the optic axes, and to accelerate it in directions lying between those axes. Hence what was a bright ring in one part of its course would become a dark ring in another part, so that each quadrant would exhibit a dislocation of half an order in the rings. In order to show this dislocation to the greatest advantage, a crystal of a certain thickness should be used. With a very thin crystal there would be no dislocation of this nature, but only a displacement like that which takes place with calcareous spar. With a very thick crystal the effect of the chromatic variation of position of the optic axes would be too much exaggerated.

It appears then that all the leading features of the abnormal rings are perfectly explicable as a result of the superposition of separately regular systems. But if known causes suffice for the explanation of phenomena, we must by no means resort to agents whose existence is purely hypothetical, such for example as invisible rays accompanying, but distinct from, visible rays of the same refrangibility. Some of the minor details of the abnormal rings may require further explanation or more precise calculation; but such calculations are of no particular interest unless the phenomena offered grounds for suspecting the agency of hitherto unrecognized causes.

The difference between the photographs taken with iodide and bromide of silver is easily explained, when we consider the manner in which those substances are respectively affected by the rays of the spectrum. With iodide of silver there is such a concentration of photographic power extending from about the fixed line G of Fraunhofer to a little beyond H, that even when white light is employed we may approximately consider that we are dealing with homogeneous rays. On this account, and not because the rays of high refrangibility are capable of producing a more extended system of rings than those of low refrangibility, the rings visible on the photograph are much more numerous than those seen directly by the eye with the same white light. Moreover, the rings do not exhibit the same abnormal character as with bromide of silver, in relation to which substance the photographic power of the rays is more diffused over the spectrum.

It is not possible to place the eye and a sensitive plate prepared with bromide of silver under the same circumstances with regard to the formation of abnormal rings. It would be easy, theoretically at least, to place the eye and the plate in the same circumstances as regards rings, by using homogeneous light; but then, I feel no doubt, the rings visible on the plate would be as regular as those seen by the eye. On the other hand, if differences of colour exist in the figure viewed by the eye, they inevitably arrest the attention, and it is impossible to get rid of them without at the same time rendering the light so nearly homogeneous that on that account nothing abnormal would be shown. Hence Mr Crookes's abnormal rings afford a very curious example of the creation, so to speak, by photography of forms which do not exist in the object as viewed by the eye.

ON THE METALLIC REFLEXION EXHIBITED BY CERTAIN NON-METALLIC SUBSTANCES.

[From the *Philosophical Magazine*, VI, *Dec.* 1853, pp. 393—403. Also *Pogg. Ann.*, XCI, 1854, pp. 300—14; *Ann. de Chimie*, XLVI, 1856, pp. 504—8.]

IN the October Number of the *Philosophical Magazine* is a translation of a paper by M. Haidinger of Vienna, containing an account of his observations relating to the optical properties of Herapathite. In this paper he refers to a communication which I made to the British Association at the meeting at Belfast*; and indeed one great object of his examination of this salt was to see whether a law which he had discovered, and already extensively verified, relating to the connexion between the reflected and transmitted tints of bodies which have the property of reflecting a different tint from that which they transmit, would be verified in this case. The report of my communication published in the Abbé Moigno's *Cosmos*† had led him to suppose that my observations were at variance with his law.

My attention was first directed to this subject while engaged in some observations on safflower-red (carthamine), which I was led to examine with reference to its fluorescence. In following out the connexion which I had observed to exist between the absorbing power of a medium and its fluorescence, I was induced to notice particularly the composition of the light transmitted by the powder; and I found that the medium, while it acted powerfully on all the more refrangible rays of the visible spectrum, absorbed green light with remarkable energy. I need not now describe the mode of absorption more particularly. During these experiments I was struck with the metallic yellowish-green reflexion which this substance exhibits. It occurred to me that the almost metallic opacity of the medium with respect to green light

[* p. 18, *ante*.]
† Vol. I. p. 574.

was connected with the reflexion of a greenish light with a metallic aspect. I found, in fact, that the medium agreed with a metal in causing a retardation in the phase of vibration of light polarized perpendicularly to the plane of incidence relatively to light polarized in that plane. The observation was made by reflecting light polarized at an azimuth of about 45° from the surface of the medium to be examined, the angle of incidence being considerable, about equal to the angle of maximum polarization, and viewing the reflected light through a Nicol's prism capped by a plate of calcareous spar cut perpendicularly to the axis. Now by using different absorbing media in succession, it was found that with red light, for which safflower-red is comparatively transparent, the reflected light was sensibly plane-polarized, while for green and blue light the ellipticity was very considerable.

In the case of a transparent medium, light would be polarized by reflexion, or at least very nearly so, at a proper angle of incidence. Hence if the light reflected by such a medium as safflower-red were decomposed into two pencils, one, which for distinction's sake may be called the ordinary, polarized in the plane of incidence, and the other, or extraordinary, polarized perpendicularly to the plane of incidence, the extraordinary pencil would vanish at the polarizing angle, except in so far as the laws of the reflexion deviate from those belonging to a transparent substance. Hence the light remaining in the extraordinary pencil might be expected to be more distinctly related to the light absorbed with such energy. Accordingly, it was found that under these circumstances the extraordinary pencil (in the case of safflower-red) was of a very rich green colour, whereas without analysis the light was of a yellowish-green colour. Similar observations were extended to specular iron.

These phenomena recalled to my mind a communication which Sir David Brewster made at the meeting of the British Association at Southampton in 1846*; and on referring to his paper, I found that the appearance of differently coloured ordinary and extraordinary pencils in the light reflected by safflower-red was the same phenomenon as he has there described with reference to chrysammate of potash.

* Notice of a new property of light exhibited in the action of chrysammate of potash upon common and polarized light. (Transactions of the Sections, p. 7.)

The observations above-mentioned were made towards the end of 1851. Accordingly, when Dr Herapath's first paper on the new salt of quinine appeared*, I was prepared to connect the metallic green reflected light with an intense absorbing action with respect to green rays. Having prepared some crystals according to his directions, I was readily able to trace the progress of the absorption in the case of light polarized in a plane perpendicular to what is usually the longest dimension of the crystalline plates, and to observe how the light passed from red to black as the thickness increased. Even the thickest of these crystals was so thin as to show hardly any colour by light polarized in the plane of the length. The result of crossing two such plates is of course obvious to any one who is conversant with physical optics. The intense absorption was readily connected with the metallic reflexion. An oral account of these observations was given at a meeting of the Cambridge Philosophical Society on March 15, 1852; but it was not till the observations were a second time described, with some slight additions, at the meeting of the British Association at Belfast, that any account of them was published. A notice of this communication appeared in the *Athenæum* of September 25, 1852, and from this the report in *Cosmos* seems to be taken, though the latter is not free from errors. In the report in the *Athenæum*, the colour of the more rapidly absorbed pencil is briefly described in these words: "But with respect to light polarized in the principal plane of the breadth, the thicker crystals are perfectly black, the thinner ones only transmitting light, which is of a deep red colour." The comparative transparency of the crystals with regard to red light is afterwards expressly connected with the green colour of the light reflected as if by a metal. But in the report in *Cosmos*, the passage just quoted is replaced by "tandis que pour le cas de la lumière polarisée dans le plan principal de la largeur ils sont opaques et noirs, quelque minces qu'ils soient d'ailleurs." This error led M. Haidinger to suppose that my observations were opposed to his law; whereas the fact is, that, without knowing at the time what he had done, I had been led independently to a similar conclusion.

In mentioning my own observations on safflower-red, Herapathite, etc., nothing is further from my wish than to neglect

* *Phil. Mag.* for March 1852 (Vol. III. p. 161).

the priority of those to whom priority belongs. M. Haidinger had several years before discovered the phenomenon of the reflexion of differently coloured oppositely polarized pencils, which Sir David Brewster shortly afterwards, and independently, discovered in the case of chrysammate of potash. M. Haidinger had from the first observed a most important character of the phenomenon in the case of many crystals, namely, the orientation of the polarization of the reflected light, which Sir David Brewster does not seem to have noticed in the case of chrysammate of potash, and which perhaps was not very evident in that salt. In a later paper M. Haidinger had announced the complementary relation of the reflected and transmitted tints*. There is nothing new in employing the rings of calcareous spar as a means of detecting elliptic polarization; and the property of producing elliptic polarization in reflecting plane-polarized light had previously been observed in substances even of vegetable origin†. I am not aware, however, that the chromatic variations of the change of phase had been experimentally connected with the chromatic variations of an intense absorbing action on the part of the medium. I have hitherto mentioned but one instance of this connexion, but I shall presently have occasion to allude to another.

* In a paper of M. Haidinger's, entitled "Über den Zusammenhang der Körperfarben, oder des farbig durchgelassenen, und der Oberflächenfarben, oder des zurückgeworfenen Lichtes gewisser Körper," from the January Number of the Proceedings of the Mathematical and Physical Class of the Academy of Sciences at Vienna for the year 1852, will be found a list of M. Haidinger's previous papers on this subject. This paper contains a methodized account of the properties, with reference to surface and substance colour, of the substances up to that time examined by the author, amounting in number to thirty. For a copy of this, as well as several others of his papers, I am indebted to the kindness of the author.

† More than twenty years ago Sir David Brewster, in his well-known paper "On the Phenomena and Laws of Elliptic Polarization, as exhibited in the action of Metals upon Light," pointed out the modification produced on the rings of calcareous spar as a character of polarized light after reflexion from a metal. (*Phil. Trans.* for 1830, p. 291.) In a communication to the British Association at the meeting at Southampton in 1846, Mr Dale mentions indigo among a set of substances in which he had detected elliptic polarization by means of the rings of calcareous spar. In this case, however, he connects the property, not with the intense absorbing power of the substance, but with its high refractive index.

I do not here mention the minute degrees of ellipticity which have been detected in polarized light reflected from transparent substances in general by the delicate researches of M. Jamin, partly because they are so small as to be widely separated from the ellipticity in the case of carthamine, etc., partly because they seem to have no relation to the present subject.

I think it but justice to myself to point out the error in *Cosmos* (from whence M. Haidinger derived his information respecting my observations), in consequence of which I would appear to have been guilty of a grievous oversight in the examination of Herapathite: but I would hardly have ventured to mention my observations on carthamine, etc., were it not that, when different persons arrive independently at a similar conclusion, it frequently happens that views present themselves to the mind of one which may not have occurred to another. In the present case, in stating in detail my own views as to the nature of the phenomenon, I hope to be able to add something to what has been already done by M. Haidinger and Sir David Brewster, and it seemed not out of place to mention the observations in which those views originated.

It appears, then, that certain substances, many of them of vegetable origin, have the property of reflecting (not scattering) light which is coloured and has a metallic aspect. The substances here referred to are observed to possess an exceedingly intense absorbing action with respect to rays of the refrangibilities of these which constitute the light thus reflected, so that for these rays the opacity of the substances is comparable with that of metals. Contrary, however, to what takes place in the case of metals, this intense absorbing action does not usually extend to all the colours of the spectrum, but is subject to chromatic variations, in some cases very rapid. The aspect of the reflected light, which itself alone would form but an uncertain indication, is not the only nor the principal character which distinguishes these substances. In the case of transparent substances, or those of which the absorbing power is not extremely intense (for example, coloured glasses, solutions, etc.), the reflected light vanishes, or almost entirely vanishes, at a certain angle of incidence, when it is analysed so as to retain only light polarized perpendicularly to the plane of incidence*, which is not the case with metals. In the case of the substances at present considered, the reflected light does not vanish, but at a considerable angle of incidence

* I do not here take into account the peculiar phenomena which have been observed by Sir David Brewster with reference to the influence of the double refraction of a medium (such as calcareous spar) on the polarization of the reflected light, which, indeed, are not very conspicuous unless the medium be placed in optical contact with a fluid having nearly the same refractive index.

the pencil polarized perpendicularly to the plane of incidence becomes usually of a richer colour, in consequence of the removal, in a great measure, of that portion of the reflected light which is independent of the metallic properties of the medium; it commonly becomes, also, more strictly related to that light which is absorbed with such great intensity. The reflexion from a transparent medium is weakened or destroyed by bringing the medium into optical contact with another having nearly or exactly the same refractive index. Accordingly, in the case of these optically metallic substances, the colours which they reflect by virtue of their metallicity* are brought out by putting the medium in optical contact with glass or water. A remarkable character of metallic reflexion consists in the circumstance, that as the angle of incidence increases from 0°, the phase of vibration of light polarized in the plane of incidence is accelerated relatively to that of light polarized in the perpendicular plane. Accordingly, the same change takes place, with the same sign, in the case of these optically metallic substances; but the amount of the change is subject to most material chromatic variations, being considerable for those colours which are absorbed with great energy, but insensible for those colours for which the medium is comparatively transparent, so that the absorption may be neglected which is produced by a stratum of the medium having a thickness amounting to a small multiple of the length of a wave of light. If the medium be crystallized, it may happen that one only of the oppositely polarized pencils which it transmits suffers, with respect to certain colours, an exceedingly intense absorption; or, if that is the case with both pencils, that the colours so absorbed are different. It may happen, likewise, that the absorption varies with the direction of the ray within the crystal. In such cases the light reflected by virtue of the metallicity of the medium will be subject to corresponding variations, so that the medium is to be regarded as not only doubly refracting and doubly absorbing, but doubly metallic.

The views which I have just explained are derived from a combination of certain theoretical notions with some experiments.

* I use this word to signify the assemblage of optical properties by which a metal differs from a transparent medium, or one moderately coloured, such as a coloured glass.

They have need of being much more extensively verified by experiment; but, so far as I at present know, they are in conformity with observation.

To illustrate the effect of bringing a transparent medium into optical contact with an optically metallic substance, I may refer to safflower-red. If a portion of this substance be deposited on glass by means of water, and the water be allowed to evaporate, a film is obtained which reflects on the upper surface a yellowish-green light, but on the surface of contact with the glass a very fine green inclining to blue. A green of the latter tint appears to be more truly related to the colours absorbed with greatest energy. Similar remarks apply to the light reflected by Herapathite, according as the crystals are in air or in the mother-liquor. If a small portion of Quadratite (platinocyanide of magnesium) be dissolved on glass in a drop or two of water, and the fluid be allowed to evaporate, the tints reflected by the upper and under surfaces of the film of crystals are related to one another much in the same way as in the case of safflower-red. For a fine specimen of the salt last mentioned I am indebted to the kindness of M. Haidinger. I may mention in passing, that the platinocyanides as a class are of extreme optical interest. The crystals are generally at the same time doubly refracting, doubly absorbing, doubly metallic, and doubly fluorescent. By the last expression I mean that the fluorescence, which the crystals generally exhibit in an eminent degree, is related to directions fixed relatively to the crystal, and to the azimuths of the planes of polarization of the incident and emitted rays.

M. Haidinger has expressed the relation between the surface and substance colours of bodies by saying that they are complementary. This expression was probably not intended to be rigorously exact; and that it cannot be so, is shown by the following simple consideration. The tint of the light transmitted across a stratum of a given substance almost always, if not always, varies more or less according to the thickness of the stratum. Now one and the same tint, namely, that of the reflected light, cannot be rigorously complementary to an infinite variety of tints of different shades, namely, those of the light transmitted across strata of different thickness. In most cases, indeed, the variation of tint may not be so great as to prevent us from regarding the reflected and transmitted tints in a general sense as complementary. But as media exist (for example, salts of sesquioxide of chromium, solutions of chlorophyll) which change their tint in a remarkable manner according to the thickness of the stratum through which the light has to pass, it is probable that instances may yet be observed in which M. Haidinger's law would appear at first sight to be violated, although in reality, when understood in the proper sense, it would be found to be obeyed. As the existence of surface-colour seems necessarily to imply a very intense absorption of those rays which are reflected according to the laws which belong to metals, it follows that it is in the very thinnest crystals or films of those which it is commonly practically possible to procure, that the transmitted tint is to be sought which is most properly complementary to the tint of the reflected light.

I will here mention another instance of the connexion between metallic reflexion and intense absorption. I choose this instance because a different explanation from that which I am about to offer has been given of a certain phenomenon observed in the substance. The instance I allude to is specular iron. As it is already known that various metallic oxides and sulphurets possess the optical properties of metals, there is nothing new in bringing forward this particular mineral as a substance of that kind. It is to the chromatic variation of the metallicity that I wish to direct attention. If light polarized at an azimuth of about 45° be reflected from a scale of this substance at about the polarizing angle, and the reflected light be viewed through a plate of calcareous spar and a Nicol's prism, it will be found, by using different absorbing media in succession, that the change of phase, as indicated by the character of the rings, while it is very evident for red light, becomes much more considerable in the highly refrangible colours. Now specular iron is almost opaque for light of all colours, but as it gives a red streak it appears that the substance-colour is red; and, in fact, it is known that very thin laminæ are blood-red by transmitted light. Accordingly, the chromatic variation of the change of phase corresponds to that of the intense absorbing power.

The light reflected by specular iron is not extinguished by analysation, whatever be the angle of incidence; but at the angle of incidence which gives the nearest approach to complete polarization, a quantity of blue light is observed to remain. This has been explained by comparing specular iron to a substance of high dispersive power, so that the polarizing angle for red light is considerably less than for blue; and accordingly on increasing the angle of incidence, the light (which is here supposed to be analysed so as to retain only the portion polarized perpendicularly to the plane of incidence), while it becomes much less copious near the polarizing angle, becomes at the same time of a decided blue colour*. I believe, however, that the blue light is mainly due to the chromatic variation of the metallicity, the medium, considered optically, being much more metallic for blue light than for red, though it may in some measure be due to the cause previously assigned.

Specular iron is a good example of a substance forming a connecting link between the true metals and substances like safflower-red. It resembles metals in the circumstance that the absorbing power, as inferred from the chromatic variation of the metallicity, and as indicated by the tint of the streak, is not subject to the same extensive chromatic variations as in the case of colouring matters like safflower-red. It resembles safflower-red in being sufficiently transparent with respect to a portion of the spectrum to allow the connexion between the metallicity and the substance-colour to be observed; whereas the substance-colour of metals is not known from direct observation, except, perhaps, in the case of gold, which in the state of gold-leaf lets through a greenish light.

I am now able to bring forward a striking confirmation of the relation which seems to exist between the light reflected as if from a metal, and that absorbed with great energy. On reading M. Haidinger's paper, of which the title has been already quoted, I was particularly interested by finding crystallized permanganate of potash mentioned among the substances which exhibit distinct surface- and substance-colours. I had previously noticed the very remarkable mode of absorption of light by red solution of mineral chameleon†. This solution, which may be regarded as an optically pure solution of permanganate of potash, inas-

* See Dr Lloyd's *Lectures on the Wave-Theory of Light*, Part II. p. 18.

† *Phil. Trans.* for 1852, p. 558. [*Ante*, Vol. III. p. 402.]

much as it is associated with only a colourless salt of potash, absorbs green light with great energy, as is indicated by the tint, even without the use of a prism. But when the light transmitted by a pale solution is analysed by a prism, it is found that there are five remarkable dark bands of absorption, or minima of transparency, which are nearly equidistant, and are situated mainly in the green region. The first is situated on the positive or more refrangible side of the fixed line D, at a distance, according to a measurement recently taken, of about four-sevenths of the interval between consecutive bands; the last nearly coincides with F. The first minimum is less conspicuous than the second and third, which are the strongest of the set. Now it occurred to me, that as the solution is so opaque for rays having the refrangibilities of these minima of transparency, corresponding maxima might be expected in the light reflected from the crystals. This expectation has since been realized by observations made on some small crystals. On analysing the reflected light by a prism, I was readily able to observe four bright bands, or maxima, in the spectrum. These, as might have been expected, were more easily seen when the light was incident nearly perpendicularly than at a large angle of incidence. The first band was yellow, the others green, passing on to bluish-green. On decomposing the light reflected at a considerable angle of incidence, in a plane parallel to the axis, into two streams, polarized respectively in, and perpendicularly to, the plane of incidence, and analysing them by a prism, the bands were hardly or not at all perceptible in the spectrum of the former pencil, while that of the latter consisted of nothing but the bright bands.

The tint alone of the first bright band already indicated that the band was more refrangible than the light lying on the negative side of the first dark band seen in the spectrum of the light transmitted by the solution, and less refrangible than the light lying between the first and second dark bands, so that its position would correspond, or nearly so, to the first dark band. However, the eye is greatly liable to be deceived, in experiments on absorption, by the effects of contrast, and therefore an observation of the nature of that just mentioned cannot, I conceive, be altogether relied on. The smallness of the crystals occasioned some difficulty; but a more trustworthy observation was made in the following manner.

The sun's light was reflected horizontally into a darkened room, and allowed to fall on a crystal. The reflected light was limited by a slit, placed at the distance of two or three feet from the crystal. This precaution was taken to ensure making the observation on the regularly reflected light. Had no slit been used, or else a slit placed close to the crystal, it might have been supposed that the light observed was not regularly reflected, but merely scattered, as it would be by a coloured powder. The appearance of a spot of green light on a screen held at the place of the slit showed that the light was really regularly reflected. The slit was also traversed by the light scattered by the support of the crystal, etc. The slit was viewed through a prism and small telescope; and the position of the dark bands, or minima of brilliancy, in the reflected light could thus be compared with the fixed lines, which were seen by means of the scattered light in the uninterrupted spectrum corresponding to that portion of the slit through which the light reflected from the crystal did not pass. The minimum situated on the positive side of the first bright band lay at something more than a band-interval on the positive side of the fixed line D; the minimum beyond the fourth bright band lay at the distance of about half a band-interval on the negative side of F. It thus appears that the minima in the light reflected by the crystal were intermediate in position between the minima seen in the light transmitted through the solution, so that the maxima of the former corresponded to the minima of the latter.

It might have been considered satisfactory to compare the reflected light with the light transmitted, not by the solution, but by the crystals themselves. But the crystals absorb light with such energy as to be opaque; and even when they are spread out on glass, the film thus obtained is too deeply coloured for the purpose. For to show the bands well, the solution must be so dilute, or else seen through so small a thickness, as to be merely pink. As M. Haidinger states that the phenomena of the reflected light are the same for all the faces in all azimuths, and for the polished surface of a mass of crushed crystals, it may be presumed that the absorption is not much affected by the crystalline arrangement, and that the composition of the light transmitted by the solution is sensibly the same as that which would be observed across a crystalline plate, were it possible to obtain one of sufficient thinness.

The first bright band in the reflected light does not usually appear to be very distinctly separated from the continuous light of lower refrangibility; but the latter may be got rid of by observing the light reflected about the polarizing angle, and analysing it so as to retain only the portion polarized perpendicularly to the plane of incidence. As the surface of the crystals is liable to become spoiled, it is safest in observations on the reflected light to make use of a crystal recently taken out of the mother-liquor. I have only observed four bright bands in the reflected light, whereas there are five distinct minima in the light transmitted by the solution. However, the extreme minima are less conspicuous than the intervening ones, besides which the fifth occurs in a comparatively feeble region of the spectrum. The fourth bright band in the reflected light was rather feeble, but with finer crystals perhaps even a fifth might have been visible. As the metallicity of the crystals is almost or perhaps quite insensible in the parts of the spectrum corresponding to the maxima of transparency, we may say, that, as regards the optical properties of the reflected light, the medium changes four or five times from a transparent substance to à metal and back again, as the refrangibility of the light changes from a little beyond the fixed line D to a little beyond F.

EXTRACTS FROM LETTER TO DR W. HAIDINGER: ON THE DIRECTION OF THE VIBRATIONS IN POLARIZED LIGHT: ON SHADOW PATTERNS AND THE CHROMATIC ABERRATION OF THE EYE: ON HAIDINGER'S BRUSHES. (*February* 9, 1854.)

Die Richtung der Schwingungen des Lichtäthers im polarisirten Lichte. Mittheilung aus einem Schreiben des Hrn. Prof. Stokes, nebst Bemerkungen von W. Haidinger.

[*Pogg. Ann.* XCVI, 1855, pp. 287—292. Mitgetheilt vom Hrn. Verf. aus d. *Sitzungsberichten d. Wiener Akademie* (*April* 1854).]

EIN Abschnitt des Schreibens vom Hrn. Prof. Stokes, den ich heute der hochverehrten mathematisch-naturwissenschaftlichen Classe vorzulegen die Ehre habe, bezieht sich auf die Richtung der Schwingungen des Lichtäthers in Bezug auf die Polarisations-Ebene, und zwar enthält er nicht nur eine Beurtheilung der Tragweite der Bemerkungen, welche ich als Beweis für die senkrechte Richtung dieser Schwingungen gegen diese Polarisations-Ebene aus den Erscheinungen an pleochromatischen Krystallen darstellen zu dürfen glaubte*, sondern auch seine Ansicht über den Gegenstand selbst, übereinstimmend mit seinen eigenen früheren Arbeiten....

"Die Thatsachen, deren ich in Bezug auf die Polarisation des Fluorescenz-Lichtes der Kalium-Platin-Cyanide (in einem andern Theile des Schreibens) gedachte, und die Art wie die Polarisirung der einfallenden Strahlen auf dieses Licht wirkt, stimmen, so viel ich glaube, viel besser mit der Annahme überein, dass die Schwingungen im polarisirten Lichte senkrecht auf der Polarisations-Ebene stehen, als mit der anderen Theorie.

* *Sitzungsberichte der kais. Akademie der Wissenschaften*, Math.-naturw. Classe, 1852, VIII, S. 52.

"Diess veranlasst mich, der Beweisgründe zu erwähnen, welche Sie anführten, um zu zeigen, dass im polarisirten Lichte die Schwingungen senkrecht auf der Polarisations-Ebene stehen. Da ich glaube, Sie würden gern meine Ansicht darüber kennen, so will ich sie ausführlich anführen. Zu allererst kann ich sagen, dass ich es nicht für *möglich* halte, durch *irgend* eine Combination von *anerkannten* Ergebnissen die Frage zu entscheiden. Unter den anerkannten Ergebnissen betrachte ich solche, wie diese — dass die Schwingungen transversal sind — dass im linear-polarisirten Lichte die Schwingungen geradlinig sind, und *symmetrisch* mit Beziehung der Polarisations-Ebene, und daher entweder parallel oder senkrecht auf diese Ebene — dass im elliptisch-polarisirten Lichte die Schwingungen elliptisch sind u. s. w. Die Entscheidung muss sich immer auf eine oder die andere Art auf dynamische oder physikalische Betrachtungen stützen, welche, mögen sie an sich noch so wahrscheinlich seyn, doch nicht zu den anerkannten Ergebnissen gezählt werden können. Es ist auch nicht schwierig zu sehen, welche die Betrachtungen dieser Art in dem Falle Ihrer Beweisführung sind. Nehmen wir den Fall eines doppelt absorbirenden einaxigen Krystalles, wie Turmalin. Es sey oC Fig. 6 parallel der Axe, oA oB zwei Richtungen senkrecht auf die Axe. Die eine Farbe (ich will sie O nennen) sieht man in der Richtung der Axe Co, und in allen Richtungen in der Ebene BoA (oder senkrecht auf die Axe) in dem oC parallel polarisirten Lichte. Die andere Farbe (E) sieht man in allen Richtungen in der Ebene BoA, wenn das Licht in dieser Ebene polarisirt ist, und man sieht sie gar nicht in der Richtung Co. 'Wenn diese Farbe nun von Transversal-Schwingungen abhängt, so sind alle solche Schwingungen, transversal oder senkrecht gegen die Axe, mit einem Male ausgeschlossen, und die einzigen Schwingungen, welche möglicherweise zu der Farbe des extraordinären Strahles, der in dem Krystall entsteht, gehören können, sind die parallel der Richtung der Axe.' Aber wenn von Schwingungen gesprochen wird, welche *zu* dieser oder jener Farbe *gehören*, so wird stillschweigend vorausgesetzt, dass in der That die Farbe abhängig ist von der Richtung der Schwingungen. Nun kann man sich allerdings ganz gut vorstellen (so unwahrscheinlich es auch immer seyn mag), dass Absorption von dem gleichzeitigen Einflusse der Richtung der Schwingungen und der Richtung der Fortpflanzung abhänge, dergestalt, dass sie als gegeben betrachtet werden kann, wenn die Richtung einer Linie gegeben ist, welche senkrecht auf den beiden vorerwähnten Richtungen steht. Nimmt man diesen Satz an in Bezug auf die Natur der Absorption, so ist vollkommen

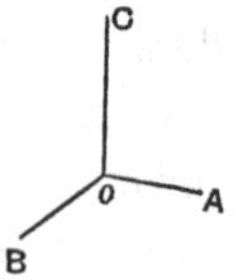

FIG. 6. 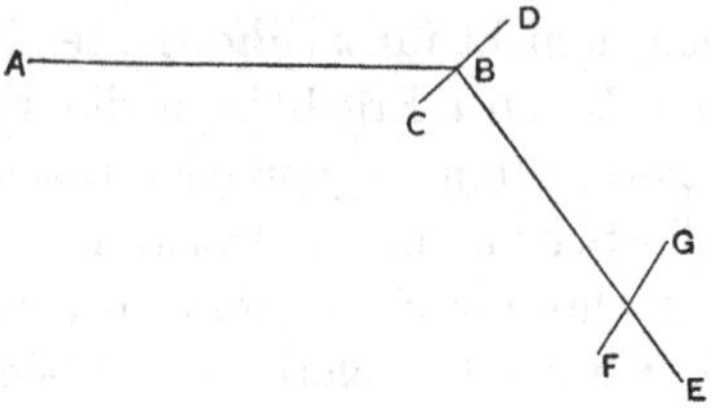 FIG. 7.

FIG. 8. 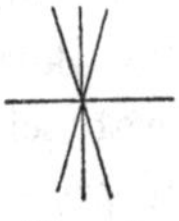 FIG. 9. 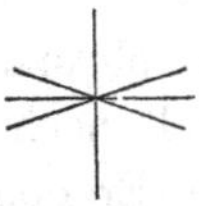 FIG. 10.

klar, dass die experimentalen Thatsachen, welche Sie in Bezug auf die doppelte Absorption anführen, zu dem Schlusse führen würde, die Schwingungen wären im polarisirten Lichte der Polarisations-Ebene parallel. Die Wahrscheinlichkeit, welche der Grund Ihrer Beweisführung der Wahrheit von Fresnel's Annahme giebt, reicht also nicht bis zur absoluten Gewissheit, sondern entspricht nur der Unwahrscheinlichkeit, dass Absorption gleichzeitig von der Richtung der Schwingungen und von der Richtung der Fortpflanzung abhängig seyn sollte, in der Art wie ich es oben erwähnte.

"Sie sagen: Wenn die Farbe *E*, welche man in der Richtung *Co* nicht sehen kann, irgendwie auf Transversal-Schwingungen beruht, so sind alle solche Schwingungen transversal oder senkrecht auf die Axe ausgeschlossen, und die einzigen Schwingungen, welche möglicherweise dem in dem Krystalle entstehenden extraordinären Strahl angehören können, sind dann der Richtung der Axe parallel. Nun könnte aber eine besondere Farbe in einer Richtung von transversalen Schwingungen abhängen, und nicht von transversalen Schwingungen abhängen in einer anderen Richtung. Sie könnte von transversalen Schwingungen in der Richtung abhängen, in welchen sie abhängig ist von der Wirkung des Mittels auf das Licht, und das Licht besteht aus transversalen Schwingungen: sie könnte nicht von transversalen Schwingungen abhängen in der Richtung, wo sie nicht bloss von der Richtung der Schwingungen bestimmt ist, sondern auch von der Richtung der Fortpflanzung abhängt.

"Daher kann ich Ihre Folgerungen nicht als einen *Beweis* in dem strengen Sinne des Wortes betrachten. Ein solcher hängt am Ende von gewissen physikalischen Betrachtungen ab, welche sich auf die Absorption beziehen. Meine eigenen Ansichten in Bezug auf die Ursache der Absorption führen mich sehr stark zu der Meinung, dass sie bloss von der Richtung der Schwingungen und der Schwingungszeit (*periodic time*) und gar nicht von der Fortpflanzungsrichtung abhängt. In meinem Sinne haben daher Ihre Gründe sehr grosses Gewicht. Aber da diess von meinen eigenen individuellen Ansichten abhängt, so betrachte ich dieselben nicht als Etwas, was *nothwendiger* Weise zu allgemeiner Beistimmung zwingen muss.

"Da ich bei diesem Gegenstande bin, so erlauben Sie mir Ihre Aufmerksamkeit auf gewisse Untersuchungen zu lenken, welche mich in einer gänzlich verschiedenen Weise zu einer ähnlichen Schlussfassung führten. Sie sind in dem 9. Band der *Cambridge Philosophical Transactions, Part I*, veröffentlicht. Eine dynamische Untersuchung des Problems der Beugung, in anderen Worten eine mathematische Untersuchung der Beugung, behandelt wie ein dynamisches Problem, führte mich zu folgendem Gesetz: Wenn linear polarisirtes Licht der Beugung unterworfen wird, so ist jeder Strahl nach der Beugung linear polarisirt, und die Schwingungsebene des gebeugten Strahles ist parallel der Schwingungsrichtung des einfallenden Strahles. Unter Schwingungsebene ist die Ebene verstanden, welche durch den Strahl und durch die Schwingungsrichtung geht. Es sey *AB* Fig. 7 in der Ebene des Papieres der einfallende Strahl, der bei *B* gebeugt wird. *BE* auch in der Ebene des Papieres ein gebeugter Strahl, der in das Auge eintritt, *CD*, in einer Ebene senkrecht auf *AB*, sey die Schwingungsrichtung des einfallenden Strahles. Man lege eine Ebene durch *BE* und *CD*, diess wird die Schwingungsebene des gebeugten Strahles seyn und wenn in dieser Ebene *FG* senkrecht auf *BE* gezogen wird, so ist *FG* die Schwingungsrichtung. Mit anderen Worten, die Schwingungsrichtung in dem gebeugten Strahle ist so nahe als möglich der Schwingungsrichtung in dem einfallenden Strahle parallel, als diess nur immer unter der Bedingung geschehen kann, dass sie senkrecht auf dem gebeugten Strahle steht. Dieses Gesetz erscheint sehr natürlich, selbst unabhängig von allem Calcül. Nun folgt aber aus demselben, dass wenn die Schwingungsebene zuerst mit der auf *ABE* senkrecht stehenden Ebene zusammenfällt, und dann allmählich durch gleiche Winkel herumgedreht wird, dass dann die Schwingungsebenen des gebeugten Strahles nicht gleichförmig ausgetheilt seyn werden, sondern sie werden mehr angehäuft gegen eine Ebene durch *BE* senkrecht auf die Ebene *ABE* erscheinen. Wenn α_i, α_d die Azimuthe der Schwingungsebene des einfallenden und des gebeugten Strahles sind, erhalten von Ebenen senkrecht auf *ABE*, und θ das Supplement des Winkels *ABE*, so haben wir $\tan \alpha_d = \cos\theta \tan \alpha_i$. Nun setzt uns aber der Versuch in den Stand die Richtung und das Maass jener Anhäufung der *Polarisationsebenen* zu bestimmen, und nach dem Ergebnisse werden wir uns geleitet finden, sie als parallel oder senkrecht auf die Schwingungsebenen zu betrachten. Wenn nun Fig. 8 die Projection der Polarisations-Ebenen des einfallenden Strahles auf einer senkrecht auf diesem Strahl stehenden Ebene in verschiedene Stellungen des Polarisirers (z. B. eines Nicol'schen Prismas in einer kreisförmig getheilten Fassung) darstellt, und Fig. 9 oder Fig. 10 dasselbe für den gebeugten Strahl vorstellt, so würden die Ebenen mehr gehäuft seyn wie in Fig. 9 und 10, je nachdem die Polarisations-Ebenen parallel oder senkrecht auf die Schwingungsebenen sind. Die Horizontallinien in Fig. 8, 9, 10 stellen die Projectionen auf der Ebene *ABE* dar. Bei einem Glasgitter geschieht die Beugung unter einem so bedeutenden Winkel, dass der theoretische Azimuth der Polarisations-Ebene des gebeugten Strahles in manchen Fällen bis zwanzig Grad variiren kann, je nachdem man voraussetzt, dass die Schwingungen des polarisirten Lichtes parallel oder senkrecht auf die Polarisations-Ebene stehen. Das Ergebniss der Versuche war vollständig zu Gunsten von Fresnel's Voraussetzung."

Mittheilung aus einem Schreiben des Hrn. Prof. Stokes, über das optische Schachbrettmuster; von W. Haidinger.

[*Pogg. Ann.* XCVI, 1855, pp. 305—312. Mitgetheilt vom Hrn. Verf. aus d. *Sitzungsberichten d. Wiener Akademie* (*April* 1854).]

......"Ich habe ähnliche Erscheinungen," wie das Interferenz-Schachbrettmuster, "auf einem Schirme dargestellt, indem ich das Sonnenlicht horizontal in ein finsteres Zimmer reflectirte, in dem Fenster, auf dem Wege des einfallenden Lichtes ein durchlöchertes Zinkblech, wie es für Fensterblenden dient, anbrachte, mit einer grossen Linse in einiger Entfernung von dem Bleche, und das Bild des Bleches nun auf einem Blatte Papier auffing, welches von dem Bilde nach beiden Seiten gegen die Linse zu und von derselben weg bewegt werden konnte. Ich überzeugte mich, dass die Erscheinung *nicht* auf Interferenz beruht, sondern einen viel einfacheren Charakter besitzt, und dass die Erklärung derselben aus der geometrischen Theorie der Schatten und Halbschatten folgt. In der That kenne ich kein Interferenz-Phänomen, das auf einer breiten Lichtfläche, wie die des Himmels ist, beruht; es ist immer erforderlich, die einfallenden Strahlen zu begränzen, indem sie etwa durch ein Loch oder einen Spalt gehen, oder indem man sich des Sonnenbildes einer Linse mit kurzer Brennweite bedient. In meinen Versuchen brachte ich nicht nur die Erscheinungen hervor, welche den von Ihnen beschriebenen ähnlich sind, indem ich den vollen Sonnenstrahl anwandte, sondern ich untersuchte auch die Interferenz-Wirkungen, indem ich das Bild der Sonne durch eine Linse von ziemlich kurzer Brennweite benützte und das Zinkblech nun in einige Entfernung vom Fenster rückte, und ich bin überzeugt, dass Interferenz-Wirkungen, selbst wenn sie nicht ganz unsichtbar seyn sollten, doch in Ihrem Phänomen so schwach sind, dass sie vernachlässigt werden können.

"Man betrachte zuerst Licht von einem einzigen Grade der Brechbarkeit (Fig. 11).

"Es seyen LL' die einfallenden Strahlen, SS' sey der durchlöcherte Schirm, ll' die Linse, $\sigma\sigma'$ das Bild der Sonne, ss' das Bild des Schirmes. Man betrachte nur den Lichtbündel, der durch eine einzige Oeffnung geht. Nach der Brechung durch die Linse werden sich diese gerade so fortpflanzen (siehe Fig. 12) als ob $\sigma\sigma'$ eine helle Scheibe wäre, von der die Strahlen ausgehen, von welchen uns aber nur diejenigen angehen, welche durch das Loch $\alpha\alpha'$ durchgehen ($\alpha\alpha'$ ist aber das Bild der Oeffnung), oder welche von $\sigma\sigma'$ in solchen Richtungen ausgehen, dass sie durch das Loch $\alpha\alpha'$ hindurchgehen würden, wenn man sie nicht durch den Schirm aufgefangen hätte. Es entsteht dadurch also, was man einen negativen Schatten $\nu\alpha\ \nu\alpha'$ und Halbschatten $A\alpha\sigma$ $A'\alpha'\sigma'$ nennen könnte, das heisst Räume, welche für Beleuchtung eben dasjenige sind, was Schatten und Halbschatten für Finsterniss. Auf einem Schirme, mit welchem man die Strahlen auffängt, würde eine Kreisfläche beleuchtet seyn, am schmälsten bei $\alpha\alpha'$ (vorausgesetzt, dass $\sigma\sigma'$ grösser ist als $\alpha\alpha'$), und in dieser Entfernung auch gleichförmig hell, während bei anderen Entfernungen die Mitte heller seyn wird als der Rand.

"Man betrachte nun die Wirkung des Uebereinanderfallens der hellen Kreise, welche den benachbarten Oeffnungen entsprechen. Um die Frage auf das Aeusserste zu vereinfachen, nehme ich die Oeffnungen sehr klein an, so dass man $\alpha\alpha'$ als Punkt betrachten kann. Ich nehme dabei die Anordnung der Oeffnungen als die nämliche an, wie in der von Ihnen gegebenen Figur*. Stellt man den Schirm in den Focus, so erscheint eine Reihe heller Flecke (Fig. 13). Bewegt man den Schirm ein wenig in die Richtung gegen die Linse oder von derselben weg, so öffnen sich die lichten Flecke zu lichten Kreisflächen (Fig. 14). Es sey d die Entfernung zwischen den Mittelpunkten zweier benachbarter Kreise, in verticaler oder horizontaler Richtung. Bewegt man den Schirm so weit, bis die Radien der Kreise grösser sind als $\frac{1}{2}d$ aber kleiner als $d/\sqrt{2}$, so werden die Ränder der Scheiben über einander fallen, etwa so wie in Fig. 15. Der Schirm wird dadurch dem grössten Theile nach beleuchtet, mit Ausnahme von dunkeln quadratartigen, regelmässig geordneten Räumen (Fig. 16). Man bewege nun den Schirm so weit, bis die Radien der vergrösserten hellen Scheiben grösser sind als $d/\sqrt{2}$, aber noch immer kleiner als d, dann ist der Mittelpunkt der bisher dunklen Räume durch vier über einander fallende Kreisscheiben beleuchtet (Fig. 17), während der Mittelpunkt jedes früher hellen Raumes immer noch nur von einer einzigen beleuchtet wird. Die am hellsten beleuchteten Räume des Schir-

* *Sitzungsberichte u. s. w.*, Bd. VII, S. 291.

mes sind also nun in regelmässiger Anordnung die Punkte wie a, welche *in ihrer Lage denjenigen Gegenden des Schirmes entsprechen, welche, wenn dieser im Focus stand, die Mittelpunkte*

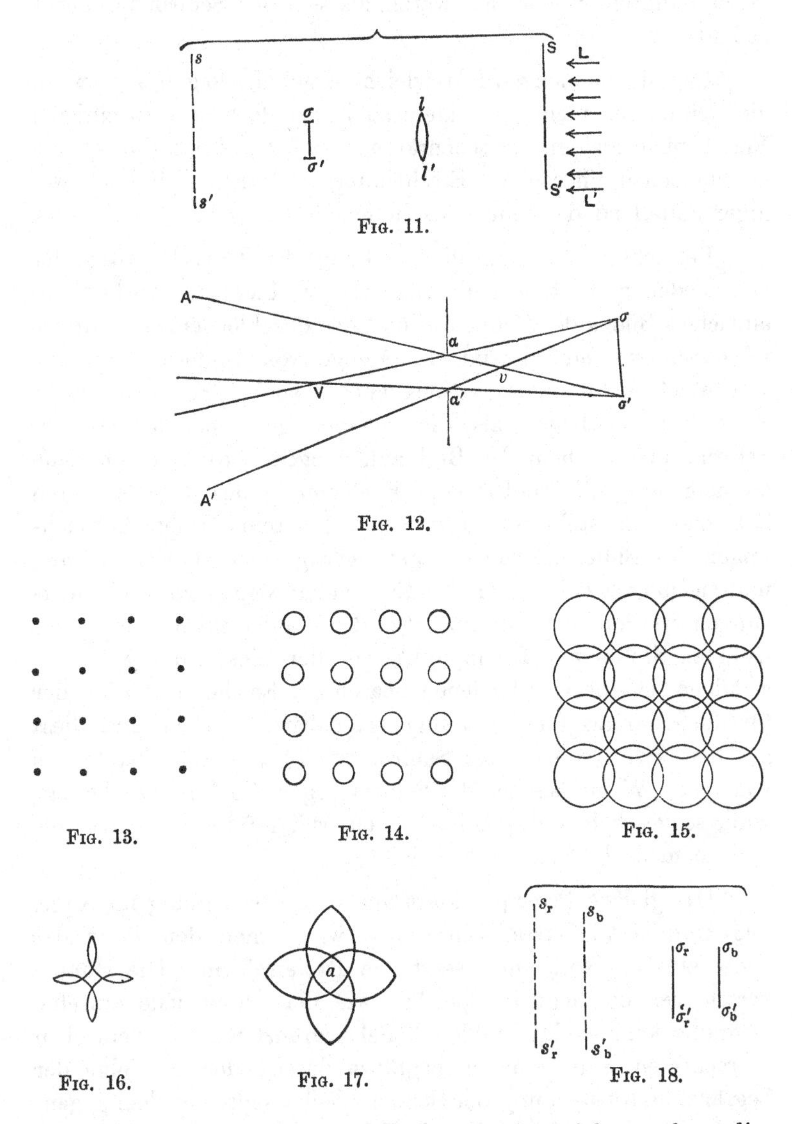

FIG. 11.

FIG. 12.

FIG. 13. FIG. 14. FIG. 15.

FIG. 16. FIG. 17. FIG. 18.

der dunkeln Zwischenräume waren. Es ist nicht nothwendig, die Details weiter zu verfolgen, nur das Eine bemerke ich noch, dass wenn der Radius nur noch um weniges kleiner ist als d, man dann nicht helle Punkte auf dunklem Felde, sondern dunkle Punkte auf hellem Felde hat, wobei die dunkeln Punkte in ihrer Lage den Mittelpunkten der Scheiben entsprechen, oder jenen Punkten, welche hell waren, als sich der Schirm im Focus befand.

"Im Allgemeinen wird das gleiche Ergebniss folgen, auch wenn die Oeffnungen nicht ganz klein sind aber doch noch in einigem Verhältnisse zu den Zwischenräumen stehen. Doch dürften die verschiedenen Phasen der Erscheinung in diesem Falle sich weniger auffallend darstellen, als in jenem.

"Betrachten wir nun die Wirkung der Ueberlagerung der verschiedenen Bestandtheile des weissen Lichtes. Anstatt des einfachen Bildes der Sonne $\sigma\sigma'$ und des durchlöcherten Schirmes ss' haben wir eine unendliche Menge von Bildern (Fig. 18), von welchen die stärker gebrochenen, wie $\sigma_b\sigma'_b$, $s_bs'_b$, näher an der Linse liegen als die weniger gebrochenen. Da der Schirm, auf welchem das Bild aufgefangen wird, ziemlich nahe an dem Orte der deutlichsten Erscheinung des durchlöcherten Schirmes aufgestellt ist, so werden die chromatischen Abweichungen des Bildes ss' viel weniger wichtig seyn, als die von $\sigma\sigma'$, und sie mögen daher hier der Einfachheit wegen gänzlich übergangen werden. Wird nun also der Papierschirm aus seiner früheren Stellung in Brennpunkte von der Linse hinweggerückt, so folgen sich die verschiedenen Phasen der Erscheinung schneller für die mehr als für die weniger brechbaren Farben, und diess aus dem Grunde, weil der Schirm von $s_bs'_b$ weiter absteht, als von $s_rs'_r$. Wenn dagegen der Schirm gegen die Linse zu bewegt wird, so geschehen die Aenderungen früher für die weniger als für die mehr brechbaren Farben.

"Das gleiche Princip erklärt auch die Erscheinung im Auge. Die Hornhaut, Krystall-Linse u. s. w. nehmen den Platz der Linse ein, die Netzhaut ersetzt den Papierschirm. Die Hauptverschiedenheit liegt in der Art, wie der durch jede einzelne Oeffnung kommende Strahlenbündel begränzt ist. In dem oben betrachteten Falle war er begränzt durch oder in Folge der begränzten Ausdehnung der Sonnenscheibe selbst, in dem gegenwärtigen Falle geschieht diess in Folge der begränzten Pupille des Auges. Ferner, anstatt dass der Schirm bewegt wird, während die Bilder $s_bs'_b$, $s_rs'_r$ eine feste Lage haben, so ist es hier der

Schirm (die Netzhaut), welcher fest steht, während die Bilder *ss'* bewegt werden, in Folge der Bewegung des Gegenstandes, der diese Bilder hervorbringt, eine Bewegung, welche in allen Fällen von Brechung eine Bewegung des Bildes in derselben Richtung zur Folge hat. Aber keiner dieser beiden Umstände hat einen Einfluss auf einen Erklärungsgrund.

"Man halte nun ein Stück durchlöcherte Karte oder durchlöchertes Papier gegen den Himmel, in der Entfernung der deutlichsten Sehweite, und nähere es dann allmählich dem Auge. Die wahren Bilder der Karte, welche den verschiedenen Farben entsprechen, fallen nun hinter die Netzhaut; da aber die mehr gebrochenen Bilder vor den weniger gebrochenen liegen, so sind sie weniger ausserhalb des Brennpunktes. Daher finden die Veränderungen der Erscheinung schneller statt für die weniger als für die mehr brechbaren Farben. Wenn daher die dunkeln Zwischenräume in die dunkeln Flecken überzugehen beginnen, so sind sie roth umsäumt, weil die rothen Kreisscheiben auf der Netzhaut grösser sind als die blauen. Diese Umsäumung durch Roth, oder vielmehr durch die mehr brechbaren Farben in ihrer Folge, könnte vielleicht zu wenig lebhaft seyn, um einen Eindruck hervorzubringen. Wenn durch das Uebereinanderfallen der Kreise die dunkeln Flecke in helle Flecke verwandelt wurden, so sind die letzteren gelblich von dem Vorwalten der weniger brechbaren Farben, während das allgemeine Feld blaulich ist, von dem Vorwalten der mehr brechbaren. Wird die durchlöcherte Karte, aus der früheren Stellung in der Entfernung des deutlichsten Sehens in eine grössere Entfernung vom Auge gerückt, so liegen die von den verschiedenen im weissen Licht enthaltenen Farben herrührenden Bilder der Karte vor der Netzhaut, zu äusserst die mehr brechbaren, und sie sind daher entfernter vom Brennpunkte als die weniger brechbaren. Daher sind die Farben der Flecken und Zwischenräume die entgegengesetzten von denen in der früheren Lage.

"Ich bemerke hier, dass das Auge nicht achromatisch ist. Schon Fraunhofer hat diess in seinen Bemerkungen über das Spectrum gezeigt. Sehr auffallend zeigt es eine Erscheinung, welche ich längst beobachtete, und welche ich später von Prof. Dove in Berlin in einer Abhandlung angegeben fand, die er mir sandte. Wenn man einen hellen, wohlbegränzten Gegenstand, wie ein Licht oder die Sonnenscheibe, durch ein tiefblaues Glas oder durch eine Verbindung mehrerer solcher Gläser betrachtet, welche keinen anderen sichtbaren Strahlen den Durchgang gestatten ausser den äussersten rothen und violetten, so sieht man die rothen und die violetten Bilder der Gegenstände nicht gleich deutlich zusammen. Wenn ich die Sonnenscheibe durch eine Combination dieser Art betrachte, was ohne die geringste Unbequemlichkeit ausführbar ist, wenn man nur ein hinlänglich dunkles Glas oder eine hinlängliche Anzahl von Gläsern anwendet, so sehe ich eine wohl begränzte rothe Scheibe und eine undeutliche violette Scheibe von etwa dem doppelten Durchmesser der ersteren. Die letztere kann durch die Anwendung einer convexen Linse deutlich gemacht werden, aber dann wird jene andere undeutliche. In der That kann ich entfernte Gegenstände deutlich vermittelst der äussersten rothen Strahlen sehen, bin aber entschieden kurzsichtig in Bezug auf die violetten Strahlen. Für mittlere Strahlen, und übereinstimmend für gewöhnliches Licht, sollte ich daher etwas weniger kurzsichtig seyn, welches auch der Fall ist."

Einige neuere Ansichten über die Natur der Polarisationsbüschel; von W. Haidinger.

[*Pogg. Ann.* XCVI, 1855, p. 314. Mitgetheilt vom Hrn. Verf. aus d. *Sitzungsberichten d. Wiener Akademie* (*Mai* 1854).]

"Ich bin keinesweges durch irgend welche der Erklärungsarten befriedigt, welche ich bisher über die Ursache Ihrer Büschel gesehen habe. Man kann allen, vorzüglich aber der des Hrn. Jamin, einen Einwurf machen, der unwiderlegbar scheint. Ich will diesen Gegenstand aber hier nicht weiter verfolgen, weil ich daran bin demnächst einen Aufsatz darüber an das *Philosophical Magazine* zu schicken. Ich bin überzeugt, dass die Erscheinung entweder in oder knapp an der Netzhaut ihren Sitz hat. Ich werde eine Muthmassung in Bezug auf die Ursache derselben aufstellen, nach welcher sie von der Art abhängen, wie die letzten Nervenfasern die Empfindung des Lichtes aufnehmen. Ich bin überzeugt, dass die sogenannte Nachahmung der Erscheinung durch Uhrgläser oder Linsen, welche Hr. Jamin vorschlug, mit derselben nichts zu thun hat."

ON THE THEORY OF THE ELECTRIC TELEGRAPH.
By Prof. W. THOMSON.* (Extract.)

[From the *Proceedings of the Royal Society*, May 1855 : also Lord KELVIN'S *Mathematical and Physical Papers*, II, pp. 61–76.]

Extract of Letter from Prof. Stokes to Prof. W. Thomson (dated Nov. 1854).

"IN working out for myself various forms of the solution of the equation $\frac{dv}{dt} = \frac{d^2v}{dx^2}$ under the conditions $v=0$ when $t=0$ from $x=0$ to $x=\infty$; $v=f(t)$, when $x=0$ from $t=0$ to $t=\infty$, I found that the solution with a single integral only (and there must necessarily be this one) was got out most easily thus:—

"Let v be expanded in a definite integral of the form

$$v = \int_0^\infty \varpi(t, \alpha) \sin \alpha x \, d\alpha,$$

which we know is possible.

"Since v does not vanish when $x=0$, $\frac{d^2v}{dx^2}$ is not obtained by differentiating under the integral sign, but the term $\frac{2}{\pi}\alpha v_{x=0}$ must be supplied†, so that (observing that $v_{x=0}=f(t)$ by one of the equations of condition) we have

$$\frac{d^2v}{dx^2} = \int_0^\infty \left\{\frac{2}{\pi}\alpha f(t) - \alpha^2\varpi\right\} \sin \alpha x \, d\alpha.$$

Hence

$$\frac{dv}{dt} - \frac{d^2v}{dx^2} = \int_0^\infty \left\{\frac{d\varpi}{dt} + \alpha^2\varpi - \frac{2}{\pi}\alpha f(t)\right\} \sin \alpha x \, d\alpha,$$

[* This investigation was commenced in consequence of a letter received by the author from Prof. Stokes dated Oct. 16, 1854, and consists mainly of two letters addressed to Prof. Stokes, followed by a reply from him as above.]

† According to the method explained in a paper "On the Critical Values of the Sums of Periodic Series," *Camb. Phil. Trans.* Vol. VIII, p. 533. [*Mathematical and Physical Papers*, G. G. Stokes, Vol. I, p. 236.]

and the second member of the equation being the direct development of the first, which is equal to zero, we must have

$$\frac{d\varpi}{dt} + \alpha^2\varpi - \frac{2}{\pi}\alpha f(t) = 0,$$

whence

$$\varpi = \epsilon^{-\alpha^2 t}\int_0^t \frac{2}{\pi}\alpha f(t)\,\epsilon^{\alpha^2 t}\,dt,$$

the inferior limit being an arbitrary function of α. But the other equation of condition gives

$$\varpi = \epsilon^{-\alpha^2 t}\int_0^t \frac{2}{\pi}\alpha f(t)\,\epsilon^{\alpha^2 t}\,dt = \left(\frac{\pi}{2}\right)^{-1}\alpha\int_0^t \epsilon^{-\alpha^2\overline{t-t'}} f(t')\,dt',$$

therefore

$$v = \left(\frac{\pi}{2}\right)^{-1}\int_0^\infty\int_0^t f(t')\,\alpha\epsilon^{-\alpha^2\overline{t-t'}}\sin\alpha x\,d\alpha\,dt'.$$

But
$$\int_0^\infty \epsilon^{-a\alpha^2}\cos b\alpha\,d\alpha = \frac{1}{2}\left(\frac{\pi}{a}\right)^{\frac{1}{2}}\epsilon^{-\frac{b^2}{4a}},$$

therefore

$$\int_0^\infty \epsilon^{-a\alpha^2}\sin b\alpha\,.\,\alpha d\alpha = -\frac{d}{db}\left\{\frac{1}{2}\left(\frac{\pi}{a}\right)^{\frac{1}{2}}\epsilon^{-\frac{b^2}{4a}}\right\}$$
$$= \frac{\pi^{\frac{1}{2}}b}{4a^{\frac{3}{2}}}\epsilon^{-\frac{b^2}{4a}},$$

whence writing $t-t'$, x, for a, b, and substituting, we have

$$v = \frac{x}{2\pi^{\frac{1}{2}}}\int_0^t (t-t')^{-\frac{3}{2}}\,\epsilon^{\frac{x^2}{4(t-t')}}\,f(t')\,dt'.$$

"Your conclusion as to the American wire follows from the differential equation itself which you have obtained. For the equation $kc\frac{dv}{dt} = \frac{d^2v}{dx^2}$ shows that two submarine wires will be similar, provided the squares of the lengths x, measured to similarly situated points, and therefore of course those of the whole lengths l, vary as the times divided by ck; or the time of any electrical operation is proportional to kcl^2.

"The equation $kc\frac{dv}{dt} = \frac{d^2v}{dx^2} - hv$ gives $h \propto l^{-2}$ for the additional condition of similarity of leakage."

On the Achromatism of a Double Object-glass.

[From the *Report of the British Association*, Glasgow, 1855, pp. 14–15.]

The general theory of the mode of rendering an object-glass achromatic by combining a flint-glass with a crown-glass lens, is well known. The achromatism is never perfect, on account of the irrationality of dispersion. The defect thence arising cannot possibly be obviated, except by altering the composition of the glass. It seemed worthy of consideration whether much improvement might not be effected in this direction; but the problem which the author proposed for consideration was only the following:—Given the kinds of glass to be employed, to find what ought to be done so as to produce the best effect; in other words, to determine the ratio of the focal lengths which gives the nearest approach to perfect achromatism. Two classes of methods may be employed for this purpose. In the one, compensations are effected by trial on a small scale; in the other, the refractive indices of each kind of glass are determined for certain well-defined objects in the spectrum, such for example as the principal fixed lines. The former has this disadvantage, that compensations on a small scale do not furnish so delicate a test as the performance of a large object-glass. The observation of refractive indices, on the other hand, admits of great precision; but it does not immediately appear what ought to be done with the refractive indices when they are obtained. After alluding to the method proposed by Fraunhofer for combining the refractive indices, which, however, as he himself remarked, did not lead to results in exact accordance with observation, the author proposed the following as the condition of nearest approach to achromatism:—that the point of the spectrum for which the focal length of the combination is a minimum shall be situated at the brightest part, namely, at about one-third of the interval DE from the fixed line D towards E.

The refractive index of the flint-glass may be regarded as a function of the refractive index of the crown-glass, and may be expressed with sufficient accuracy by a series with three terms only. The three arbitrary constants may be determined by the values of three refractive indices determined for each kind of glass. The result is as follows:—Let μ_1, μ_2, μ_3 be the refractive indices for the crown-glass; μ_1', μ_2', μ_3' the same for the flint-glass; μ, μ' the refractive indices of the two glasses for any arbitrary ray; m the value of μ for the point at which the focal length is to be made a minimum; r the ratio of $\Delta\mu'$ to $\Delta\mu$ to be employed in the ordinary formula for achromatism. Then having calculated numerically

$$r_{1,2} = \frac{\mu_2' - \mu_1'}{\mu_2 - \mu_1}, \qquad r_{2,3} = \frac{\mu_3' - \mu_2'}{\mu_3 - \mu_2},$$

we shall have*

$$r = r_{1,2} + \frac{2m - \mu_1 - \mu_2}{\mu_3 - \mu_1}(r_{2,3} - r_{1,2}).$$

For the value of m it will be sufficient to take

$$\mu_D + \tfrac{1}{3}(\mu_E - \mu_D).$$

On applying this formula to calculate r for the object-glass for which Fraunhofer has given both the refractive indices of the component glasses and the value of r, which, as observation showed, gave the best results, and taking in succession various combinations of three lines each out of the seven used by Fraunhofer, the author found that whenever the combination was judiciously chosen, the resulting value of r was the same, whatever might have been the combination, and equal to 1·980, which is precisely the value determined by Fraunhofer from observation, as giving the best effect†.

[* In fact the value of the relative dispersion r for the ray of index μ may be obtained from $r = d\mu'/d\mu$, where $\mu' = A + Bx + Cx^2$, in which x represents $\mu - \mu_2$.]

[† The subject of this note is resumed and amplified in a lecture "On the Principles of the Chemical Correction of Object-glasses" delivered to the Photographic Society, and published in the *Photographic Journal*, Feb. 15, 1873; reprinted *infra*.]

Remarks on Professor Challis's paper, entitled "A Theory of the Composition of Colours etc."

[From the *Philosophical Magazine*, xii, 1856, pp. 421–5.]

My object in the present communication is not to discuss Professor Challis's theory, but to rectify some statements as to the experimental facts of the case, as well as one relating to the extent of some researches of my own. I have, however, on some points expressed opinions, respecting the justice of which it is only one who is familiar with certain classes of optical experiments who can feel the confidence that I entertain.

From the paragraph commencing at the foot of page 330, it is plain that Professor Challis has made some confusion between three perfectly distinct things: Sir David Brewster's controverted analysis of the solar spectrum by means of absorbing media*; his discovery of the phenomenon of internal dispersion†; and my own discovery, that a beam of rays of prismatic purity (whether belonging to the visible or invisible portion of the spectrum is indifferent) may, by their action on certain media, produce light which may be decomposed by the prism into portions extending over a wide range of refrangibility, and having colours answering to their refrangibilities‡.

As to the first, it was asserted by Sir David Brewster that light of prismatic purity may have its colour changed by passing through absorbing media. This has nothing to do with "internal" or "epipolic" dispersion, or "fluorescence." Glass coloured blue by cobalt, for instance, has none of these properties, although it is one of the media which exhibit most strikingly the phenomena adduced by Sir David Brewster. Were such a change of colour

* *Edinburgh Transactions*, Vol. xii, p. 123.

† *Edinb. Trans.* Vol. xvi, p. 111; and *Phil. Mag.* Vol. xxxii, 1848, p. 401.

‡ *Philosophical Transactions* for 1852, p. 463.

made out, it would be a point of the utmost importance to consider in reference to any physical theory of light. But while none deny that the appearances are as stated by Sir David Brewster, the inference to be drawn from those appearances remains open to discussion. Airy*, Helmholtz†, and Bernard‡, by operating in a different manner, have come to the conclusion that the colour is not changed; and Helmholtz has attributed the apparent change partly to the mixture of a very small quantity of stray light, partly to the effects of contrast. Having been much in the habit of analysing the light transmitted by coloured solutions, and having repeatedly seen the phenomena on which Sir David Brewster relies, I may be permitted to express my belief that the change of colour is only apparent, being an illusion depending upon contrast, and that this is one of the cases in which the direct evidence of the senses must be controlled. Were the change of colour real, Prof. Challis's statement (p. 330), that "experiment has proved that both the colour and the angle of refraction for a given angle of incidence depend, the substance being given, only on the value of λ," would cease to be true.

As to the second, the principal phenomenon consists in this: that when a beam of sunlight, condensed by a lens, is admitted into certain perfectly clear (*i.e.* not muddy) media, the path of the rays is marked by light, of different colours in different cases, which emanates in all directions. As the real nature of this remarkable phenomenon was not at the time understood, and the phenomenon itself was confounded with the effects of mere suspended particles, it is needless to discuss its possible bearing on any theory of the sensation of colour under this head.

As to the third, the new light emanating from the media which possess the property in question is just like any other light of the same prismatic composition. In its physical properties it retains no traces of its parentage, and its colour depends simply upon its new refrangibility, having nothing to do with that of the producing rays, nor to the circumstance of their belonging

* *Phil. Mag.* Vol. XXX, p. 73.

† *Poggendorff's Annalen*, Vol. LXXXVI, p. 501.

‡ Report of the Meeting of the British Association at Liverpool in 1854, 2nd Part, p. 5.

to the visible or the invisible part of the spectrum. Hence, in speculating on the sensation of colour, this phenomenon may be set aside as not bearing upon the question. I may remark, however, that with regard to the sensation of colour, an analogy has often struck me between the retina and a fluorescent substance, or rather a mixture of three or more fluorescent substances: but this is only an analogy.

It is not true, as Professor Challis seems to suppose (p. 332), that absorption is always, or even generally, accompanied by epipolic dispersion. Among the great variety of coloured metallic solutions, I have hitherto found that property only in solutions of salts of sesquioxide of uranium. I make this remark merely by the way, to prevent misconception: I perfectly agree with Professor Challis in believing that a ray of definite refrangibility is uncompounded; in fact, it was my firm belief in that doctrine which led me to make out the phenomenon of the change of refrangibility of light.

The superposition of two coloured glasses or ribbons by no means gives the effect of the mixture of the two colours. Various methods of mixing colours are enumerated by Mr Maxwell at the end of his paper, entitled "Experiments on Colour, etc.," in the twenty-first volume of the *Edinburgh Transactions*, p. 275. The production of white by a mixture of blue and yellow is by no means confined to prismatic blue and yellow, but takes place just as well with the colours of coloured bodies. In making experiments with the spectrum, in order to neutralize, when possible, a prismatic colour of given intensity by another prismatic colour, so as to produce white, *two* points must be attended to: the place of the second colour in the spectrum must be properly chosen, and the intensity of the light properly regulated. Hence any speculations as to the cause of the variations of intensity in the solar spectrum can have no bearing on the subject before us, seeing that the relation between the intensities of the mixed colours necessary for the production of whiteness is a matter of experimental adjustment.

The reason why the superposition of two coloured bodies does not give the mixture of the colours is known, and is very simple. The composition of the light transmitted through a coloured glass may very conveniently be represented by a curve, in the manner of Sir John Herschel, in which the abscissa x denotes refrangibility, measured, suppose, by the distance from the extreme red in some standard spectrum, and the ordinate y denotes the intensity; so that ydx is the quantity of light between the refrangibilities x and $x+dx$, the intensity in the incident light being taken equal to unity, for simplicity's sake, whatever be the value of x, as we only care to compare intensities for the same value of x. Let y, y' be the ordinates in the curves belonging to two glasses, y_s the ordinate belonging to the tint obtained by superposing the glasses, y_m the ordinate belonging to the mixed tint, as procured, for instance, by a double-image prism, in which case each of the superposed differently coloured images has half the brightness of the original. Then $y_s = yy'$, but $y_m = \frac{1}{2}(y+y')$; and it is easy to see how different may be the curves whose ordinates are y_s, y_m respectively. Thus, let the scale of abscissæ be such that the spectrum extends from $x=0$ to $x=\pi$, and let $y=\frac{1}{4}(1-\cos x)^2$, $y'=\frac{1}{4}(1+\cos x)^2$. In this case $y_s=\frac{1}{16}\sin^4 x$, which vanishes at the extremities, and is a maximum in the middle; whereas $y_m=\frac{1}{4}(1+\cos^2 x)$, which is a maximum at the two extremities, and a minimum in the middle. In the former case, the tint would be a sort of green, a pretty full colour; in the latter, a sort of dilute purple. The colours of two ribbons may very conveniently be mixed in equal proportion by placing them side by side, and viewing them through a double-image achromatic prism; and it will be seen how different the mixed colour is from that seen on superposing the ribbons and holding them up to the light.

I cannot agree with Professor Challis, that "the coloured light of substances, though derived from sunlight, is in fact new light," except so far as relates to that portion which arises from fluorescence. But fluorescence is often absent altogether; and even when it exists, the colour thence arising must in most cases be but a small fraction of the whole colour observed when the substance is freely exposed to white light, not viewed under absorbing media. I think that any one who has been in the constant habit of analysing by the prism the light transmitted through clear coloured fluids or solids, and the light transmitted through or reflected from dyed or other coloured substances, must be forced to admit, that, setting aside a comparatively small number of cases in which the colour observed is referable to other causes, the colours of natural bodies are due to absorption. The exceptions are colours due to fluorescence, as in the case of solutions of quinine, or to regular chromatic reflexion, as in the case of gold, copper, platino-cyanide of magnesium, murexide, etc., not to mention such colours as those of the rainbow, etc., which result from the general properties of bodies with regard to their action on light, not from any speciality of the substance by which the colours happen to be produced. The mode in which I conceive absorption to operate in occasioning the colours observed in dyed ribbons, flowers, coloured powders, etc., I have more fully explained elsewhere*. Now absorption is best studied in clear solids or solutions, where it is not complicated by irregular reflexions or refractions. But when such media are studied by the aid of a pure spectrum, there cannot be a moment's hesitation that the colour of the transmitted light is due to the abstraction from the incident white light of some of the component rays, as explained by Newton. The colour results, not from the light acted on by the medium, but precisely from the portion left unaffected. Hence its origin is celestial (supposing the sun to be the source of the light employed), not terrestrial. But if the colours of natural bodies arise from absorption, the origin of those colours must be deemed celestial too. To make the origin of the green colour of a leaf terrestrial, but that of the green colour of the light transmitted through an alcoholic solution of the colouring matter celestial, notwithstanding that the two greens agree in their very remarkable prismatic composition, would be needlessly and most capriciously to multiply the causes of natural phenomena. The light which gives us the sensation of greenness when we look at a leaf is, I conceive, no more terrestrial in its origin than the sun's light reflected from a mirror is terrestrial, as not retaining the direction which it had in travelling to us from the sun. It is only in the phenomenon of fluorescence, and the closely allied phenomenon of phosphorescence, that the light emitted can be considered as new light having a terrestrial origin.

* *Philosophical Transactions* for 1852, p. 527. [*Ante*, Vol. III, p. 356; cf. also p. 396. See also p. 153, *infra*.]

SUPPLEMENT TO THE "ACCOUNT OF PENDULUM EXPERIMENTS UNDERTAKEN IN THE HARTON COLLIERY...". BY G. B. AIRY, Esq., Astronomer Royal.

[From the *Philosophical Transactions* for 1856. Received *Feb.* 13, read *Mar.* 6, 1856.]

ADDENDUM (pp. 353–355).

ON communicating with Professor Stokes, in reference to the effect of the Earth's rotation and ellipticity in modifying the numerical results of the Harton Experiment, I was favoured by that gentleman with an investigation, which, with his permission, I subjoin as a valuable addition to my own paper.

"I shall suppose the surface of the Earth to be an ellipsoid of revolution, and will employ the notation made use of in my paper on Clairaut's Theorem, published in the fourth volume of the *Cambridge and Dublin Mathematical Journal* *. In this,

V is the potential of the Earth's mass.

r, θ are the polar coordinates of any point in or exterior to the Earth's surface; r being measured from the centre, and θ from the axis of rotation.

a is the equatorial radius.

ϵ the ellipticity.

ω the angular velocity.

m the ratio of the centrifugal force to gravity at the equator.

E the mass of the Earth.

ν the angle between the normal and radius vector at any point of the surface.

In the following investigation, small quantities of the second order are neglected, ϵ and m being regarded as small quantities of the first order.

[* *Ante*, Vol. II, p. 104. See also p. 153, *infra*.]

If
$$U = V + \frac{\omega^2}{2} r^2 \sin^2 \theta,$$
the differential coefficients
$$\frac{dU}{dr}, \quad \frac{1}{r}\frac{dU}{d\theta}$$
will give the components of the force along and perpendicular to the radius vector; and, g being the force of gravity,
$$g = -\cos\nu \frac{dU}{dr} + \sin\nu \frac{1}{r}\frac{dU}{d\theta};$$
which becomes, since ν and $\dfrac{dU}{d\theta}$ are small quantities of the first order,
$$g = -\frac{dU}{dr}.$$
Let v be measured along the vertical; then
$$\frac{dg}{dv} = \cos\nu \frac{dg}{dr} - \sin\nu \frac{1}{r}\frac{dg}{d\theta},$$
or, to the first order,
$$\frac{dg}{dv} = \frac{dg}{dr} = -\frac{d^2U}{dr^2}.$$

Let c be the depth of the mine; then if $(c/a)^2$ be neglected, we shall have for the value of the fraction $\dfrac{\text{gravity below}}{\text{gravity above}}$ (which I will call F), calculated on the supposition that all the attracting mass is internal to both stations,
$$F = 1 - \frac{c}{g}\frac{dg}{dv},$$
where, after differentiation, r is to be put equal to the radius vector of the surface, namely $a(1 - \epsilon\cos^2\theta)$. Now the value of V (Article 5 of the paper referred to) is
$$V = \frac{E}{r} - \left(\frac{E\epsilon}{a} - \frac{1}{2}\omega^2 a^2\right)\frac{a^3}{r^3}\left(\cos^2\theta - \frac{1}{3}\right),$$
which is true, independently of any particular hypothesis respecting the distribution of matter in the interior of the Earth; so that
$$U = \frac{E}{r} - \left(\frac{E\epsilon}{a} - \frac{1}{2}\omega^2 a^2\right)\frac{a^3}{r^3}\left(\cos^2\theta - \frac{1}{3}\right) + \frac{\omega^2}{2} r^2 \sin^2\theta$$

and $$g = -\frac{dU}{dr}$$

$$= \frac{E}{r^2} - 3\left(\frac{E\epsilon}{a} - \frac{1}{2}\omega^2 a^2\right)\frac{a^3}{r^4}\left(\cos^2\theta - \frac{1}{3}\right) - \omega^2 r \sin^2\theta,$$

whence $$-\frac{dg}{dv} = -\frac{dg}{dr}$$

$$= \frac{2E}{r^3} - 12\left(\frac{E\epsilon}{a} - \frac{1}{2}\omega^2 a^2\right)\frac{a^3}{r^5}\left(\cos^2\theta - \frac{1}{3}\right) + \omega^2 \sin^2\theta.$$

Putting now $r = a(1 - \epsilon\cos^2\theta)$, $\omega^2 = m\frac{E}{a^3}$, we find

$$g = \frac{E}{a^2}(1 + 2\epsilon\cos^2\theta) - \frac{3E}{a^2}\left(\epsilon - \frac{m}{2}\right)\left(\cos^2\theta - \frac{1}{3}\right) - m\frac{E}{a^2}(1 - \cos^2\theta)$$

$$= \frac{E}{a^2}\left\{1 + \left(\frac{5m}{2} - \epsilon\right)\cos^2\theta + \epsilon - \frac{3m}{2}\right\},$$

$$-\frac{dg}{dv} = \frac{2E}{a^3}(1 + 3\epsilon\cos^2\theta) - \frac{12E}{a^3}\left(\epsilon - \frac{m}{2}\right)\left(\cos^2\theta - \frac{1}{3}\right) + \frac{mE}{a^3}(1 - \cos^2\theta)$$

$$= \frac{2E}{a^2}\left\{1 + \left(\frac{5m}{2} - 3\epsilon\right)\cos^2\theta + 2\epsilon - \frac{m}{2}\right\}.$$

Whence

$$-\frac{1}{g}\frac{dg}{dv} = \frac{2}{a}\left\{\begin{array}{l} 1 + \left(\frac{5m}{2} - 3\epsilon\right)\cos^2\theta + 2\epsilon - \frac{m}{2} \\ \quad - \left(\frac{5m}{2} - \epsilon\right)\cos^2\theta - \epsilon + \frac{3m}{2} \end{array}\right\}$$

$$= \frac{2}{a}\{1 - 2\epsilon\cos^2\theta + \epsilon + m\};$$

and therefore

$$F = 1 + \frac{2c}{a}\{1 - 2\epsilon\cos^2\theta + \epsilon + m\}.$$

Now the method adopted in the 'Account of Experiments,' etc., Article 57, gives

$$-\frac{1}{g}\frac{dg}{dr} = \frac{2c}{r} = \frac{2c}{a}(1 + \epsilon\cos^2\theta),$$

whence $$F = 1 + \frac{2c}{a}(1 + \epsilon\cos^2\theta).$$

Therefore, if R be the ratio of the value of $F-1$ given above, to $F-1$ as calculated by the method of the 'Account of Experiments,'

$$R = \frac{1-2\epsilon\cos^2\theta+\epsilon+m}{1+\epsilon\cos^2\theta} = 1-3\epsilon\cos^2\theta+\epsilon+m.$$

If l be the geocentric latitude of the place, we may in tł small term replace θ by $90°-l$; and since

$$\cos^2\theta = \sin^2 l = \tfrac{1}{2}(1-\cos 2l),$$

we find

$$R = 1+m-\frac{\epsilon}{2}+\frac{3\epsilon}{2}\cos 2l.$$

Now
$$m = \tfrac{1}{289} = 0{\cdot}00346$$

$$\epsilon = \frac{1}{300{\cdot}8} = 0{\cdot}00333$$

$$l, \text{ for Harton}, = 54° \; 48';$$

$$R = 1+0{\cdot}00346-0{\cdot}00334 = 1{\cdot}00012.$$

That R should have been so very nearly equal to unity, depends upon an accidental numerical relation between the values of m, ϵ, and l. At the equator, $R-1$ would have been as great as 0·00679.

In Article 60 of the 'Account,' $F-1$ was found $=$ ·00012032; whence $R(F-1) =$ ·00012033; which only alters the final value of the mean density in the ratio of 6836 to 6835, giving for result

6·565.

At the equator, the correction to the deduced value 6·566 would have been − ·077."

On the Polarization of Diffracted Light.

[From the *Philosophical Magazine*, XIII, 1857, pp. 159–61:
also *Pogg. Ann.* CI, 1857, pp. 154–7.]

On considering the recent interesting experimental researches of M. Holtzmann on this subject*, I am induced to make the following remarks†.

In the more common phenomena of diffraction, in which the angle of diffraction is but small, we know that the character of the diffracting edge, and the nature of the body by which the light is obstructed, are matters of indifference. This was made the object of special experimental investigation by Fresnel; and its truth is further confirmed by the wonderful accordance which he found between the results of the most careful measurements and the predictions of a theory in which it is assumed that the office of the opaque body is merely to stop a portion of the incident light. But when diffraction is produced by a fine grating, the angle of diffraction is no longer restricted to be small; and it becomes an open question whether the precise circumstances of the diffraction may not have to be taken into account, and not merely the form and dimensions of the apertures through which the light passes. If so, the problem becomes one of extreme complexity. In my memoir on the Dynamical Theory of Diffraction, published in the ninth volume of the *Cambridge Philosophical Transactions,* I investigated the problem on the hypothesis that in diffraction at a large angle, as we know to be the case in diffraction at a small one, the office of the opaque body is *merely to stop* a portion of the incident light. I distinctly stated this as a hypothesis, and I always regarded it as rather precarious. I was guided by the following consideration. Let AB

[* *Pogg. Ann.* 1856; also *Phil. Mag.* XIII, 1857, pp. 125–9.]

[† Cf. also the author's remarks of date 1882, *ante* Vol. II, p. 327, in which Lorenz's experiments of 1860, confirming his own results, are referred to.]

be the section of a transparent interval by the plane of diffraction, supposing for simplicity the diffraction to take place in air or in a homogeneous medium, and not at the confines of two different media; let $AB = b$; let β be the angle of diffraction, and λ the wave-length in the medium. Supposing the light to be incident perpendicularly on the grating, the difference of phase of the secondary waves which started from A, B, respectively, will be determined by the length of path $b \sin \beta$ within the medium. In experiment this will usually be a considerable multiple of λ. In the line AB take two points, A', B', equidistant from A, B, respectively, and comprising between them as large a multiple as possible of $\lambda \operatorname{cosec} \beta$. If we suppose the influence of the opaque body insensible at the distance AA' or BB' from A or B, the secondary waves which start from all points in the interval $A'B'$ will neutralize each other by interference, so that the whole effect will be due to the secondary waves which start from AA' and BB'. Suppose the angle β to belong to the brightest part of a "spectrum of the first class" (Fraunhofer); then $AA' + BB' = \frac{1}{2}\lambda \operatorname{cosec} \beta$, λ referring to mean rays, so that AA' or BB' is only equal to $\frac{1}{4}\lambda \operatorname{cosec} \beta$. If, for example, $\beta = 30°$, AA' is only equal to $\frac{1}{2}\lambda$. At such very small distances it may well be doubted whether the influence of the opaque body may not have to be taken into account.

When diffraction takes place at the confines of two different media, suppose air and glass, the problem is still further complicated. We may, however, apply the theory to which reference has been made on the two extreme suppositions, first, that the diffraction takes place wholly in the first, secondly, that it takes place wholly in the second medium. The results of my own experiments were very fairly represented by theory, the vibrations being supposed perpendicular to the plane of polarization, provided the diffraction be conceived to take place in the *first* medium, or in other words, just *before* the light reaches the grating; but they would not at all fit the hypothesis of vibrations parallel to the plane of polarization. I put forth some considerations, founded on probable reasoning, to show that the supposition of diffraction taking place in the *first* medium was in accordance with the physical circumstances of the case. So decided was the result obtained, that it seemed to me a strong argument in favour of the hypothesis that the vibrations are perpendicular to the plane of polarization, though I still felt the necessity of repeating the experiments under varied circumstances.

But since the appearance of M. Holtzmann's researches the state of the question is changed. I have no reason to doubt the correctness of his results, while on the other hand the result I myself obtained was far too decided to be passed by. The conclusion which, in the present state of the question, seems to me most probable is, that the polarization of light diffracted at a large angle is, in fact, influenced by the nature of the diffracting body. The subject demands a much more extensive experimental investigation, in which the circumstances of diffraction shall be varied as much as possible. I hope to have leisure to undertake such an investigation: meanwhile it would be premature to offer any decided opinion. It seems to me, however, worthy of attentive consideration, whether a glass grating may not offer a fairer experiment for the decision of the question as to the direction of vibration in polarized light than a smoke grating, inasmuch as in the former we have to do with an uninterrupted medium, glass, the surface of which is merely rendered irregular, whereas in the latter the problem is complicated by the existence of two distinct media, glass and soot, placed alternately. I call the layer of soot a medium, for though no light can pass through any sensible thickness of it, we must not conclude from that that it is without influence on the light which passes excessively close to it.

I have not mentioned the effect of oblique refraction in the experiments of M. Holtzmann, because if it were allowed for, the character of the results obtained would remain unchanged, the magnitude of the observed effect would only be somewhat diminished.

ON THE DISCONTINUITY OF ARBITRARY CONSTANTS WHICH APPEAR IN DIVERGENT DEVELOPMENTS.

[From the *Transactions of the Cambridge Philosophical Society*, Vol. x, pp. 106–128. Read *May* 11, 1857.]

[Abstract. *Proc. Camb. Phil. Soc.* Vol. I, pp. 181–2.]

IN a paper "On the Numerical Calculation of a class of Definite Integrals and Infinite Series" printed in the ninth volume of the *Cambridge Philosophical Transactions*, the author succeeded in putting the integral

$$\int_0^\infty \cos\frac{\pi}{2}\,(w^3 - mw)\,dw$$

under a form which admits of extremely easy numerical calculation when m is large, whether positive or negative. The integral is obtained in the first instance under the form of circular functions for m positive, or an exponential for m negative, multiplied by series according to descending powers of m. These series, which are at first convergent, though ultimately divergent, have arbitrary constants as coefficients, the determination of which is all that remains to complete the process. From the nature of the series, which are applicable only when m is large, or when it is an imaginary quantity with a large modulus, the passage from a large positive to a large negative value of m cannot be made through zero, but only by making m imaginary and altering its amplitude by π. The author succeeded in determining directly the arbitrary constants for m positive, but not for m negative. It was found that if, in the analytical expression applicable in the case of m positive, $-m$ were written for m, the result would become correct on throwing away the part involving an exponential with a positive index. There was nothing however to show *à priori* that this process was legitimate, nor, if it were, at what value of the amplitude of m a change in the analytical expression ought to be made, although the occurrence of radicals in the descending and ultimately divergent series, which did not occur in ascending convergent series by which the function might always be expressed, showed that some change analogous to the change of sign of a radical ought to be made in passing through some values of the amplitude of the variable m. The method which the author applied to this function is of very general application, but is subject throughout to the same difficulty.

In the present paper the author has resumed the subject, and has pointed out the character by which the liability to discontinuity in the arbitrary constants may be ascertained, which consists in this, that the terms of an associated divergent series come to be regularly positive. It is thus found that, notwithstanding the discontinuity, the complete integrals, by means of divergent series, of the differential equations which the functions treated of satisfy, are expressed in such a manner as to involve only as many unknown constants as correspond to the degree of the equation.

Divergent series are usually divided into two classes, according as the terms are regularly positive, or alternately positive and negative. But according to the view here taken, series of the former kind appear as singularities of the general case of divergent series proceeding according to powers of an imaginary variable, as indeterminate forms in passing through which a discontinuity of analytical expression takes place, analogous to a change of sign of a radical.

In a paper "On the Numerical Calculation of a class of Definite Integrals and Infinite Series," printed in the ninth volume of the *Transactions* of this Society, I succeeded in developing the integral $\int_0^\infty \cos\frac{\pi}{2}(w^3 - mw)\,dw$ in a form which admits of extremely easy numerical calculation when m is large, whether positive or negative, or even moderately large. The method there followed is of very general application to a class of functions which frequently occur in physical problems. Some other examples of its use are given in the same paper; and I was enabled by the application of it to solve the problem of the motion of the fluid surrounding a pendulum of the form of a long cylinder, when the internal friction of the fluid is taken into account*.

These functions admit of expansion, according to ascending powers of the variables, in series which are always convergent, and which may be regarded as defining the functions for all values of the variable real or imaginary, though the actual numerical calculation would involve a labour increasing indefinitely with the magnitude of the variable. They satisfy certain linear differential equations, which indeed frequently are what present themselves in the first instance, the series, multiplied by arbitrary constants, being merely their integrals. In my former paper, to which the present may be regarded as a supplement, I have employed these equations to obtain integrals in the form of descending series multiplied by exponentials. These integrals, when once the arbitrary constants are determined, are exceedingly convenient for numerical calculation when the variable is large, notwithstanding that the series involved in them, though at first rapidly convergent, become ultimately rapidly divergent.

* *Cambridge Philosophical Transactions*, Vol. IX, Part II. [*Ante*, Vol. II, p. 329. Further papers on this subject of dates 1868, 1889, 1902, are reprinted *infra*.]

The determination of the arbitrary constants may be effected in two ways, numerically or analytically. In the former, it will be sufficient to calculate the function for one or more values of the variable from the ascending and descending series separately, and equate the results. This method has the advantage of being generally applicable, but is wholly devoid of elegance. It is better, when possible, to determine analytically the relations between the arbitrary constants in the ascending and descending series. In the examples to which I have applied the method, with one exception, this was effected, so far as was necessary for the physical problem, by means of a definite integral, which either was what presented itself in the first instance, or was employed as one form of the integral of the differential equation, and in either case formed a link of connexion between the ascending and the descending series. The exception occurs in the case of Mr Airy's integral for m negative. I succeeded in determining the arbitrary constants in the divergent series for m positive; but though I was able to obtain the correct result for m negative, I had to profess myself (p. 177, [343]) unable to give a satisfactory demonstration of it.

But though the arbitrary constants which occur as coefficients of the divergent series may be completely determined for real values of the variable, or even for imaginary values with their amplitudes lying between restricted limits, something yet remains to be done in order to render the expression by means of divergent series analytically perfect. I have already remarked in the former paper (p. 176, [342]) that inasmuch as the descending series contain radicals which do not appear in the ascending series, we may see, *a priori*, that the arbitrary constants must be discontinuous. But it is not enough to know that they must be discontinuous; we must also know where the discontinuity takes place, and to what the constants change. Then, and not till then, will the expressions by descending series be complete, inasmuch as we shall be able to use them for all values of the amplitude of the variable.

I have lately resumed this subject, and I have now succeeded in ascertaining the character by which the liability to discontinuity in these arbitrary constants may be ascertained. I may mention at once that it consists in this; that an associated divergent series comes to have all its terms regularly positive. The expression becomes thereby *to a certain extent* illusory; and thus it is that analysis gets over the apparent paradox of furnishing a discontinuous expression for a continuous function. It will be found that the expressions by divergent series will thus acquire all the requisite generality, and that though applied without any restriction as to the amplitude of the variable they will contain only as many unknown constants as correspond to the degree of the differential equation. The determination, among other things, of the constants in the development of Mr Airy's integral will thus be rendered complete *.

---

1. Before proceeding to more difficult examples, it will be well to consider a comparatively simple function, which has been already much discussed. As my object in treating this function is to facilitate the comprehension of methods applicable to functions of much greater complexity, I shall not take the shortest course, but that which seems best adapted to serve as an introduction to what is to follow.

Consider the integral

$$u = 2\int_0^\infty e^{-x^2} \sin 2ax\, dx = \frac{2a}{1} - \frac{(2a)^3}{2 \,.\, 3} + \frac{(2a)^5}{3 \,.\, 4 \,.\, 5} \;\ldots\ldots (1).$$

[* The considerations developed in this memoir form the complement of Section III of the memoir "On the Critical Values of the Sums of Periodic Series" (*Camb. Phil. Trans.* 1847; *ante* Vol. I, p. 279), in which the cause of discontinuity in the values of series and integrals was traced to infinitely slow convergence at the place of the sudden change. Here the converse phenomenon of discontinuous changes in the constants multiplying the semi-convergent series, which together represent a continuous function, is traced to the same kind of origin, in the domain however of the complex variable,—namely (p. 103) the multiplier of a series can change discontinuously only in crossing places at which one of the other series involved in the expression of the function loses its semi-convergence.

The formulae for the Bessel functions, employed here (pp. 100—103) as an illustration, were again obtained eleven years later by H. Hankel, in his memoir on the cylinder functions, *Math. Annalen*, Vol. I, Dec. 1868, p. 498, with the same explanation of the discontinuities of their constants. At that time the method of complex contour integration, developed by Riemann in his fundamental memoir on the hypergeometric series, a few months before the date of the present paper, afforded the natural means of procedure. It is of great interest to trace the approaches (cf. p. 91) toward that general method in analysis, also probably the result of physical considerations, to which the solution of the present most intricate problem of discontinuous representation had independently given rise. The procedure of the present paper also applies when no expression of the form of a complex integral is available.]

The integral and the series are both convergent for all values of a, and either of them completely defines u for all values real or imaginary of a. We easily find from either the integral or the series

$$\frac{du}{da} + 2au = 2 \quad \ldots\ldots\ldots\ldots\ldots\ldots\ldots(2).$$

This equation gives, if we observe that $u = 0$ when $a = 0$,

$$u = 2e^{-a^2}\int_0^a e^{a^2}da = 2e^{-a^2}\left\{a + \frac{a^3}{1.3} + \frac{a^5}{1.2.5} + \frac{a^7}{1.2.3.7} + \ldots\right\}\ldots(3).$$

This integral or series like the former gives a determinate and unique value to u for any assigned value of a real or imaginary. Both series, however, though ultimately convergent, begin by diverging rapidly when the modulus of a is large. For the sake of brevity I shall hereafter speak of an imaginary quantity simply as large or small when it is meant that its modulus is large or small.

2. In order to obtain u in a form convenient for calculation when a is large, let us seek to express u by means of a descending series. We see from (2) that when the real part of a^2 is positive, the most important terms of the equation are $2au$ and 2, and the leading term of the development is a^{-1}. Assuming a series with arbitrary indices and coefficients, and determining them so as to satisfy the equation, we readily find

$$u = \frac{1}{a} + \frac{1}{2a^3} + \frac{1.3}{2^2a^5} + \ldots$$

This series can be only a particular integral of (2), since it wants an arbitrary constant. To complete the integral we must add the complete integral of

$$\frac{du}{da} + 2au = 0,$$

whence we get for the complete integral of (2)

$$u = Ce^{-a^2} + \frac{1}{a} + \frac{1}{2a^3} + \frac{1.3}{2^2a^5} + \frac{1.3.5}{2^3a^7} + \ldots \ \ldots\ldots\ldots(4).$$

This expression might have been got at once from (3) by integration by parts. It remains to determine the arbitrary constant C.

3. The expression (1) or (3) shows that u is an odd function of a, changing sign with a. But according to (4) u is expressed as the sum of two functions, the first even, the second odd, unless

$C=0$, in which case the even function disappears. But since, as we shall presently see, the value of C is not zero, it must change sign with a. Let

$$a=\rho(\cos\theta+\sqrt{-1}\sin\theta).$$

Since in the application of the series (4) it is supposed that ρ is large, we must suppose a to change sign by a variation of θ, which must be increased or diminished (suppose increased) by π. Hence, if we knew what C was for a range π of θ, suppose from $\theta=\alpha$ to $\theta=\alpha+\pi$, we should know at once what it was from $\theta=\alpha+\pi$ to $\theta=\alpha+2\pi$, which would be sufficient for our purpose, since we may always suppose the amplitude of a included in the range α to $\alpha+2\pi$, by adding, if need be, a positive or negative multiple of 2π, which as appears from (1) or (3) makes no difference in the value of u.

4. When ρ is large the series (4) is at first rapidly convergent, but be ρ ever so great it ends by diverging with increasing rapidity. Nevertheless it may be employed in calculation provided we do not push the series too far, but stop before the terms get large again. To show in a general way the legitimacy of this, we may observe that if we stop with the term

$$\frac{1.3.5\ldots(2i-1)}{2^i a^{2i+1}},$$

the value of u so obtained will satisfy exactly, not (2), but the differential equation

$$\frac{du}{da}+2au=2-\frac{1.3\ldots(2i+1)}{2^i a^{2i+2}} \quad\ldots\ldots\ldots\ldots\ldots(5).$$

Let u_0 be the true value of u for a large value of a_0 of a, and suppose that we pass from a_0 to another large value of a keeping the modulus of a large all the while. Since u ought to satisfy (2), we ought to have

$$u=u_0+2e^{-a^2}\int_{a_0}^{a} e^{a}\,da,$$

whereas since our approximate expression for u actually satisfies (5) we actually have, putting A_i for the last term,

$$u=u_0+e^{-a^2}\int_{a_0}^{a}(2-A_i)\,e^{a^2}da \quad\ldots\ldots\ldots\ldots\ldots(6).$$

If a be very large, and in using the series (4) we stop about where the moduli of the terms are smallest, the modulus of A_i will be very small. Hence in general A_i may be neglected in comparison with (2), and we may use the expression (4), though we stop after $i+1$ terms of the series, as a near approximation to u.

5. But to this there is an important restriction, to understand which more readily it will be convenient to suppose the integration from a_0 to a performed, first by putting

$$da = (\cos\theta + \sqrt{-1}\sin\theta)\,d\rho,$$

and integrating from ρ_0 to ρ, θ remaining equal to θ_0, and then

$$da = \rho(-\sin\theta + \sqrt{-1}\cos\theta)\,d\theta,$$

and integrating from θ_0 to θ, ρ remaining unchanged. This is allowable, since u is a finite, continuous, and determinate function of a, and therefore the mode in which ρ and θ vary when a passes from its initial value a_0 to its final value a is a matter of indifference. The modulus of e^{a^2} will depend on the real part $\rho^2\cos 2\theta$ of the index. Now should $\cos 2\theta$ become a maximum within the limits of integration, we can no longer neglect A_i in the integration. For however great may be the value previously assigned to i, the quantity $\rho^{-2i-1}e^{\rho^2\cos 2\theta}$ will become, for values of θ comprised within the limits of integration, infinitely great, when ρ is infinitely increased, compared with the value of $e^{\rho^2\cos 2\theta}$ at either limit. And though the modulus of the quantity $2e^{a^2}$ under the integral sign will become far greater still, inasmuch as it does not contain the factor ρ^{-2i-1}, yet as the mutual destruction of positive and negative parts may take place quite differently in the two integrals $\int 2e^{a^2}da$ and $\int A_i e^{a^2}da$, we can conclude nothing as to their relative importance.

6. Now $\cos 2\theta$ will continually increase or decrease from one limit to the other, or else will become a maximum, according as the two limits θ_0 and θ lie in the same interval 0 to π or π to 2π, or else lie one in one of the two intervals and the other in the other. Hence we may employ the expression (4), with an invariable value of C yet to be determined, so long as $0 < \theta < \pi$, and we may employ the expression obtained by writing C' for C so long as $\pi < \theta < 2\pi$, but we must not pass from one interval to the other, retaining the same expression. Now we have seen (Art. 3) that the constant changes sign when θ is increased by π, and therefore $C' = -C$. And since u is unchanged when θ is increased by any multiple of 2π, we readily see that in order to make the

expression (4) generally applicable, it will be sufficient to change the sign of the constant whenever θ passes through zero or a multiple of π.

7. We may arrive at the same conclusion in another way, which will be of more general or at least easier application, as not involving the integration of the differential equation.

The modulus of the general term (Art. 4) of the series (4), expressed by means of the function Γ, is

$$\frac{\Gamma(i+\frac{1}{2})}{\Gamma(\frac{1}{2})\,\rho^{2i+1}}.$$

Suppose i very large. Employing the formula

$$\Gamma(x+1)=\sqrt{2\pi x}\left(\frac{x}{e}\right)^x, \text{ nearly, when } x \text{ is large,}$$

observing that $\Gamma(\frac{1}{2})=\pi^{\frac{1}{2}}$, and calling the modulus μ_i, we find

$$\mu_i=2^{\frac{1}{2}}(i-\tfrac{1}{2})^i e^{-i+\frac{1}{2}}\rho^{-2i-1},$$

which, since $(i+c)^i=i^ie^c$, nearly, becomes

$$\mu_i=2^{\frac{1}{2}}i^ie^{-i}\rho^{2i-1} \text{ ....(7).}$$

We easily get, either from this expression or from the general term,

$$\frac{\mu_{i+1}}{\mu_i}=\frac{i}{\rho^2}, \text{ nearly. ....(8).}$$

Hence when ρ is large the ratio of consecutive moduli becomes very nearly equal to unity for a great number of terms together, about where the modulus is a minimum. To find approximately the minimum modulus μ, we must put $i=\rho^2$ in (7), which gives

$$\mu=2^{\frac{1}{2}}\rho^{-1}e^{-\rho^2} \text{ ....(9).}$$

If we knew precisely at what term it would be best to stop, the expression for μ would be a measure of the uncertainty to which we were liable in using the series (4) directly, that is, without any transformation. For although it is clear that we must stop *somewhere about* the term with a minimum modulus, in order that the differential equation (5) which our function really satisfies may be as good an approximation as can be had to the true differential equation (2), the number of terms comprised in this *about* will increase with i, the order of the term of minimum modulus. If we suppose that we are uncertain to the extent of

n terms, the sum of the moduli of these n nearly equal terms will be

$$2^{\frac{1}{2}} n\rho^{-1} e^{-\rho^2},$$

nearly. It seems as if n must increase with i, but not so fast as i. If we suppose that it is of the form $ki^{\frac{1}{2}}$ or $k\rho$, the sum of the n terms will be a quantity of the order $e^{-\rho^2}$. But even if n increased as any power p of i, however great, still the sum of the n terms would be a quantity of the order $\rho^{2p-1}e^{-\rho^2}$, which when ρ was infinitely increased would become infinitely small in comparison with the modulus $e^{-\rho^2 \cos 2\theta}$ of the term multiplied by C in (4), provided θ had any given value differing from zero or a multiple of π. Hence if θ have any value lying between α and $\pi - \alpha$, or else between $\pi + \alpha$ and $2\pi - \alpha$, where α is a small positive quantity which in the end may be made as small as we please, the quantity C in (4) cannot pass from one of its values to another without rendering the function u discontinuous, which it is not. But when $\theta = 0$ or $= \pi$, the term $Ce^{-\alpha^2}$ becomes merged in the vagueness with which, in this case, the divergent series defines the function. Hence we arrive in a way quite different from that of Art. 5 at the conclusions enunciated in Art. 6.

8. Nor is this all. When the terms of a regular series are alternately positive and negative, the series may be converted by the formulæ of finite differences into others which converge rapidly. In the present case the terms are not simply positive and negative alternately, except when θ is an odd multiple of $\frac{1}{2}\pi$, but the same methods will apply with the proper modification. Suppose that we sum the series (4) directly as far as terms of the order $i-1$ inclusive. Omitting the common factor $e^{-(2i+1)\theta\sqrt{-1}}$, which may be restored in the end, we have for the rest of the series

$$\mu_i + e^{-2\theta\sqrt{-1}}\mu_{i+1} + e^{-4\theta\sqrt{-1}}\mu_{i+2} + \dots$$

If we denote by D or $1+\Delta$ the operation of passing from μ_i to μ_{i+1}, and separate symbols of operation, this becomes

$$(1 + e^{-2\theta\sqrt{-1}}D + e^{-4\theta\sqrt{-1}}D^2 + \dots)\,\mu_i,$$

or

$$\{1 - (1+\Delta)\,e^{-2\theta\sqrt{-1}}\}^{-1}\mu_i.$$

Now $1 - e^{-2\theta\sqrt{-1}} = 1 - \cos 2\theta + \sqrt{-1}\sin 2\theta = 2\sin\theta e^{\left(\frac{\pi}{2}-\theta\right)\sqrt{-1}}$, which reduces the expression to

$$(2\sin\theta)^{-1}e^{\left(\theta-\frac{\pi}{2}\right)\sqrt{-1}}\{1 - (2\sin\theta)^{-1}e^{-\left(\frac{\pi}{2}+\theta\right)\sqrt{-1}}\Delta\}^{-1}\mu_i,$$

or, putting q for $(2 \sin \theta)^{-1}$, to

$$qe^{\left(\theta-\frac{\pi}{2}\right)\sqrt{-1}}\mu_i + q^2e^{-\pi\sqrt{-1}}\Delta\mu_i + q^3e^{-\left(\frac{3\pi}{2}+\theta\right)\sqrt{-1}}\Delta^2\mu_i + \ldots$$

Now if ρ be very large, and μ_i belong to the part of the series where the moduli of consecutive terms are nearly equal, the successive differences $\Delta\mu_i, \Delta^2\mu_i, \ldots$ will decrease with great rapidity. Hence if θ have any given value different from zero or a multiple of π, by taking ρ sufficiently great, we may transform the series about where it ceases to converge into one which is at first rapidly convergent, and thus a quantity which may be taken as a measure of the remaining uncertainty will become incomparably smaller even than μ, much more, incomparably smaller than the modulus of e^{-a^2}. But if $\theta = 0$ or $= \pi$, the above transformation fails, since q becomes infinite. In this case if we want to calculate u closer than to admit of the uncertainty to which we are liable, knowing only that we must stop *somewhere about* the place where the series begins to diverge after having been convergent, we must have recourse to the ascending series (1) or (3), or to some perfectly distinct method. The usual method by which Σu_x is made to depend on $\int u_x dx$ would evidently fail, in consequence of the divergence of the integral.

9. In applying *practically* the transformation of the last article to the summation of the series (4), it would not usually, when ρ was very large, be necessary to go as far as the part of the series where the moduli of consecutive terms are nearly equal. It would be sufficient to deduct $l, 2l \ldots$ from the logarithms of μ_{i+1}, $\mu_{i+2} \ldots$, where l is nearly equal to the mean increment of the logarithms at that part of the series, to associate the factor f whose logarithm is l with the symbol D, and take the differences of the numbers

$$\mu_i, \; f^{-1}\mu_{i+1}, \; f^{-2}\mu_{i+2}, \text{ etc.}$$

However, my object leads me to consider, not the actual summation of the series, but the theoretical possibility of summation, and consequent interpretation of the equation (4).

10. The mode of discontinuity of the constant C having been now ascertained, nothing more remains except to determine that constant, which is done at once. Writing $\sqrt{-1}a$ for a in (4) after having put for u its first expression in (3), we have

$$2e^{a^2}\int_0^a e^{-a^2}da = -\sqrt{-1}\,Ce^{a^2} - \frac{1}{a} + \frac{1}{2a^3} - \ldots,$$

whence, putting $a=\infty$, we have $C=\sqrt{-1}\pi^{\frac{1}{2}}$. Hence we get for the general expression for C in (4),

$$\left.\begin{aligned} C &= \sqrt{-1}\pi^{\frac{1}{2}}, \text{ when } 0<\theta<\pi, \\ C &= -\sqrt{-1}\pi^{\frac{1}{2}} \text{ when } \pi<\theta<2\pi; \end{aligned}\right\} \quad \text{.............(10)},$$

and therefore from (3) and (4)

$$2e^{a^2}\int_0^a e^{a^2}da = \pm\sqrt{-1}\pi^{\frac{1}{2}}e^{a^2} + \frac{1}{a} + \frac{1}{2a^3} + \frac{1.3}{2^2a^5} + \frac{1.3.5}{2^3a^7} + \ldots\ldots (11),$$

the sign being + or − according as θ, the amplitude of a, is comprised within the limits 0 and π, or π and 2π.

Writing $a\sqrt{-1}$ for a in (11), which comes to altering the origin of θ by $\frac{1}{2}\pi$, we find

$$2e^{a^2}\int_0^a e^{-a^2}da = \pm\pi^{\frac{1}{2}}e^{a^2} - \frac{1}{a} + \frac{1}{2a^3} - \frac{1.3}{2^2a^5} + \frac{1.3.5}{2^3a^7} - \ldots\ldots (12),$$

the sign being + or − according as the amplitude of a lies within the limits $-\frac{1}{2}\pi$ and $\frac{1}{2}\pi$, or $\frac{1}{2}\pi$ and $\frac{3}{2}\pi$. It is worthy of remark that in this expression the transcendental quantity $\pi^{\frac{1}{2}}$ appears as a true radical, admitting of the double sign.

Two cases of the integral $\int_0^a e^{a^2}da$ occur in actual investigations, namely when $\theta=\frac{1}{2}\pi$, when the integral leads to $\int_0^t e^{t^2}dt$, which occurs in the theory of probabilities, and when $\theta=\frac{1}{4}\pi$, when it leads to Fresnel's integrals $\int_0^s \cos(\frac{1}{2}\pi s^2)\,ds$ and $\int_0^s \sin(\frac{1}{2}\pi s^2)\,ds$. In the latter case the expression (11) is equivalent to the development of these integrals which has been given by M. Cauchy.

11. If in equation (11) we put $a=\rho(\cos\theta\pm\sqrt{-1}\sin\theta)$, where θ is a small positive quantity, and after equating the real parts of both sides of the equation make θ vanish, we find, whichever sign be taken,

$$\frac{1}{\rho} + \frac{1}{2\rho^3} + \frac{1.3}{2^2\rho^5} + \frac{1.3.5}{2^3\rho^7} + \ldots\ldots = 2e^{-\rho^2}\int_0^\rho e^{\rho^2}d\rho \quad \ldots (13).$$

The expression which appears on the second side of this equation may be regarded as a *singular value* of the sum of the series

$$\frac{1}{a}+\frac{1}{2a^3}+\frac{1.3}{2^2a^5}+\frac{1.3.5}{2^3a^7}+\dots \quad\dots\dots\dots\dots(14),$$

a series which when θ vanishes *takes the form* of the first member of the equation. The equivalent of the series for general values of the variable is given, not by (13), but by (11). It may be remarked that the singular value is the mean of the general values for two infinitely small values of θ, one positive and the other negative.

These results, to which we are led by analysis, may be compared with the known theory of periodic series. If $f(x)$ be a finite function of x, the value of which changes abruptly from a to b as x increases through the value c, a quantity lying between 0 and π, and $f(x)$ be expanded between the limits 0 and π in a series of sines of multiples of x, and if $\phi(n, x)$ be the sum of n terms of the series, the value of $\phi(n, x)$ for an infinitely large value of n and a value of x infinitely near to c is indeterminate, like that of the fraction

$$\frac{(x+y)^2+x-y}{(x-y)^2+x+y},$$

which takes the form $\frac{0}{0}$ when x and y vanish, but of which the limiting value is wholly indeterminate if x and y are independent. We may enquire, if we please, what is the limit of the fraction when x first vanishes and then y, or the limit when y first vanishes and then x, for each of these has a perfectly clear and determinate signification. In the former case we have, calling the fraction $\psi(x, y)$,

$$\lim._{y=0}\lim._{x=0}\psi(x, y)=\lim._{y=0}\frac{y^2-y}{y^2+y}=-1;$$

in the latter

$$\lim._{x=0}\lim._{y=0}\psi(x, y)=\lim._{x=0}\frac{x^2+x}{x^2+x}=1.$$

So in the case of the periodic series if we denote by ξ a small positive quantity

$$\lim._{\xi=0}\lim._{n=\infty}\phi(n, c-\xi)=\lim._{\xi=0}f(c-\xi)=a,$$
$$\lim._{\xi=0}\lim._{n=\infty}\phi(n, c+\xi)=\lim._{\xi=0}f(c+\xi)=b;$$

but we know that

$$\lim._{n=\infty} \lim._{\xi=0} \phi(n, c \pm \xi) = \lim._{n=\infty} \phi(n, c) = \tfrac{1}{2}(a+b).$$

Similarly in the case of the series (14) if we denote its sum by $\chi(a) = \varpi(\rho, \theta)$, and use the term *limit* in an extended sense, so as to understand by $\lim._{\rho=\infty} F(\rho)$ a function of ρ to which $F(\rho)$ may be regarded as equal when ρ is large enough, and if we suppose θ to be a small positive quantity, we have from (11)

$$\lim._{\theta=0} \lim._{\rho=\infty} \varpi(\rho, \theta) = \lim._{\theta=0} \{2e^{-a^2} \int_0^a e^{a^2} da - \sqrt{-1}\, \pi^{\frac{1}{2}} e^{-a^2}\}$$

$$= 2e^{-\rho^2} \int_0^\rho e^{\rho^2} d\rho - \sqrt{-1}\, \pi^{\frac{1}{2}} e^{-\rho^2};$$

$$\lim._{\theta=0} \lim._{\rho=\infty} \varpi(\rho, -\theta) = \lim._{\theta=0} \{2e^{-a^2} \int_0^a e^{a^2} da + \sqrt{-1}\, \pi^{\frac{1}{2}} e^{a^2}\}$$

$$= 2e^{-\rho^2} \int_0^\rho e^{\rho^2} d\rho + \sqrt{-1}\, \pi^{\frac{1}{2}} e^{-\rho^2},$$

whereas equation (13) may be expressed by

$$\lim._{\rho=\infty} \lim._{\theta=0} \varpi(\rho, \pm\theta) = \lim._{\rho=\infty} \chi(\rho) = 2e^{-\rho^2} \int_0^\rho e^{\rho^2} d\rho.$$

There is however this difference between the two cases, that in the case of the periodic series the series whose general term is $\Delta\phi(n, c)$ is convergent, and may be actually summed to any assigned degree of accuracy, whereas the series (13), though at first convergent, is ultimately divergent; and though we know that we must stop somewhere about the least term, that alone does not enable us to find the sum, except subject to an uncertainty comparable with $e^{-\rho^2}$. Unless therefore it be possible to apply to the series (13) some transformation rendering it capable of summation to a degree of accuracy incomparably superior to this, the equation (13) must be regarded as a mere symbolical result. We might indeed *define* the sum of the ultimately divergent series (13) to mean the sum taken to as many terms as should *make* the equation (13) true, and express that condition in a manner which would not require the quantity taken to denote the number of terms to be integral; but then equation (13) would become a mere truism. However I shall not pursue this subject further, as these singular values of divergent series appear to be merely matters of curiosity.

12. In order still further to illustrate the subject, before going on to the actual application of the principles here established, let us consider the function defined by the equation

$$u = 1 + \frac{1}{2}x - \frac{1.1}{2.4}x^2 + \frac{1.1.3}{2.4.6}x^3 - \dots\dots\dots\dots (15).$$

Suppose that we have to deal with such values only of the imaginary variable x as have their moduli less than unity. For such values the series (15) is convergent, and the equation (15) assigns a determinate and unique value to u. Now we happen to know that the series is the development of $(1+x)^{\frac{1}{2}}$. But this function admits of one or other of the following developments according to descending powers of x:—

$$u = x^{\frac{1}{2}} + \frac{1}{2}x^{-\frac{1}{2}} - \frac{1.1}{2.4}x^{-\frac{3}{2}} + \frac{1.1.3}{2.4.6}x^{-\frac{5}{2}} - \dots\dots\dots (16),$$

$$u = -x^{\frac{1}{2}} - \frac{1}{2}x^{-\frac{1}{2}} + \frac{1.1}{2.4}x^{-\frac{3}{2}} - \frac{1.1.3}{2.4.6}x^{-\frac{5}{2}} + \dots\dots\dots (17).$$

Let $x = \rho(\cos\theta + \sqrt{-1}\sin\theta)$, and let $x^{\frac{1}{2}}$ denote that square root of x which has $\frac{1}{2}\theta$ for its amplitude. Although the series (16), (17) are divergent when $\rho < 1$, they may in general, for a given value of θ, be employed in actual numerical calculation, by subjecting them to the transformation of Art. 8, provided ρ do not differ too much from 1. The greater be the accuracy required, θ being given, the less must ρ differ from 1 if we would employ the series (16) or (17) in place of (15). It remains to be found which of these series must be taken.

If θ lie between $(2i-1)\pi + \alpha$ and $(2i+1)\pi - \alpha$, where i is any positive or negative integer or zero, and α a small positive quantity which in the end may be made as small as we please, either series (16) or (17) may by the method of Art. 8 be converted into another, which is at first sufficiently convergent to give u with a sufficient degree of accuracy by employing a finite number only of terms. If m terms be summed directly, and in the formula of Art. 8 the nth difference be the last which yields significant figures, the number of terms actually employed in some way or other in the summation will be $m+n+1$. And in this case we cannot pass from one to the other of the two series (16), (17) without rendering u discontinuous. But when θ passes through an odd multiple of π we may have to pass from one of the two series to the other. Now when θ is increased by 2π the series (16) or (17) changes sign, whereas (15) remains unchanged. Therefore in calculating u for two values of θ differing by 2π we must employ the two series (16) and (17), one in each case. Hence we must employ one of the series from $\theta = -\pi$ to $\theta = \pi$, the other from $\theta = \pi$ to $\theta = 3\pi$, and so on; and therefore if we knew which series to take for some one value of x everything would be determined.

Now when $\rho = 1$ the series (15) becomes identical with (16) when θ has the particular value 0. Hence (16) and not (17) gives the true value of u when $-\pi < \theta < \pi$.

13. Let ρ, θ be the polar co-ordinates of a point in a plane, O the origin, C a circle described round O with radius unity, S the point determined by $x = -1$, that is, by $\rho = 1$, $\theta = \pi$. To each value of x corresponds a point in the plane; and the restriction laid down as to the moduli of x confines our attention to points within the circle, to each of which corresponds a determinate value of u. If P_0 be any point in the plane, either within the circle or not, and a moveable point P start from P_0, and after making any circuit, without passing through S, return to P_0 again, the function $(1+x)^{\frac{1}{2}}$ will regain its primitive value u_0, or else become equal to $-u_0$, according as the circuit excludes or includes the point S, which for the present purpose may be called a *singular point.* Suppose that we wished to tabulate u, using when possible the divergent series (16) in place of the convergent series (15). For a given value of θ, in commencing with small values of ρ we should have to begin with the series (15), and when ρ became large enough we might have recourse to (16). Let OP be the smallest value of ρ for which the series (16) may be employed; for which, suppose, it will give u correctly to a certain number of decimal places. The length OP will depend upon θ, and the locus of P will be some curve, symmetrical with respect to the diameter through S. As θ increases the curve will gradually approach the circle C, which it will run into at the point S. For points lying between the curve and the circle we may employ the series (16), but we cannot, keeping within this space, make θ pass through the value π. The series (16), (17) are convergent, and their sums vary continuously with x, when $\rho > 1$; and if we employed the

same series (16) for the calculation of u for values of x having amplitudes $\pi - \beta$, $\pi + \beta$, corresponding to points P, P', we should get for the value of u at P' that into which the value of u at P passes continuously when we travel from P to P' *outside* the point S, which as we have seen is *minus* the true value, the latter being defined to be that into which the value of u at P passes continuously when we travel from P to P' *inside* the point S.

In the case of the simple function at present under consideration, it would be an arbitrary restriction to confine our attention to values of x having moduli less than unity, nor would there be any advantage in using the divergent series (16) rather than the convergent series (15). But in the example first considered we have to deal with a function which has a perfectly determinate and unique value for all values of the variable a, and there is the greatest possible advantage in employing the descending series for large values of ρ, though it is ultimately divergent. In the case of this function there are (to use the same geometrical illustration as before) as it were two singular points at infinity, corresponding respectively to $\theta = 0$ and $\theta = \pi$.

14. The principles which are to guide us having been now laid down, there will be no difficulty in applying them to other cases, in which their real utility will be perceived. I will now take Mr Airy's integral, or rather the differential equation to which it leads, the treatment of which will exemplify the subject still better. This equation, which is No. 11 of my paper "On the Numerical Calculation, etc.," becomes on writing u for U, $-3x$ for n

$$\frac{d^2u}{dx^2} - 9xu = 0 \quad \text{..................(18).}$$

The complete integral of this equation in ascending series, obtained in the usual way, is

$$\left.\begin{aligned} u = A &\left\{1 + \frac{9x^3}{2.3} + \frac{9^2x^6}{2.3.5.6} + \frac{9^3x^9}{2.3.5.6.8.9} + \dots\right\} \\ + B &\left\{x + \frac{9x^4}{3.4} + \frac{9^2x^7}{3.4.6.7} + \frac{9^3x^{10}}{3.4.6.7.9.10} + \dots\right\} \end{aligned}\right\} \text{...(19).}$$

These series are always convergent, and for any value of x real or imaginary assign a determinate and unique value to u.

The integral in a form adapted for calculation when x is large, obtained by the method of my former paper, is

$$u = Cx^{-\frac{1}{4}}e^{-2x^{\frac{3}{2}}}\left\{1 - \frac{1.5}{1.144x^{\frac{3}{2}}} + \frac{1.5.7.11}{1.2.144^2x^3} - \frac{1.5.7.11.13.17}{1.2.3.144^3x^{\frac{9}{2}}} + \ldots\right\}$$
$$+ Dx^{-\frac{1}{4}}e^{2x^{\frac{3}{2}}}\left\{1 + \frac{1.5}{1.144x^{\frac{3}{2}}} + \frac{1.5.7.11}{1.2.144^2x^3} + \frac{1.5.7.11.13.17}{1.2.3.144^3x^{\frac{9}{2}}} + \ldots\right\} \quad \ldots\ldots\ldots(20).$$

The constants C, D must however be discontinuous, since otherwise the value of u determined by this equation would not recur, as it ought, when the amplitude of x is increased by 2π. We have now first to ascertain the mode of discontinuity of these constants, secondly, to find the two linear relations which connect A, B with C, D.

Let the equation (20) be denoted for shortness by

$$u = Cx^{-\frac{1}{4}}f_1(x) + Dx^{-\frac{1}{4}}f_2(x) \quad \ldots\ldots\ldots\ldots\ldots(21);$$

and let $f(x)$, when we care only to express its dependence on the amplitude of x, be denoted by $F(\theta)$. We may notice that

$$F_1(\theta + \tfrac{2}{3}\pi) = F_2(\theta); \qquad F_2(\theta + \tfrac{2}{3}\pi) = F_1(\theta) \ldots\ldots(22).$$

15. In equation (21), let that term in which the real part of the index of the exponential is positive be called the *superior*, and the other the *inferior* term. In order to represent to the eye the existence and progress of the functions $f_1(x)$, $f_2(x)$ for different values of θ, draw a circle with any radius, and along a radius vector inclined to the prime radius at the variable angle θ take two distances, measured respectively outwards and inwards from the circumference of the circle, proportional to the real part of the index of the exponential in the superior and inferior terms, θ alone being supposed to vary,

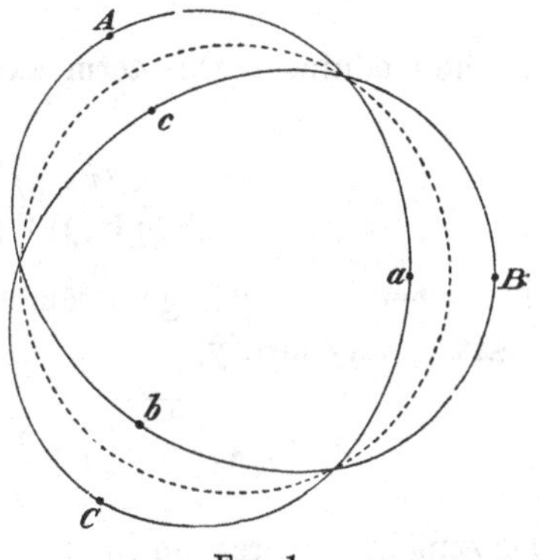

Fig. 1.

or in other words proportional to $\cos \frac{3}{2}\theta$. For greater convenience suppose these distances moderately small compared with the radius. Consider first the function $F_1(\theta)$ alone. The curve will evidently have the form represented in the figure, cutting the circle at intervals of 120°, and running into itself after two complete revolutions. The equations (22) show that the curve corresponding to $F_2(\theta)$ is already traced, since $F_2(\theta) = F_1(\theta + 2\pi)$. If now we conceive the curve marked with the proper values of the constants C, D, it will serve to represent the complete integral of equation (18).

In marking the curve we may either assume the amplitude θ of x to lie in the interval 0 to 2π, and determine the values of C, D accordingly, or else we may retain the same value of C or D throughout as great a range as possible of the curve, and for that purpose permit θ to go beyond the above limits. The latter course will be found the more convenient.

16. We must now ascertain in what cases it is possible for the constant C or D to alter discontinuously as θ alters continuously. The tests already given will enable us to decide.

The general term of either series in (20), taken without regard to sign, is

$$\frac{1 . 5 \ldots (6i-5)(6i-1)}{1 . 2 \ldots i\,(144x^{\frac{3}{2}})^i};$$

and the modulus of this term, expressed by means of the function Γ, is

$$\frac{\Gamma(i+\frac{1}{6})\,\Gamma(i+\frac{5}{6})}{\Gamma(\frac{1}{6})\,\Gamma(\frac{5}{6})\,\Gamma(i+1)(4\rho^{\frac{3}{2}})^i},$$

which when i is very large becomes by the transformations employed in Art. 7, very nearly,

$$\sqrt{\frac{2\pi}{i}}\left(\frac{i}{e}\right)^i \div \Gamma(\tfrac{1}{6})\,\Gamma(\tfrac{5}{6})\,(4\rho^{\frac{3}{2}})^i.$$

Denoting this expression by μ_i, and putting for $\Gamma(\frac{1}{6})\,\Gamma(\frac{5}{6})$ its value $\pi \operatorname{cosec} \frac{1}{6}\pi$ or 2π, we have

$$\mu_i = (2\pi i)^{-\frac{1}{2}} \frac{i^i}{(4\rho^{\frac{3}{2}}e)^i} \quad \ldots\ldots\ldots\ldots\ldots\ldots\ldots(23);$$

whence for very large values of i

$$\frac{\mu_{i+1}}{\mu_i} = \frac{i}{4\rho^{\frac{3}{2}}} \qquad\qquad (24).$$

For large values of ρ the moduli of several consecutive terms are nearly equal at the part of the series where the modulus is a minimum, and for the minimum modulus μ we have very nearly from (24), (23)

$$i = 4\rho^{\frac{3}{2}}, \qquad \mu = (2\pi i)^{-\frac{1}{2}} e^{-i} = (2\pi i)^{-\frac{1}{2}} e^{-4\rho^{\frac{3}{2}}}.$$

If the exponential in the expression for μ be multiplied by the modulus of the exponential in the superior term, the result will be

$$e^{-(4\mp 2\cos\frac{3}{2}\theta)\,\rho^{\frac{3}{2}}},$$

the sign $-$ or $+$ being taken according as $\cos\frac{3}{2}\theta$ is positive or negative. Hence even if the terms of the divergent series were all positive, the superior term would be defined by means of its series within a quantity incomparably smaller, when ρ is indefinitely increased, than the inferior term, except only when $\pm\cos\frac{3}{2}\theta = 1$, and in this case too and this alone are the terms of the divergent series in the superior term regularly positive. In no other case then can the coefficient of the inferior term alter discontinuously, and the coefficient of the other term cannot change so long as that term remains the superior term. Referring for convenience to the figure (Fig. 1), we see that it is only at the points a, b, c, at the middle of the portions of the curve which lie within the circle, that the coefficient belonging to the curve can change.

It might appear at first sight that we could have three distinct coefficients, corresponding respectively to the portions aAb, bBc, cCa of the curve, which would make three distinct constants occurring in the integral of a differential equation of the second order only. This however is not the case; and if we were to assign in the first instance three distinct constants to those three portions of the curve, they would be connected by an equation of condition.

To show this assume the coefficient belonging to the part of the curve about B to be equal to zero. We shall thus get an integral of our equation with only one arbitrary constant. Since there is no superior term from $\theta = -\frac{1}{3}\pi$ to $\theta = +\frac{1}{3}\pi$, the coefficient of the other term cannot change discontinuously at a (*i.e.* when θ passes through the value zero); and by what has been already shown the coefficient must remain unchanged throughout the portion bBc of the curve, and therefore be equal to zero; and again the coefficient must remain unchanged throughout the portion $cCaAb$, and therefore have the same value as at a; but these two portions between them take in the whole curve. The integral at present under consideration is represented by Fig. 2, the coefficient having the same value throughout the portion of the curve there drawn, and being equal to zero for the remainder of the course*.

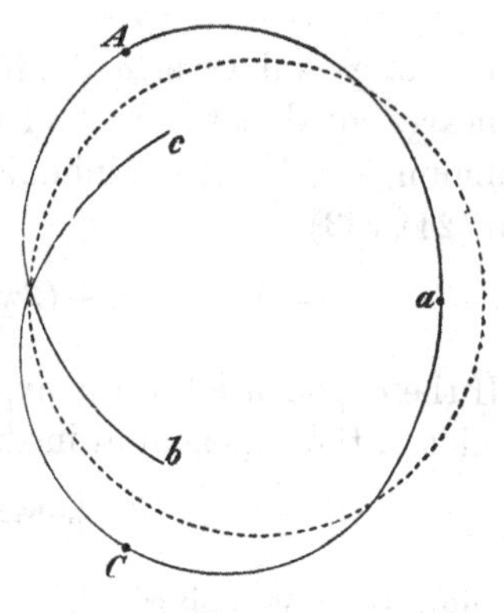

Fig. 2.

The second line on the right-hand side of (20) is what the first becomes when the origin of θ is altered by $\pm \frac{2}{3}\pi$, and the arbitrary constant changed. Hence if we take the term corresponding to the curve represented in Fig. 3, and having a constant coefficient throughout the portion there represented, we shall get another particular integral with one arbitrary constant, and the sum of these two particular integrals will be the complete integral.

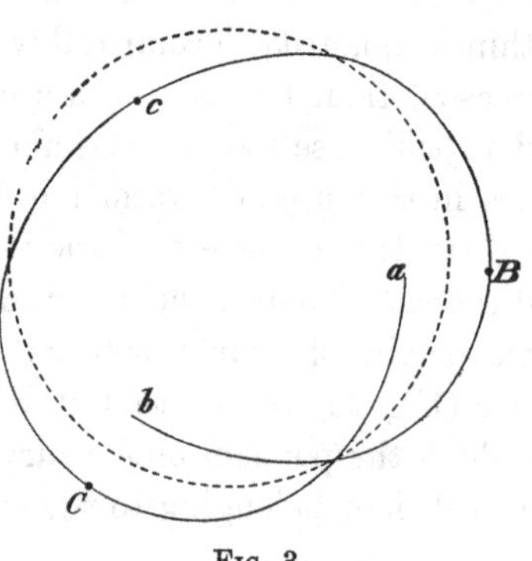

Fig. 3.

In Fig. 3 the uninterrupted interior branch of the curve is made to lie in the interval $\frac{1}{3}\pi$ to π. It would have done equally well to make it lie in the interval $-\frac{1}{3}\pi$ to $-\pi$; we should thus in fact obtain the same complete integral merely somewhat differently expressed.

* A numerical verification of the discontinuity here represented is given as an Appendix to this paper.

The integral (20) may now be conveniently expressed in the following form, in which the discontinuity of the constants is exhibited:

$$\left.\begin{aligned} u = \left(-\frac{4\pi}{3} \text{ to } +\frac{4\pi}{3}\right) C x^{-\frac{1}{4}} e^{2x^{\frac{3}{2}}} \left\{1 - \frac{1.5}{1.144 x^{\frac{3}{2}}} + \frac{1.5.7.11}{1.2.144^2 x^3} - \ldots\right\} \\ + \left(-\frac{2\pi}{3} \text{ to } +2\pi\right) D x^{-\frac{1}{4}} e^{2x^{\frac{3}{2}}} \left\{1 + \frac{1.5}{1.144 x^{\frac{3}{2}}} + \frac{1.5.7.11}{1.2.144^2 x^3} + \ldots\right\} \end{aligned}\right\} \quad \ldots\ldots\ldots(25).$$

In this equation the expression $(-\frac{4}{3}\pi \text{ to } +\frac{4}{3}\pi)$ denotes that the function written after it is to be taken whenever an angle in the indefinite series

$$\ldots\, \theta - 4\pi, \quad \theta - 2\pi, \quad \theta, \quad \theta + 2\pi, \quad \theta + 4\pi, \ldots$$

falls within the specified limits, which will be either once or twice according to the value of θ.

17. If we put $D = 0$ in (25), the resulting value of u will be equal to Mr Airy's integral, multiplied by an arbitrary constant, x being equal to $-\left(\frac{\pi}{2}\right)^{\frac{2}{3}} \frac{m}{3}$. When $\theta = 0$ we have the integral belonging to the dark side of the caustic, when $\theta = \pi$ that belonging to the bright side. We easily see from (25), or by referring to Fig. 2, in what way to pass from one of these integrals to the other, the integrals being supposed to be expressed by means of the divergent series. If we have got the analytical expression belonging to the dark side we must add $+\pi$, $-\pi$ in succession to the amplitude of x, and take the sum of the results. If we have got the analytical expression belonging to the bright side, we must alter the amplitude of x by π, and reject the superior function in the resulting expression. It is shown in Art. 9 of my paper "On the Numerical Calculation, etc." that the latter process leads to a correct result, but I was unable then to give a demonstration. This desideratum is now supplied.

18. It now only remains to connect the constants A, B with C, D in the two different forms (19) and (25) of the integral of (18). This may be done by means of the complete integral of (18) expressed in the form of definite integrals.

Let $$v = \int_0^\infty e^{-\lambda^3 - cx\lambda} d\lambda,$$

then $$\frac{d^2v}{dx^2} = \frac{c^2}{3}\int_0^\infty e^{-\lambda^3 - cx\lambda}\{(3\lambda^2 + cx) - cx\}\, d\lambda$$

$$= \frac{c^2}{3} - \frac{c^3}{3}xv;$$

whence

$$\frac{d^2 \, . \, cv}{dx^2} + \frac{c^3}{3}x \, . \, cv = \frac{c^3}{3} \quad \text{.................(26).}$$

In order to make the left-hand member of this equation agree with (18), we must have $c^3 = -27$, and therefore

$$c = -3, \text{ or } 3\alpha, \text{ or } 3\beta,$$

α, β being the imaginary cube roots of -1, of which α will be supposed equal to

$$\cos \tfrac{1}{3}\pi + \sqrt{-1}\sin \tfrac{1}{3}\pi.$$

Whichever value of c be taken, the right-hand member of equation (26) will be equal to -9, and therefore will disappear on taking the difference of any two functions cv corresponding to two different values of c. This difference multiplied by an arbitrary constant will be an integral of (18), and accordingly we shall have for the complete integral

$$u = E\int_0^\infty e^{-\lambda^3}(e^{3x\lambda} + \alpha e^{-3\alpha x\lambda})\, d\lambda + F\int_0^\infty e^{-\lambda^3}(e^{3x\lambda} + \beta e^{-3\beta x\lambda})\, d\lambda \text{...(27).}$$

That this expression is in fact equivalent to (19) might be verified by expanding the exponentials within parentheses, and integrating term by term.

To find the relations between E, F and A, B, it will be sufficient to expand as far as the first power of x, and equate the results. We thus get

$$A + Bx = \int_0^\infty e^{-\lambda}\{(1+\alpha)E + (1+\beta)F$$
$$+ 3[(1-\alpha^2)E + (1-\beta^2)F]x\lambda\}\, d\lambda$$

which gives, since

$$\alpha^2 = -\beta,\ \beta^2 = -\alpha,\ \int_0^\infty e^{-\lambda^3}d\lambda = \tfrac{1}{3}\Gamma(\tfrac{1}{3}),\ \int_0^\infty e^{-\lambda^3}\lambda d\lambda = \tfrac{1}{3}\Gamma(\tfrac{2}{3}),$$

$$\left.\begin{aligned} A &= \tfrac{1}{3}\Gamma(\tfrac{1}{3})\{(1+\alpha)E + (1+\beta)F\} \\ B &= \Gamma(\tfrac{2}{3})\{(1+\beta)E + (1+\alpha)F\} \end{aligned}\right\} \quad \text{..............(28).}$$

19. We have now to find the relations between E, F and C, D, for which purpose we must compare the expressions (25), (27), supposing x indefinitely large.

In order that the exponentials in (25), may be as large as possible, we must have $\theta = \frac{2}{3}\pi$ in the term multiplied by C, and $\theta = 0$ in the term multiplied by D. We have therefore for the leading term of u

$$Ce^{-\frac{\pi}{2}\sqrt{-1}}\rho^{-\frac{1}{4}}e^{2\rho^{\frac{3}{2}}}, \text{ when } \theta = \frac{2\pi}{3};$$

$$D\rho^{-\frac{1}{4}}e^{2\rho^{\frac{3}{2}}}, \qquad \text{when } \theta = 0.$$

Let us now seek the leading term of u from the expression (27), taking first the case in which $\theta = 0$. It is evident that this must arise from the part of the integral which involves $e^{3x\lambda}$ or in this case $e^{3\rho\lambda}$, which is

$$(E+F)\int_0^\infty e^{-\lambda^3+3\rho\lambda}d\lambda.$$

Now $3\rho\lambda - \lambda^3$ is a maximum for $\lambda = \rho^{\frac{1}{2}}$. Let $\lambda = \rho^{\frac{1}{2}} + \zeta$; then

$$3\rho\lambda - \lambda^3 = 2\rho^{\frac{3}{2}} - 3\rho^{\frac{1}{2}}\zeta^2 - \zeta^3,$$

and our integral becomes

$$e^{2\rho^{\frac{3}{2}}}\int_{-\rho^{\frac{1}{2}}}^{\infty} e^{-3\rho^{\frac{1}{2}}\zeta^2-\zeta^3}d\zeta.$$

Put $\zeta = 3^{-\frac{1}{2}}\rho^{-\frac{1}{4}}\xi$; then the integral becomes

$$3^{-\frac{1}{2}}\rho^{-\frac{1}{4}}e^{2\rho^{\frac{3}{2}}}\int_{-3^{\frac{1}{2}}\rho^{\frac{3}{4}}}^{\infty} e^{-\xi^2-3^{-\frac{3}{2}}\rho^{-\frac{3}{4}}\xi^3}d\xi.$$

Let now ρ become infinite; then the last integral becomes $\int_{-\infty}^{\infty} e^{-\xi^2}d\xi$ or $\pi^{\frac{1}{2}}$. For though the index $-\xi^2 - 3^{-\frac{3}{2}}\rho^{-\frac{3}{4}}\xi^3$ becomes positive for a sufficient large negative value of ξ, that value lies far beyond the limits of integration, within which in fact the index continually decreases with ξ, having at the inferior limit the value $-2\rho^{\frac{3}{2}}$. Hence then for $\theta = 0$, and for very large values of ρ, we have ultimately

$$u = 3^{-\frac{1}{2}}\pi^{\frac{1}{2}}(E+F)\rho^{-\frac{1}{4}}e^{2\rho^{\frac{3}{2}}}.$$

Next let $\theta = \frac{2}{3}\pi$. In this case $\alpha x = -\rho$, and we get for the leading part of u

$$\alpha E \int_0^\infty e^{-\lambda^3 + 3\rho\lambda} d\lambda,$$

which when ρ is very large becomes, as before,

$$3^{-\frac{1}{2}}\pi^{\frac{1}{2}}\alpha E \rho^{-\frac{1}{4}} e^{2\rho^{\frac{3}{2}}}.$$

Comparing the leading terms of u both for $\theta = \frac{2}{3}\pi$ and for $\theta = 0$, we find, observing that $\alpha = e^{\frac{\pi}{3}\sqrt{-1}}$

$$\left.\begin{aligned} C &= \sqrt{-1}\, 3^{-\frac{1}{2}}\pi^{\frac{1}{2}} E \\ D &= 3^{-\frac{1}{2}}\pi^{\frac{1}{2}} (E + F) \end{aligned}\right\} \dots\dots\dots\dots\dots\dots\dots (29).$$

Eliminating E, F between (28) and (29) we have finally

$$\left.\begin{aligned} A &= \pi^{-\frac{1}{2}}\Gamma(\tfrac{1}{3})\{\ C + e^{-\frac{\pi}{6}\sqrt{-1}}\ D\} \\ B &= 3\pi^{-\frac{1}{2}}\Gamma(\tfrac{2}{3})\{-C + e^{\frac{\pi}{6}\sqrt{-1}}\ D\} \end{aligned}\right\} \dots\dots\dots (30).$$

20. As a last example of the principles of this paper, let us take the differential equation

$$\frac{d^2u}{dx^2} + \frac{1}{x}\frac{du}{dx} - u = 0 \dots\dots\dots\dots\dots\dots\dots (31).$$

The complete integral of this equation in series according to ascending powers of x involves a logarithm. If the arbitrary constant multiplying the logarithm be equated to zero we shall obtain an integral with only one arbitrary constant. This integral, or rather what it becomes when $\sqrt{-1}x$ is written for x, occurs in many physical investigations, for example the problem of annular waves in shallow water, and that of diffraction in the case of a circular disk. I had occasion to employ the integral with a logarithm in determining the motion of a fluid about a long cylindrical rod oscillating as a pendulum, the internal friction of the fluid itself being taken into account*. In that paper the integral of (31) both in ascending and in descending series was employed, but the discussion of the equation was not quite completed, one of the arbitrary constants being left undetermined. A knowledge of the value of this constant was not required for

* *Camb. Phil. Trans.* Vol. IX, Part II, [1850], p. [38]. [*Ante*, Vol. III, p. 38.]

determining the resultant force of the fluid on the pendulum, which was the great object of the investigation, but would have been required for determining the motion of the fluid at a great distance from the pendulum.

21. The three forms of the integral of (31) which we shall require are given in Arts. 28 and 29 of my paper on pendulums. The complete integral according to ascending series is

$$\left.\begin{array}{l} u=(A+B\log x)\left(1+\dfrac{x^3}{2^2}+\dfrac{x^4}{2^2.4^2}+\dfrac{x^6}{2^2.4^2.6^2}+\dots\right) \\ \qquad -B\left(\dfrac{x^2}{2^2}S_1+\dfrac{x^4}{2^2.4^2}S_2+\dfrac{x^6}{2^2.4^2.6^2}S_3+\dots\right) \end{array}\right\}\dots(32)$$

where $$S_i=1^{-1}+2^{-1}+3^{-1}\dots+i^{-1}.$$

The series contained in this equation are convergent for all real or imaginary values of x, but the value of u determined by the equation is not unique, inasmuch as $\log x$ has an infinite number of values. To pass from one of these to another comes to the same thing as changing the constant A by some multiple of $2\pi B\sqrt{-1}$. If ρ, θ, the modulus and amplitude of x, be supposed to be polar co-ordinates, and the expression (32) be made to vary continuously by giving continuous variations to ρ and θ without allowing the former to vanish, the value of $\log x$ will increase by $2\pi\sqrt{-1}$ in passing from any point in the positive direction once round the origin so as to arrive at the starting point again. In order to render everything definite we must specify the value of the logarithm which is supposed to be taken.

The complete integral of (31) expressed by means of descending series is

$$\left.\begin{array}{l} u=Cx^{-\frac{1}{2}}e^{-x}\left\{1-\dfrac{1^2}{2.4x}+\dfrac{1^2.3^2}{2.4(4x)^2}-\dfrac{1^2.3^2.5^2}{2.4.6(4x)^3}+\dots\right\} \\ \quad +Dx^{-\frac{1}{2}}e^{x}\left\{1+\dfrac{1^2}{2.4x}+\dfrac{1^2.3^2}{2.4(4x)^2}+\dfrac{1^2.3^2.5^2}{2.4.6(4x)^3}+\dots\right\} \end{array}\right\}\dots(33).$$

These series are ultimately divergent, and the constants C, D are discontinuous. It may be shown precisely as before that the values of θ for which the constants are discontinuous are

$$\dots-4\pi,\quad -2\pi,\quad 0,\quad 2\pi,\quad 4\pi\dots \text{ for } C,$$
$$\dots-3\pi,\quad -\pi,\quad \pi,\quad 3\pi,\dots \quad \text{ for } D.$$

Hence the equation (33) may be written, according to the notation employed in Art. 16, as follows:

$$u = (0 \text{ to } 2\pi)\, Cx^{-\frac{1}{2}}e^{-x}\left(1 - \frac{1^2}{2.4x} + \ldots\right)$$
$$+ (-\pi \text{ to } +\pi)\, Dx^{-\frac{1}{2}}e^{x}\left(1 + \frac{1^2}{2.4x} + \ldots\right) \ldots\ldots(34).$$

22. It remains to connect A, B with C, D. For this purpose we shall require the third form of the integral of (31), namely

$$u = \int_0^{\frac{\pi}{2}} \{E + F \log (x \sin^2 \omega)\} (e^{x \cos \omega} + e^{-x \cos \omega})\, d\omega \ldots (35).$$

As to the value of $\log x$ to be taken, it will suffice for the present to assume that whatever value is employed in (32), the same shall be employed also in (35).

To connect A, B with E, F, it will suffice to compare (32) and (35), expanding the exponentials, and rejecting all powers of x. We have

$$A + B \log x = 2 \int_0^{\frac{\pi}{2}} \{E + F \log (x \sin^2 \omega)\}\, d\omega$$
$$= \pi (E + F \log x) + 2\pi \log (\tfrac{1}{2}) . F;$$

whence

$$\left.\begin{aligned} A &= \pi E - 2\pi \log 2 . F \\ B &= \pi F \end{aligned}\right\} \ldots\ldots\ldots\ldots\ldots\ldots(36).$$

To connect C, D with E, F, we must seek the ultimate value of u when ρ is infinitely increased. It will be convenient to assume in succession $\theta = 0$ and $\theta = \pi$. We have ultimately from (34)

$$u = D\rho^{-\frac{1}{2}}e^{\rho} \text{ when } \theta = 0; \quad u = -\sqrt{-1}\, C\rho^{-\frac{1}{2}}e^{\rho} \text{ when } \theta = \pi \ldots(37).$$

It will be necessary now to specify what value of $\log x$ we suppose taken in (35). Let it be $\log \rho + \sqrt{-1}\,\theta$, θ being supposed reduced within the limits 0 and 2π by adding or subtracting if need be $2i\pi$, where i is an integer.

The limiting value of u for $\theta = 0$ from (35) may be found as in Art. 29 of my paper on pendulums, above referred to. In fact, the reasoning of that Article will apply if the imaginary quantity there denoted by m be replaced by unity. The constants

$$C,\ D,\ C',\ D',\ C'',\ D'',$$

of the former paper correspond to

$$A,\ B,\ C,\ D,\ E,\ F,$$

of the present. Hence we have for the ultimate value of u for $\theta = 0$

$$u = \left(\frac{\pi}{2\rho}\right)^{\frac{1}{2}} e^{\rho} \{E + (\pi^{-\frac{1}{2}}\Gamma'(\tfrac{1}{2}) + \log 2) F\} \ldots\ldots\ldots(38).$$

For $\theta = \pi$, (35) becomes

$$u = \int_0^{\frac{\pi}{2}} \{E + \pi F \sqrt{-1} + F \log(\rho \sin^2 \omega)\} (e^{-\rho\cos\omega} + e^{\rho\cos\omega})\, d\omega\,;$$

and to find the ultimate value of u we have merely to write $E + \pi F\sqrt{-1}$ for E in the above, which gives ultimately for $\theta = \pi$

$$u = \left(\frac{\pi}{2\rho}\right)^{\frac{1}{2}} e^{\rho} [E + \pi F \sqrt{-1} + \{\pi^{-\frac{1}{2}}\Gamma'(\tfrac{1}{2}) + \log 2\} F] \ldots(39).$$

Comparing the equations (38), (39) with (37), we get

$$\left.\begin{aligned} C &= \left(\frac{\pi}{2}\right)^{\frac{1}{2}} [E\sqrt{-1} - \pi F + \{\pi^{-\frac{1}{2}}\Gamma'(\tfrac{1}{2}) + \log 2\}\sqrt{-1}F] \\ D &= \left(\frac{\pi}{2}\right)^{\frac{1}{2}} [E + \{\pi^{-\frac{1}{2}}\Gamma'(\tfrac{1}{2}) + \log 2\} F] \end{aligned}\right\} \ldots(40).$$

Eliminating E, F between (36) and (40), we get finally

$$\left.\begin{aligned} C &= (2\pi)^{-\frac{1}{2}} [\sqrt{-1}A + \{(\pi^{-\frac{1}{2}}\Gamma'(\tfrac{1}{2}) + \log 8)\sqrt{-1} - \pi\} B] \\ D &= (2\pi)^{-\frac{1}{2}} [A + \{\pi^{-\frac{1}{2}}\Gamma'(\tfrac{1}{2}) + \log 8\} B] \end{aligned}\right\} \ldots(41).$$

Conclusion.

23. It has been shown in the foregoing paper,

First, That when functions expressible in convergent series according to ascending powers of the variable are transformed so as to be expressed by exponentials multiplied by series according to descending powers, applicable to the calculation of the functions for large values of the variable, and ultimately divergent, though at first rapidly convergent, the series contain in general discontinuous constants, which change abruptly as the amplitude of the imaginary variable passes through certain values.

Secondly, That the liability to discontinuity in one of the constants is pointed out by the circumstance, that for a particular value of the amplitude of the variable, all the terms of an associated divergent series become regularly positive.

Thirdly, That a divergent series with all its terms regularly positive is in many cases a sort of indeterminate form, in passing through which a discontinuity takes place.

Fourthly, That when the function may be expressed by means of a definite integral, the constants in the ascending and descending series may usually be connected by one uniform process. The comparison of the leading terms of the ascending series with the integral presents no difficulty. The comparison of the leading terms of the descending series with the integral may usually be effected by assigning to the amplitude of the variable such a value, or such values in succession, as shall render the real part of the index of the exponential a maximum, and then seeking what the integral becomes when the modulus of the variable increases indefinitely. The leading term obtained from the integral will be found within a range of integration comprising the maximum value of the real part of the index of the exponential under the integral sign, and extending between limits which may be supposed to become indefinitely close after the modulus of the original variable has been made indefinitely great, whereby the integral will be reduced to one of a simpler form. Should a definite integral capable of expressing the function not be discovered, the relations between the constants in the ascending and descending series may still be obtained numerically by calculating from the ascending and descending series separately and equating the results*.

APPENDIX.

[*Added since the reading of the Paper.*]

On account of the strange appearance of figures 2 and 3, the reader may be pleased to see a numerical verification of the discontinuity which has been shown to exist in the values of the arbitrary constants. I subjoin therefore the numerical calculation of the integral to which Fig. 2 relates, for two values of x, from the ascending and descending series separately. For this integral $D=0$, and I will take $C=1$, which gives (equations 30),

$$A=\pi^{-\frac{1}{2}}\Gamma\left(\tfrac{1}{3}\right);\quad B=-3\pi^{-\frac{1}{2}}\Gamma\left(\tfrac{2}{3}\right);$$

and $$\log A=0\cdot1793878;\ \log(-B)=0\cdot3602028.$$

[* Cf. also *Proc. Camb. Phil. Soc.* VI, 1889, pp. 362–6, reprinted *infra*.]

The two values of x chosen for calculation have 2 for their common modulus, and 90°, 150°, respectively, for their amplitudes, so that the corresponding radii in Fig. 2 are situated at 30° on each side of the radius passing through the point of discontinuity c. The terms of the descending series are calculated to 7 places of decimals. As the modulus of the result has afterwards to be multiplied by a number exceeding 40, it is needless to retain more than 6 decimal places in the ascending series. In the multiplications required after summation, 7-figure logarithms were employed. The results are given to 7 significant figures, that is, to 5 places of decimals.

The following is the calculation by ascending series for the amplitude 90° of x. By the first and second series are meant respectively those which have A, B for their coefficients in equation (19).

	First Series.		*Second Series.*	
Order of term	Real part	Coefficient of $\sqrt{-1}$	Real part	Coefficient of $\sqrt{-1}$
0	+ 1·000000			+ 2·000000
1		−12·000000	+12·000000	
2	−28·800000			−20·571429
3		+28·800000	−16·457143	
4	+15·709091			+ 7·595605
5		− 5·385974	+ 2·278681	
6	− 1·267288			− 0·479722
7		+ 0·217249	− 0·074762	
8	+ 0·028337			+ 0·008971
9		− 0·002906	+ 0·000855	
10	− 0·000240			− 0·000066
11		+ 0·000016	− 0·000004	
12	+ 0·000001			
Sum	−13·330099	$+11·628385\sqrt{-1}$	− 2·252373	$-11·446641\sqrt{-1}$

Sum multiplied

by A, $-20·14750 + 17·57548\sqrt{-1}$; by B, $+ 5·16230 + 26·23499\sqrt{-1}$.

When the amplitude of x becomes 150° in place of 90°, the amplitude of x^3 is increased by 180°. Hence in the first series it will be sufficient to change the sign of the imaginary part. To see what the second series becomes, imagine for a moment the factor x put outside as a coefficient. In the reduced series it would be sufficient to change the sign of the imaginary part; and to correct for the change in the factor x it would be sufficient to multiply by

$\cos 60° + \sqrt{-1} \sin 60°$. But since the amplitude of x was at first 90°, the real and imaginary parts of the series calculated correspond respectively to the imaginary and real parts of the reduced series. Hence it will be sufficient to change the sign of the real part in the product of the sum of the second series by B, and multiply by $\frac{1}{2}(1 + \sqrt{3}\sqrt{-1})$, which gives the result

$$-25{\cdot}30132 + 8{\cdot}64681\sqrt{-1}.$$

Hence we have for the result obtained from the ascending series:

	for amp. $x = 90°$,	for amp. $x = 150°$,
From first series	$-20{\cdot}14750 + 17{\cdot}57548\sqrt{-1}$	$-20{\cdot}14750 - 17{\cdot}57548\sqrt{-1}$
From second series	$+\ 5{\cdot}16230 + 26{\cdot}23499\sqrt{-1}$	$-25{\cdot}30132 +\ 8{\cdot}64681\sqrt{-1}$
Total	$-14{\cdot}98520 + 43{\cdot}81047\sqrt{-1}$	$-45{\cdot}44882 -\ 8{\cdot}92867\sqrt{-1}$

On account of the particular values of amp. x chosen for calculation, the terms in the ascending series were either wholly real or wholly imaginary. In the case of the descending series this is only true of every second term, and therefore the values of the moduli are subjoined in order to exhibit their progress. The following is the calculation for amp. $x = 90°$, in which case there is no inferior term.

Order	Modulus	Real part	Coefficient of $\sqrt{-1}$
0	1·0000000	+1·0000000	
1	0·0122762	+0·0086806	+0·0086806
2	0·0011604		+0·0011604
3	0·0002099	−0·0001484	+0·0001484
4	0·0000563	−0·0000563	
5	0·0000200	−0·0000142	−0·0000142
6	0·0000089		−0·0000089
7	0·0000047	+0·0000033	−0·0000033
8	0·0000029	+0·0000029	
9	0·0000021	+0·0000015	+0·0000015
10	0·0000017		+0·0000017
11	0·0000015	−0·0000010	+0·0000010
Remainder		−0·0000007	−0·0000017
Sum		+1·0084677	$+0{\cdot}0099655\sqrt{-1}$.

The modulus of the term of the order 12 is 14 in the seventh place, and is the least of the moduli. Those of the succeeding terms are got by multiplying the above by the factors 1·0616, 1·2208, 1·5116, 2·0053, etc., and the successive differences of the series of factors headed by unity are

$$\Delta^1 = +0{\cdot}0616,\ \ \Delta^2 = +0{\cdot}0976,\ \ \Delta^3 = +0{\cdot}0340,\ \ \Delta^4 = +0{\cdot}0373, \text{ etc.}$$

These differences when multiplied by 14 are so small that in the application of the transformation of Art. 8, for which in the present case $q = 1$, the differences may be neglected, and the series there given reduced to its first term. It is thus that the remainder given above was calculated.

The sum of the series is now to be reduced to the form $\rho(\cos\theta + \sqrt{-1}\sin\theta)$, and thus multiplied by $e^{-2x^{\frac{3}{2}}}$ and by $x^{-\frac{1}{4}}$. We have

for series	log. mod. = 0·0036832	amp. = + 0° 33′ 58″·21
for exponential	log. mod. = 1·7371779	amp. = + 130° 49′ 0″·78
for $x^{-\frac{1}{4}}$	log. mod. = $\bar{1}$·9247425	amp. = − 22° 30′
	1·6656036	+ 108° 52′ 58″·99

When the amplitude of x is 150°, there are both superior and inferior terms in the expression of the function by means of descending series. It will be most convenient, as has been explained, to put in succession, in the function multiplied by C in equation (20), amp. $x = 150°$ and amp. $x = -210°$, and to take the sum of the results. The first will give the superior, the second the inferior term.

For the amplitudes 90°, 150° of x, or more generally for any two amplitudes equidistant from 120°, the amplitudes of $x^{\frac{3}{2}}$ will be equidistant from 180°, so that for any rational and real function of $x^{\frac{3}{2}}$ we may pass from the result in the one case to the result in the other by simply changing the sign of $\sqrt{-1}$, or, which comes to the same, changing the sign of the amplitude of the result. The series and the exponential are both such functions, and for the factor $x^{-\frac{1}{4}}$ we have simply to replace the amplitude $-22° 30'$ by $-37° 30'$. Hence we have for the superior term

$$\text{log. mod.} = 1{\cdot}6656036\,;\ \ \text{amp.} = -168°\ 52'\ 58''{\cdot}99.$$

When amp. x is changed from 150° to $-210°$, amp. $x^{\frac{3}{2}}$ is altered by $3 \times 180°$, and therefore the sign of $x^{\frac{3}{2}}$ is changed. Hence the log. mod. of the exponential is less than it was by $2 \times 1·737...$ or by more than 3. Hence 4 decimal places will be sufficient in calculating the series, and 4-figure logarithms may be employed in the multiplications. The terms of the series will be obtained from those already calculated by changing first the signs of the imaginary parts, and secondly the sign of every second term, or, which comes to the same, by changing the signs of the real parts in the terms of the orders 1, 3, 5 ..., and of the imaginary parts in the terms of the orders 0, 2, 4 ... Hence we have

Real part	Coefficient of $\sqrt{-1}$
+1·0000	
−0·0087	+0·0087
	−0·0012
+0·0001	+0·0001
+0·9914	+0·0076 $\sqrt{-1}$.

log. mod. $= \bar{1}·9963$; amp. $= +26'·5$.

Hence we have altogether for the inferior term,

$$\text{log. mod.} = \bar{2}·1838; \text{ amp.} = +183° \, 45'·5.$$

Hence reducing each imaginary result from the form $\rho(\cos\theta + \sqrt{-1}\sin\theta)$ to the form $a + \sqrt{-1}\,b$, we have for the final result, obtained from the descending series:

	For amp. $x = 90°$.	For amp. $x = 150°$.
From superior term	$-14·98520 + 43·81046\sqrt{-1}$;	$-45·43360 - 8·92767\sqrt{-1}$
From inferior term		$-\ \ 0·01524 - 0·00100\sqrt{-1}$
		$-45·44884 - 8·92867\sqrt{-1}$

Had the asserted discontinuity in the value of the arbitrary constant not existed, either the inferior term would have been present for amp. $x = 90°$, or it would have been absent for amp. $x = 150°$, and we see that one or other of the two results would have been wrong in the second place of decimals.

In considering the relative difficulty of the calculation by the ascending and descending series, it must be remembered that the blanks only occur in consequence of the special values of the amplitude of x chosen for calculation: for general values they would have been all filled up by figures. Hence even for so low a value of the modulus of x as 2 the descending series have a decided advantage over the ascending.

On the Effect of Wind on the Intensity of Sound.

[From the *Report of the British Association*, Dublin, 1857, p. 22.]

The remarkable diminution in the intensity of sound, which is produced when a strong wind blows in a direction from the observer towards the source of sound, is familiar to everybody, but has not hitherto been explained, so far as the author is aware. At first sight we might be disposed to attribute it merely to the increase in the radius of the sound-wave which reaches the observer. The whole mass of air being supposed to be carried uniformly along, the time which the sound would take to reach the observer, and consequently the radius of the sound-wave, would be increased by the wind in the ratio of the velocity of sound to the sum of the velocities of sound and of the wind, and the intensity would be diminished in the inverse duplicate ratio. But the effect is much too great to be attributable to this cause. It would be a strong wind, whose velocity was a twenty-fourth part of that of sound; yet even in this case the intensity would be diminished by only about a twelfth part. The first volume of the *Annales de Chimie* (1816) contains a paper by M. Delaroche, giving the results of some experiments made on this subject. It appeared from the experiments,—first, that at small distances the wind has hardly any perceptible effect, the sound being propagated almost equally well in a direction contrary to the wind and in the direction of the wind; secondly, that the disparity between the intensity of the sound propagated in these two directions becomes proportionately greater and greater as the distance increases; thirdly, that sound is propagated rather better in a direction perpendicular to the wind than even in the direction of the wind. The explanation offered by the author of the present communication is as follows*. If we imagine the whole mass of air in the neighbourhood of the source of disturbance divided into horizontal strata, these strata do not all move with the same velocity. The lower

[* Cf. Osborne Reynolds, *Roy. Soc. Proc.* XXII, 1874, p. 531; Lord Rayleigh, *Theory of Sound*, 1878, Vol. II, §§ 289–290.]

strata are retarded by friction against the earth, and by the various obstacles they meet with; the upper by friction against the lower, and so on. Hence the velocity increases from the ground upwards, conformably with observation. This difference of velocity disturbs the spherical form of the sound-wave, tending to make it somewhat of the form of an ellipsoid, the section of which by a vertical diametral plane parallel to the direction of the wind is an ellipse meeting the ground at an obtuse angle on the side towards which the wind is blowing, and an acute angle on the opposite side. Now, sound tends to propagate itself in a direction perpendicular to the sound-wave; and if a portion of the wave is intercepted by an obstacle of large size, the space behind is left in a sort of sound-shadow, and the only sound there heard is what diverges from the general wave after passing the obstacle. Hence, near the earth, in a direction contrary to the wind, the sound continually tends to be propagated upwards, and consequently there is a continual tendency for an observer in that direction to be left in a sort of sound-shadow. Hence, at a sufficient distance, the sound ought to be very much enfeebled; but near the source of disturbance this cause has not yet had time to operate, and therefore the wind produces no sensible effect, except what arises from the augmentation in the radius of the sound-wave, and this is too small to be perceptible. In the contrary direction, that is, in the direction towards which the wind is blowing, the sound tends to propagate itself downwards, and to be reflected from the surface of the earth; and both the direct and reflected waves contribute to the effect perceived. The two waves assist each other so much the better, as the angle between them is less, and this angle vanishes in a direction perpendicular to the wind. Hence, in the latter direction the sound ought to be propagated a little better than even in the direction of the wind, which agrees with the experiments of M. Delaroche. Thus the effect is referred to two known causes,—the increased velocity of the air in ascending, and the diffraction of sound.

On the Existence of a Second Crystallizable Fluorescent Substance (Paviin) in the Bark of the Horse-Chestnut.

[From the *Quarterly Journal of the Chemical Society*, xi, 1859, pp. 17–21: also *Pogg. Ann.* cxiv, 1861, pp. 646–51.]

On examining, a good while ago, infusions of the barks of various species of Æsculus, and the closely allied genus Pavia, I found that the remarkably strong fluorescence shown by the horse-chestnut ran through the whole family. The tint of the fluorescent light was, however, different in different cases, being as a general rule blue throughout the genus Æsculus, and a blue-green throughout Pavia. This alone rendered it evident, either that there were at least two fluorescent substances present, one in one bark and another in another, or, which appeared more probable, that there were two (or possibly more) fluorescent substances present in different proportions in different barks.

On examining, under a deep violet glass, a freshly cut section of a young shoot, of at least two years' growth, of these various trees, the sap which oozed out from different parts of the bark or pith was found to emit a differently coloured fluorescent light. Hence, even the same bark must have contained more than one fluorescent substance; and as the existence of two would account for the fluorescent tints of the whole family, a family so closely allied botanically, the second of the suppositions mentioned above appeared by far the more probable.

I happened to put some small pieces of horse-chestnut bark with a little ether into a bottle, which was laid aside, imperfectly corked. On examining the bottle after some time, the ether was found to have evaporated, and had left behind a substance crystallized in delicate radiating crystals. This substance, which I will call paviin, when dissolved in water, yields, like æsculin, a highly fluorescent solution, and the fluorescence is in both cases destroyed (comparatively speaking) by acids, and restored by alkalies. The tint, however, of the fluorescent light is decidedly different from that given by pure æsculin, for a specimen of which I am indebted to the kindness of the Prince of Salm-Horstmar, being a blue-green in place of a sky-blue. The fluorescent tint of an infusion of horse-chestnut bark is intermediate between the two, but much nearer to æsculin than to paviin.

In all probability, the fluorescence of the infusions of barks from the closely allied genera Æsculus and Pavia, is due to æsculin and paviin present in different proportions, æsculin predominating generally in the genus Æsculus, and paviin in Pavia.

Æsculin and paviin are extremely similar in their properties, so far as they have yet been observed. They are most easily distinguished by the different colour of the fluorescent light of their solutions, a character which is especially trustworthy, as it does not require for its observation that the solutions should be pure. Paviin, as appears from the way in which it was first obtained, must be much more soluble than æsculin in ether. Æsculin is indeed described as insoluble in ether, but it is sufficiently soluble to render the ether fluorescent. Paviin, like æsculin, is withdrawn from its ethereal solution by agitation with water. Though of feeble affinities, it is rather more disposed than æsculin to combine with oxide of lead. If a decoction of horse-chestnut bark be purified by adding a sufficient quantity of a salt of peroxide of iron or of alumina, precipitating by ammonia, and filtering, and the ammoniacal filtrate be partially precipitated by very dilute acetate of lead, the whole redissolved by acetic acid, reprecipitated by ammonia, and filtered, the fluorescent tint of the filtrate will be found to be a deeper blue than that of the original solution; while, if the fluorescent substances combined with oxide of lead (the compound itself is not fluorescent) be again obtained in alkaline solution, the tint, as compared with the original, will be found to verge towards green. The required solution is most easily obtained from the lead-compounds by means of an alkaline bicarbonate, which plays the double part of an acid and an alkali, yielding carbonic acid to the oxide of lead, and ensuring the alkalinity of the filtrate from carbonate of lead. It is very easy in this way, by repeating the process, if necessary, on the filtrate from the first precipitate, to obtain a solution which will serve as a standard for the fluorescent tint of pure æsculin. A solution, serving nearly enough as a standard of comparison in this respect for pure paviin, may be had by making a decoction of a little ash bark, adding a considerable quantity of a salt of alumina, precipitating by ammonia, and filtering. By partial precipitation in the manner explained, it is very easy to prove a mixture of æsculin and paviin to be a mixture, even when operating on extremely small quantities.

It must be carefully borne in mind, that the characteristic fluorescent tint of a solution is that of the fluorescent light coming from the solution *directly* to the eye. Even should a solution of the pure substance be nearly colourless by transmitted light, though strong enough to develope the fluorescence to perfection, if the solution be impure· it is liable to be coloured, most commonly yellow of some kind, which would make a blue seen through it appear green. To depend upon the fluorescent tint, as seen through and modified by a coloured solution, would be like depending on the analysis, not of the substance to be investigated, but of a mixture containing it. Yet in solutions obtained from the horse-chestnut, and in similar cases, the true fluorescent tint can be observed very well, in spite of considerable colour in the solution.

The best method of observing the true fluorescent tint is to dilute the fluid greatly, and to pass into it a beam of sunlight, condensed by a lens fixed in a board, in such a manner that as small a thickness of the fluid as may be shall intervene between the fluorescent beam and the eye. If a stratum of this thickness of the dilute solution be sensibly colourless, the tint of the fluorescent light will not be sensibly modified by subsequent absorption. This, however, requires sunlight, which is not always to be had. Another excellent method, requiring only daylight, and capable of practically superseding the former in the examination of horse-chestnut bark, is the following, in using which it is best that the solutions should be pretty strong, or at least not extremely dilute.

A glass vessel with water is placed at a window, the vessel being blackened internally at the bottom by sinking a piece of black cloth or velvet in the water, or otherwise. The solutions to be compared as to their fluorescent tint are placed in two test-tubes, which are held nearly vertically in the water, their tops slightly inclining from the window, and the observer regards the fluorescent light from above, looking outside the test-tubes. Since by far the greater part of the fluorescent light comes from a very thin stratum of fluid next the surface by which the light enters, the fluorescent rays have mostly to traverse only a very small thickness of the coloured fluid before reaching the eye; the water permits the escape of those fluorescent rays which would otherwise be internally reflected at the external surface of the test tubes; and the intensity of the light of which the tint is to be observed is increased by foreshortening. The observer would do well to practise with a fluorescent fluid purposely made yellow by introducing some non-fluorescent indifferent substance; thus, a portion of the standard solution of æsculin mentioned above may be rendered yellow by ferrid-cyanide of potassium. The more completely the fluorescent tints of the yellow and the nearly colourless solution agree, the more nearly perfect is the method of observation. If ferro-cyanide of potassium be used in the experiment suggested, instead of ferrid-cyanide, the most marked effect is a diminution in the intensity of the fluorescent light, the cause of which is that the absorption by this salt takes place more upon the active, or fluorogenic, than upon the fluorescent rays. Since substances of a similar character may be present in an impure solution, the observer must not always infer poverty with regard to fluorescent substances from a want of brilliancy in the fluorescent light.

The existence of paviin may perhaps account for the discrepancies between the analyses of æsculin given by different chemists. I should mention, however, that I have met with three specimens of æsculin, and they all appeared to be free from paviin. The reason why æsculin was obtained pure from a decoction containing paviin also, is probably that the former greatly preponderates over the latter in the bark of the horse-chestnut. A decoction of this bark yielded to me a copious crop of crystals of æsculin, while the paviin, together with a quantity of æsculin still apparently in excess, remained in the mother-liquor. I may, perhaps, on some future occasion communicate to the Society the method employed, when I have leisure to examine it further; I will merely state for the present that it enabled me to obtain crystallized æsculin in a few hours, without employing any other solvent than water. In the method commonly employed, the first crystallization of æsculin is described as requiring some fourteen days.

On account of the small quantity, apparently, of paviin, as compared with æsculin, present in the bark of the horse-chestnut, a chemist who wished to obtain the substance for analysis would probably do well to examine a bark from the genus Pavia, if such could be procured. The richness of the bark in paviin, as compared with æsculin, may be judged of by boiling a small portion with water in a test tube; those barks in which the substance presumed to be paviin abounds yield a decoction having almost exactly the same fluorescent tint as that of a decoction of ash bark.

A crystallizable substance, giving a highly fluorescent solution, has been discovered in the bark of the ash, by the Prince of Salm-Horstmar*, who has favoured me with a specimen. This substance, which has been named *fraxin* by its discoverer, is so similar in its optical characters to paviin that the two can hardly, if at all, be distinguished thereby; but as fraxin is stated to be insoluble in ether, it can hardly be identical with paviin, which was left in a crystallized state by that solvent. I find, however, that fraxin is sufficiently soluble in ether to render the fluid fluorescent, so that after all it is only a question of degree, which cannot be satisfactorily settled till paviin shall have been prepared in greater quantity.

* *Poggendorff's Annalen*, Vol. c, (1857), p. 607.

On the bearing of the Phenomena of Diffraction on the Direction of the Vibrations of Polarized Light, with Remarks on the Paper of Professor F. Eisenlohr.

[From the *Philosophical Magazine*, xviii, 1859, pp. 426–7.]

The appearance in the *Philosophical Magazine* for September of a translation of Professor F. Eisenlohr's paper in the 104th volume of *Poggendorff's Annalen*, induces me to offer some remarks on the subject there treated of.

Had my paper "On the Dynamical Theory of Diffraction*" been accessible to M. Eisenlohr at the time when he wrote, he would have seen that I did not content myself with merely resolving the vibrations of the incident light in directions parallel and perpendicular to the diffracted ray, and neglecting the former component, as competent to produce only normal vibrations, but that I gave a rigorous dynamical solution of the problem, in which the normal vibrations, or their imaginary representatives, as well as the transversal vibrations, were fully taken into account, though the result of the investigation showed that, in case of diffraction in one and the same medium (the only case investigated), the state of polarization of the diffracted ray was independent of the normal vibrations. M. Eisenlohr's result, on the other hand, confessedly rests on the assumption that the diffracted ray may be regarded as produced by an incident ray agreeing in *direction of propagation* with an incident ray which would produce the diffracted ray by regular refraction, but in *direction of vibration* (in the immediate neighbourhood of the surface at which the diffraction takes place) with the actual incident ray. This assumption, though plausible at first sight, is altogether precarious; and since in the particular case of diffraction in one and the same medium it leads to a result at variance with that of a rigorous investigation, it cannot be admitted.

* *Cambridge Philosophical Transactions*, Vol. ix, p. 1. [*Ante*, Vol. ii, p. 243.]

M. Eisenlohr's formula agrees no doubt very well with M. Holtzmann's experiments; but then it must be recollected that the formula contains a disposable constant, whereby such an agreement can in good measure be brought about. But in agreeing with these experiments, it is necessarily at variance with mine, in passing to which it is not allowable to change the value of the disposable constant. I can no more ignore the uniform result of my own experiments, than I am disposed to dispute the accuracy of M. Holtzmann's, made under different experimental circumstances. Whether the circumstances of his experiments or of mine made the nearer approach to the simplicity assumed in theory, or whether in both there did not exist experimental conditions sensibly influencing the result, but of such a nature that it would be impracticable to take account of them in theory, is a question which at present I think it would be premature to discuss. I still adhere to the opinion I formerly expressed*, that the whole question must be subjected to a thoroughly searching experimental investigation before physical conclusions can safely be drawn from the phenomena.

* *Phil. Mag.* Ser. 4, Vol. XIII, p. 159. [*Ante*, p. 74; see also footnote.]

Note on Paviin.

[From the *Quarterly Journal of the Chemical Society*, xii, 1860, pp. 126--128.]

The crystallizable substance, the existence of which in the bark of the horse-chestnut I noticed in a former communication to this Society*, and to which I gave the name paviin, together with æsculin which it closely resembles in its properties, may be thus prepared.

A decoction of the bark having been made, and allowed to grow cold, there is added a persalt of iron, such as pernitrate, until on testing a sample by the addition of ammonia, the precipitate separates at once in distinct flocks, leaving a bright pale yellow highly fluorescent fluid, giving a mixture which filters readily. The whole is then precipitated by ammonia and filtered; about one-fourth of the ammoniacal filtrate is precipitated with acetate of lead, avoiding an excess, ammonia being added if required; the mixture is restored to solution by the addition of acetic or dilute nitric acid, added to the remainder of the first filtrate previously acidulated; the whole precipitated by ammonia and filtered; and the filtrate precipitated by ammoniacal acetate of lead and filtered. The two precipitates are collected separately, and treated with acetic acid until they are dissolved, or at least wholly broken up, filtered if need be, and set aside in a cool place. The second precipitate yields æsculin, which makes its appearance as a light precipitate, appearing under the microscope to consist wholly of minute needles. The first precipitate yields paviin, which ordinarily crystallizes in tufts of very long slender silky crystals. When the solution is comparatively impure, it sometimes crystallizes very slowly in shorter and far thicker crystals. The crystallization of paviin is sometimes remarkably

* *Chem. Soc. Quarterly Journal*, Vol. xi, p. 17. [*Ante*, p. 112.]

facilitated by dropping in a very minute portion of the substance from a previous preparation. When the crystallization, whether of æsculin or of paviin, ceases to progress, the mass of crystals is thrown on a filter, drained, and pressed out. The mother-liquors being similarly treated by partial precipitation, yield additional quantities of the substances.

The æsculin thus obtained, after merely pressing out the crystals, without washing, is usually snow-white. The paviin, which appears to be naturally slightly yellow, is often mixed with a brown product of decomposition of other substances, which however, being insoluble in water, is of little consequence.

The properties, especially the optical properties, of paviin, so far as I have observed, appear to be absolutely identical with those of fraxin, a crystallizable substance discovered in the bark of the ash by the Prince of Salm-Horstmar*, who has favoured me with a specimen. The substances would seem to be identical†, but analyses are yet wanting. Accordingly, it is unnecessary to describe the properties of paviin. I will limit myself to two observations relative to horse-chestnut bark, in which optics come in aid of chemistry.

The quantity of paviin contained in horse-chestnut bark is larger than I had at first imagined. I find that in order to match the tint of the fluorescent light of a decoction of horse-chestnut bark by a mixed solution of æsculin and paviin, the substances must be present in the proportion by weight of about 3 to 2. The relative proportion of paviin as compared with æsculin actually obtained is liable to be much less than this, which arises, I believe, from the circumstance that paviin, from its somewhat stronger affinities, is less easily separated than æsculin from certain readily decomposed substances present in the decoction.

The production of æsculetin from æsculin may be elegantly followed by combining the optical and chemical properties of the substances. A solution liable to contain æsculetin is acidulated, if not already acid, and agitated with about an equal volume of ether. The ether withdraws the æsculetin, and when the whole is examined by daylight transmitted through a deep manganese

* *Poggendorff's Annalen*, Vol. c, p. 607.

† I find fraxin, like paviin, to be slightly soluble in ether, at least washed ether.

purple glass (*Phil. Trans.*, 1853, p. 385*), the æsculetin shows itself by the strong fluorescence of the ethereal solution. In this way it is easy to tell what acids give rise to the formation of æsculetin by boiling a teaspoonful of water containing, say, the hundredth of a grain of æsculin with the acid to be tried. It is easy, too, to demonstrate the absence of æsculetin in the bark itself, and its presence in a decoction which has stood some time.

In describing an analysis of horse-chestnut bark not yet complete, Rochleder mentions a *nearly neutral*† crystallizable substance which *accompanies æsculin in small quantity*. If this be paviin, as seems probable, the subject is at present in the hands of Rochleder, who has also undertaken the analysis of fraxin.

[* *Ante*, p. 1.]

† Gmelin's *Handbuch*, Vol. VIII, (1858), p. 25.

ON THE COLOURING MATTERS OF MADDER.
By DR E. SCHUNCK. (Extract.)

[From the *Quarterly Journal of the Chemical Society*, XII, 1860, pp. 218–222.]

BEFORE concluding, it remains for me to say a few words in regard to purpurine, the colouring matter which has by some chemists been supposed to be, in addition to alizarine, essential to the production of madder colours. The two chief properties, whereby, according to those chemists who have examined it, it is distinguished from alizarine are: 1. That it dissolves in alkalies with a cherry-red or bright red colour, alizarine giving with alkalies beautiful violet solutions. 2. That it is entirely soluble in boiling alum-liquor, forming a solution of a beautiful pink colour with a yellow fluorescence, whereas alizarine is almost insoluble in the same menstruum. These properties are, however, not of a sufficiently decided character to entitle it to rank as a distinct substance, as they might possibly be produced by an admixture of alizarine with some foreign substance. Having convinced myself by numerous experiments that almost all madder colours may be produced by means of alizarine only, and that the finer madder colours of the dyer contain little besides alizarine in combination with the mordants, I made some attempts to prove that purpurine contains ready-formed alizarine, to which its tinctorial power may be supposed to be due. The optical phenomena exhibited by solutions of purpurine are however so peculiar, as to lead Professor Stokes, who has carefully examined them, to the conclusion that they cannot be produced by any compound of alizarine, or by a mixture of alizarine with any other substance hitherto obtained from madder. Nevertheless it is certain that alizarine and purpurine are nearly allied substances, since both of them yield phthalic acid when decomposed by nitric acid, a property which belongs, as far as is known, to no other substance with the exception of naphthaline. There is one property by which purpurine may be easily distinguished from alizarine, viz. that of being decomposed when its solution in caustic alkali is exposed to the air. The bright red colour of the solution, when left to stand in an open vessel, soon changes to reddish-yellow, and at length almost the whole of the colour disappears, after which the purpurine can no longer be discovered in the solution. This is probably the cause of the disappearance of purpurine, when the method given by me for the preparation of alizarine from madder and its separation from the impurities with which it is associated, is adopted. This method, which depends on the employment of caustic alkalies, is an imitation of that to which dyers have recourse for the purpose of improving and beautifying ordinary madder colours, and it is certain that during this process the purpurine is either decomposed or by some means disappears. The only advantage which purpurine presents over alizarine in dyeing is that it imparts to the alumina-mordant a fiery red tint which in some cases is preferred to the purplish-red colour from alizarine. To the iron mordant it communicates a very unsightly reddish-purple colour, presenting a disagreeable contrast with the lovely purple from alizarine. In all madder colours which have been subjected to a long course of after-treatment, the purpurine is found to have almost entirely disappeared.

Professor Stokes has had the kindness to draw up, for the purpose of being appended to this paper, the following account of the optical characters of purpurine and alizarine, containing the results obtained by him on a renewed examination of their action on light.

Optical Characters of Purpurine and Alizarine.

The optical characters of purpurine are distinctive in the very highest degree; those of alizarine are also very distinctive. The characters here referred to consist in the mode of absorption of light by certain solutions of the bodies, and occasionally in the powerful fluorescence of a solution. They are specially valuable because their observation is independent of more than a moderate degree of purity of the specimens, and requires no apparatus beyond a test-tube, a slit, and a small prism, a little instrument which ought to be in the hands of every chemist.

Alkaline solution of purpurine.—If purpurine be dissolved in a solution of carbonate of potash or soda, (it is easily decomposed by caustic alkalies,) the solution obtained absorbs with greatest energy the green part of the spectrum. In this and similar cases it is necessary to take care either to use a sufficiently small quantity of the substance, or else to dilute sufficiently the solution, or view it through a sufficiently small thickness; otherwise a broad region of the spectrum is absorbed, and the peculiar characters of the substance depending on its mode of absorbing light are not perceived. If the solution be contained in a wedge-shaped vessel, the effect of different thicknesses is seen at a glance; but a test-tube will answer perfectly well if two or three different degrees of dilution be tried in succession. When the light transmitted through an alkaline solution of purpurine of suitable

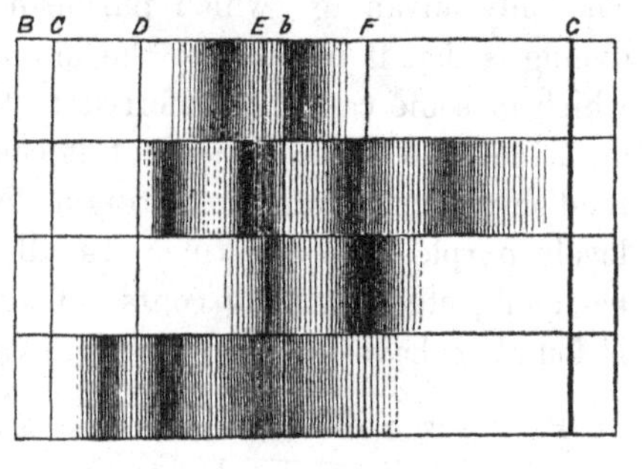

Fig. 1.—Solution of purpurine in carbonate of soda or potash, or in alum-liquor.

Fig. 2.—Solution of purpurine in bisulphide of carbon.

Fig. 3.—Solution of purpurine in ether.

Fig. 4.—Alkaline solution of alizarine.

strength, after being limited by a slit, is viewed through a prism, two remarkable dark bands of absorption (Fig. 1) are seen about the green part of the spectrum, comprising between them a band of green light, which, though much weakened in comparison with the same part of the unabsorbed spectrum, is bright compared with the two dark bands, which latter in a sufficiently strong solution appear perfectly black. The places of the dark bands, estimated with reference to the principal fixed lines of the spectrum, are given in the figure.

Solution in a solution of alum.—This solution has the same peculiar mode of absorption, and (Fig. 1) will serve equally well for it. But it has the further property of being eminently fluorescent, which the alkaline solution is not at all. The fluorescent light is yellow, but ordinarily appears orange from being seen through the fluid. The difference between the alkaline and alum-liquor solutions as to fluorescence does not depend on the acid reaction of the latter, but on the alumina. A solution exhibiting to perfection the peculiar properties of the alum-liquor solution may be obtained by adding to a solution of purpurine in carbonate of soda a solution of alum to which enough tartaric acid to prevent precipitation, and then carbonate of soda, has previously been added; and in this case the fluorescent solution is obtained at once and in the cold. This forms a very striking reaction in a dark room, according to the method described in the *Philosophical Transactions* for 1853, p. 385*, with the combination, solution of nitrate of copper and a red (Cu_2O) glass. Some other colourless oxides besides alumina develop in this manner fluorescence, though to a less degree.

Solution in bisulphide of carbon.—This solution gives the highly characteristic spectrum (Fig. 2) exhibiting four bands of absorption, of which the first is narrower than the others, and the fourth is very inconspicuous, hardly standing out from the general absorption which takes place in that region of the spectrum. The second and third bands are the most conspicuous of the set.

Solution in ether.—This gives the characteristic spectrum (Fig. 3) exhibiting two bands of absorption. The solution is fluorescent, but not enough so to be perceptible by common observation.

The spectra of the solutions of purpurine in other solvents might be mentioned, but these are more than sufficient. In an optical point of view purpurine is remarkable for the general similarity of character, combined with diversity as to detail, which its various solutions exhibit as to their mode of absorbing light.

Alkaline solution of alizarine.—The solution of alizarine in caustic or carbonate of potash or soda, or in ammonia, exhibits on analysis the characteristic spectrum (Fig. 4) having a band of absorption in the yellow, and another narrower one between the red and the orange. There is a third very inconspicuous band at E, almost lost in the general darkening of that part of the spectrum.

[* *Ante*, p. 1.]

Other solutions.—The solution of alizarine in ether or in bisulphide of carbon shows nothing particular. There is a general absorption of the more refrangible part of the spectrum, but there are none of those remarkable alternations of comparative transparency and opacity which characterize purpurine. Alizarine is hardly soluble in alum-liquor; and in the case of the red solution of mixed alizarine and verantine mentioned by Dr Schunck at page 451 of the *Philosophical Transactions* for 1851, the absence of the remarkable absorption-bands (Fig. 1) and the absence of fluorescence, show instantly and independently of each other that it is distinct from purpurine.

Optical detection of purpurine and alizarine.—The characters of these substances are so marked that I do not know any substance with which either of them could be confounded, even if we restricted ourselves to *any one* of the solutions yielding the peculiar spectra. Not only so, but these properties enable us to detect small quantities, in the case of purpurine the merest trace, of the substance, present in the midst of a quantity of impurities. In the case of purpurine, a solution of alum is specially convenient for use, because the impurities liable to be present do not with this solvent absorb the part of the spectrum in which the bands occur. In this way I was able, though operating on only a very minute quantity of the root, to detect purpurine in more than twenty species of the family *Rubiaceae* which were examined with this view, comprising the genera *Rubia, Asperula, Galium, Crucianella,* and *Sherardia.* The detection of alizarine by means of the characters of its alkaline solution is much less delicate, because many of the impurities liable to be present absorb the part of the spectrum in which all but the least refrangible of the absorption-bands occur; and as this band is not that which corresponds to the most intense absorption, a larger quantity of the substance must be present in order that the band may be perceived.

EXTRACTS RELATING TO THE EARLY HISTORY OF SPECTRUM ANALYSIS.

On the Simultaneous Emission and Absorption of Rays of the same definite Refrangibility; being a translation of a portion of a paper by M. LÉON FOUCAULT, *and of a paper by* Professor KIRCHHOFF.

[From the *Philosophical Magazine*, XIX, 1860, pp. 196–7.]

GENTLEMEN,

Some years ago M. Foucault mentioned to me in conversation a most remarkable phænomenon which he had observed in the course of some researches on the voltaic arc, but which, though published in *L'Institut*, does not seem to have attracted the attention which it deserves. Having recently received from Prof. Kirchhoff a copy of a very important communication to the Academy of Sciences at Berlin, I take the liberty of sending you translations of the two, which I doubt not will prove highly interesting to many of your readers.

I am, Gentlemen, Yours sincerely,

G. G. STOKES.

M. Foucault's discovery is mentioned in the course of a paper published in *L'Institut* of Feb. 7, 1849, having been brought forward at a meeting of the Philomathic Society on the 20th of January preceding. In describing the result of a prismatic analysis of the voltaic arc formed between charcoal poles, M. Foucault writes as follows (p. 45):—

"Its spectrum is marked, as is known, in its whole extent by a multitude of irregularly grouped luminous lines; but among these may be remarked a double line situated at the boundary of the yellow and orange. As this double line recalled by its form and situation the line *D* of the solar spectrum, I wished to try if it corresponded to it; and in default of instruments for measuring the angles, I had recourse to a particular process.

"I caused an image of the sun, formed by a converging lens, to fall on the arc itself, which allowed me to observe at the same time the electric and the solar spectrum superposed; I convinced myself in this way that the double bright line of the arc coincides exactly with the double dark line of the solar spectrum.

"This process of investigation furnished me matter for some unexpected observations. It proved to me in the first instance the extreme transparency of the arc, which occasions only a faint shadow in the solar light. It showed me that this arc, placed in the path of a beam of solar light, absorbs the rays *D*, so that the above-mentioned line *D* of the solar light is considerably strengthened when the two spectra are exactly superposed. When, on the contrary, they jut out one beyond the other, the line *D* appears darker than usual in the solar light, and stands out bright in the electric spectrum, which allows one easily to judge of their perfect coincidence. Thus the arc presents us with a medium which emits the rays *D* on its own account, and which at the same time absorbs them when they come from another quarter.

"To make the experiment in a manner still more decisive, I projected on the arc the reflected image of one of the charcoal points, which, like all solid bodies in ignition, gives no lines; and under these circumstances the line *D* appeared to me as in the solar spectrum."

Professor Kirchhoff's communication "On Fraunhofer's Lines," dated Heidelberg, 20th of October, 1859, was brought before the Berlin Academy on the 27th of that month, and is printed in the *Monatsbericht*, p. 662.

"On the occasion of an examination of the spectra of coloured flames not yet published, conducted by Bunsen and myself in common, by which it has become possible for us to recognise the qualitative composition of complicated mixtures from the appearance of the spectrum of their blowpipe-flame, I made some observations which disclose an unexpected explanation of the origin of Fraunhofer's lines, and authorise conclusions therefrom respecting the material constitution of the atmosphere of the sun, and perhaps also of the brighter fixed stars.

"Fraunhofer had remarked that in the spectrum of the flame of a candle there appear two bright lines, which coincide with the two dark lines *D* of the solar spectrum. The same bright lines are obtained of greater intensity from a flame into which some common salt is put. I formed a solar spectrum by projection, and allowed the solar rays concerned, before they fell on the slit, to pass through a powerful salt-flame. If the sunlight were sufficiently reduced, there appeared in place of the two dark lines *D* two bright lines; if, on the other hand, its intensity surpassed a certain limit, the two dark lines *D* showed themselves in much greater distinctness than without the employment of the salt-flame.

"The spectrum of the Drummond light contains, as a general rule, the two bright lines of sodium, if the luminous spot of the cylinder of lime has not long been exposed to the white heat; if the cylinder remains unmoved these lines become weaker, and finally vanish altogether. If they have vanished, or only faintly appear, an alcohol flame into which salt has been put, and which is placed between the cylinder of lime and the slit, causes two dark lines of remarkable sharpness and fineness, which in that respect agree with the lines *D* of the solar spectrum, to show themselves in their stead. Thus the lines *D* of the solar spectrum are artificially evoked in a spectrum in which naturally they are not present.

"If chloride of lithium is brought into the flame of Bunsen's gas-lamp, the spectrum of the flame shows a very bright sharply defined line, which lies midway between Fraunhofer's lines *B* and *C*. If, now, solar rays of moderate intensity are allowed to fall through the flame on the slit, the line at the place pointed out is seen bright on a darker ground; but with greater strength of sunlight there appears in its place a dark line, which has quite the same character as Fraunhofer's lines. If the flame be taken away, the line disappears, as far as I have been able to see, completely.

"I conclude from these observations, that coloured flames, in the spectra of which bright sharp lines present themselves, so weaken rays of the colour of these lines, when such rays pass through the flames, that in place of the bright lines dark ones appear as soon as there is brought behind the flame a source of light of sufficient intensity, in the spectrum of which these lines are otherwise wanting. I conclude further, that the dark lines of the solar spectrum which are not evoked by the atmosphere of the earth, exist in consequence of the presence, in the incandescent atmosphere of the sun, of those substances which in the spectrum of a flame produce bright lines at the same place. We may assume that the bright lines agreeing with *D* in the spectrum of a flame always arise from sodium contained in it; the dark line *D* in the solar spectrum allows us, therefore, to conclude that there exists sodium in the sun's atmosphere. Brewster has found bright lines in the spectrum of the flame of saltpeter at the place of Fraunhofer's lines *A*, *a*, *B*; these lines point to the existence of potassium in the sun's atmosphere*. From my observation, according to which no dark line in the solar spectrum answers to the red line of lithium, it would follow with probability that in the atmosphere of the sun lithium is either absent, or is present in comparatively small quantity.

"The examination of the spectra of coloured flames has

[* See however p. 135 *infra*.]

accordingly acquired a new and high interest; I will carry it out in conjunction with Bunsen as far as our means allow. In connexion therewith we will investigate the weakening of rays of light in flames that has been established by my observations. In the course of the experiments which have at present been instituted by us in this direction, a fact has already shown itself which seems to us to be of great importance. The Drummond light requires, in order that the lines D should come out in it dark, a salt-flame of lower temperature. The flame of alcohol containing water is fitted for this, but the flame of Bunsen's gas-lamp is not. With the latter the smallest mixture of common salt, as soon as it makes itself generally perceptible, causes the bright lines of sodium to show themselves. We reserve to ourselves to develope the consequences which may be connected with this fact."

Note.—The remarkable phenomenon discovered by Foucault, and rediscovered and extended by Kirchhoff, that a body may be at the same time a source of light giving out rays of a definite refrangibility, and an absorbing medium extinguishing rays of that same refrangibility which traverse it, seems readily to admit of a dynamical illustration borrowed from sound.

We know that a stretched string which on being struck gives out a certain note (suppose its fundamental note) is capable of being thrown into the same state of vibration by aërial vibrations corresponding to the same note. Suppose now a portion of space to contain a great number of such stretched strings, forming thus the analogue of a "medium." It is evident that such a medium on being agitated would give out the note above mentioned, while on the other hand, if that note were sounded in air at a distance, the incident vibrations would throw the strings into vibration, and consequently would themselves be gradually extinguished, since otherwise there would be a creation of *vis viva*. The optical application of this illustration is too obvious to need comment

---

On the Relation between the Radiating and Absorbing Powers of different Bodies for Light and Heat. By G. KIRCHHOFF. (*Postscript.*)

[*Pogg. Ann.* CIX, 1860, p. 275: trans. by F. Guthrie, *Philosophical Magazine*, XX, July 1860. This postscript communicated by author to *Phil. Mag.*, pp. 19–21.]

1. Since the appearance of the above paper in *Poggendorff's Annalen*, I have received information of a prior communication closely related to mine. The communication in question is by Mr Balfour Stewart, and appeared in the *Transactions of the Royal Society of Edinburgh*, Vol. XXII, 1858. Mr Stewart has made the interesting observation that a plate of rock-salt is much less diathermic for rays emitted by a mass of the same substance heated to 100° C., than for rays emitted by any other black body of the same temperature. From this circumstance he draws certain conclusions, and is led to a result similar to that which I have established concerning the connexion between the powers of absorption and emission. The principle enunciated by Mr Stewart is, however, less distinctly expressed, less general, and not altogether so strictly proved as mine. It is as follows:—"The absorption of a plate equals its radiation, and that for every description of heat."

2. The fact that the bright lines of the spectra of sodium- and lithium-flames may be reversed, was first published by me in a communication to the Berlin Academy, October 27, 1859. This communication is noticed by M. Verdet in the February Number of the *Ann. de Chim. et de Phys.* of the following year, and is translated by Professor Stokes in the March Number of the *Philosophical Magazine.* The latter gentleman calls attention to a similar observation made by M. Léon Foucault eleven years ago, and which was unknown to me, as it seems to have been to most physicists. This observation was to the effect that the electric arch between charcoal points behaves, with respect to the emission and absorption of rays of refrangibility answering to Fraunhofer's line *D*, precisely as the sodium-flame does according to my experiments. The communication made on this subject by M. Foucault to the Soc. Philom. in 1849 is reproduced by M. Verdet, from the *Journal de l'Institut,* in the April Number of the *Ann. de Chim. et de Phys.*

M. Foucault's observation appears to be regarded as essentially the same as mine; and for this reason I take the liberty of drawing attention to the difference between the two. The observation of M. Foucault relates to the electric arch between charcoal points, a phenomenon attended by circumstances which are in many respects extremely enigmatical. My observation relates to ordinary flames into which vapours of certain chemical substances have been introduced. By the aid of my observation, the other may be accounted for on the ground of the presence of sodium in the charcoal, and indeed might even have been foreseen. M. Foucault's observation does not afford any explanation of mine, and could not have led to its anticipation. My observation leads necessarily to the law which I have announced with reference to the relation between the powers of absorption and emission; it explains the existence of Fraunhofer's lines, and leads the way to the chemical analysis of the atmosphere of the sun and the fixed stars. All this M. Foucault's observation did not and could not accomplish, since it related to a too complicated phenomenon, and since there was no means of determining how much of the result was due to electricity, and how much to the presence of sodium. If I had been earlier acquainted with this observation, I should not have neglected to introduce some notice of it into my communication, but I should nevertheless have considered myself justified in representing my observation as essentially new.

3. Since the above communication was printed in *Poggendorff's Annalen*, I have learned in the course of a written correspondence with Professor Thomson, that the idea was some years ago thrown out, if not published, that it might be possible, by comparing the spectra of various chemical flames with that of the sun and fixed stars, in the manner I have described, to become acquainted with the chemical constitution of the latter bodies (an idea now demonstrated to be correct by the observations and theoretical considerations above set forth). Prof. Thomson writes:—

"Professor Stokes mentioned to me at Cambridge some time ago, probably about ten years, that Professor Miller* had made an experiment testing to a very high degree of accuracy the agreement of the double dark line D of the solar spectrum with the double bright line constituting the spectrum of the spirit-lamp burning with salt. I remarked that there must be some physical connexion between two agencies presenting so marked a characteristic in common. He assented, and said he believed a mechanical explanation of the cause was to be had on some such principles as the following:—Vapour of sodium must possess by its molecular structure a tendency to vibrate in the periods corresponding to the degrees of refrangibility of the double

[* W. H. Miller, of Cambridge.]

line *D*. Hence the presence of sodium in a source of light must tend to originate light of that quality. On the other hand, vapour of sodium in an atmosphere round a source, must have a great tendency to retain in itself, *i.e.* to absorb and to have its temperature raised by, light from the source, of the precise quality in question. In the atmosphere around the sun, therefore, there must be present vapour of sodium, which, according to the mechanical explanation thus suggested, being particularly opaque for light of that quality, prevents such of it as is emitted from the sun from penetrating to any considerable distance through the surrounding atmosphere. The test of this theory must be had in ascertaining whether or not vapour of sodium has the special absorbing power anticipated. I have the impression that some Frenchman did make this out by experiment, but I can find no reference on the point.

"I am not sure whether Professor Stokes's suggestion of a mechanical theory has ever appeared in print. I have given it in my lectures regularly for many years, always pointing out along with it that solar and stellar chemistry were to be studied by investigating terrestrial substances giving bright lines in the spectra of artificial flames corresponding to the dark lines of the solar and stellar spectra."

On the Early History of Spectrum Analysis.

[From *Nature*, Vol. XII, 1876, pp. 188-9.]

The following extract from a letter*, relating to the early history of spectrum analysis, from our highest English authority on physical optics, cannot fail to interest, apart from its intrinsic importance, a wide circle of readers. I have therefore obtained permission from Professor Stokes to forward it to *Nature*.

C. T. L. WHITMELL.

"CAMBRIDGE, *Dec.* 23, 1875.

"......I felt that the coincidence between the dark *D* of the solar spectrum and the bright *D* of a spirit-lamp with salted wick could not be a matter of chance; and knowing as I did that the latter was specially produced by salts of soda, and believing as

[* This letter was elicited by a syllabus of a course of optical lectures sent by Mr Whitmell to Prof. Stokes.]

I did that even when such were not ostensibly present, they were present in a trace (thus alcohol burnt on a watch-glass and a candle snuffed close, so that the wick does not project into the incandescent envelope, do not show bright *D*), I concluded in my own mind that dark *D* was due to absorption by sodium in some shape. In what shape? I knew that such narrow absorption-bands were only observed in vapours; I knew that as a rule vapours agree in a general way with their liquids or solutions as to absorption, save that in lieu of the capricious absorption of the vapour, we have a general absorption attacking those regions of the spectrum in which the vapour-bands are chiefly found. Hence as the sodium *compounds*, chloride, oxide, etc., are transparent, I concluded that the absorbing vapour was that of sodium itself. Knowing the powerful affinities of sodium, I did not dream of its being present in a *free* state in the flame of a spirit-lamp; and so I supposed that the emitting body in the case of a spirit-lamp with salted wick was volatilised chloride of sodium, capable of vibrating in a specific time, or rather two specific and nearly equal periods, by virtue of its sodium constituent; but that to produce absorption the sodium must be free. I never thought of the extension of Prevost's law of exchanges from radiation as a whole to radiation of each particular refrangibility by itself, afterwards made by B. Stewart; and so I failed to perceive that a soda flame which emits bright *D* must *on that very account* absorb light of the same refrangibility.

"When Foucault, whom I met at dinner at Dr Neil Arnott's when he came to receive the Copley Medal in 1855, told me of his discovery of the absorption and emission of *D* by a voltaic arc, I was greatly struck with it. But though I had pictured to my mind the possibility of emitting and absorbing light of the same refrangibility by the mechanism of a system of piano strings tuned to the same pitch, which would, if struck, give out a particular note, or would take it up from the air at the expense of the aërial vibrations, I did not think of the extension of Prevost's theory, afterwards discovered by Stewart*, nor perceive that the emission of light of definite refrangibility *necessitated* (and not merely *permitted*) absorption of light of the same refrangibility.

[* *Trans. R. S. Edin.*, XXII, March 1858; cf. Lord Rayleigh, *Phil. Mag.* I, 1901, pp. 98–100, or *Scientific Papers*, Vol. IV, p. 494.]

"Reviewing my then thoughts by the light of our present knowledge, I see that my error lay in the erroneous chemical assumption that sodium could not be free in the flame of a spirit-lamp; I failed to perceive the extension of Prevost's theory, which would have come in conflict with that error*.—Yours sincerely,

(*Signed*) "G. G. STOKES.

"To CHAS. WHITMELL, Esq."

"P.S., Dec. 31.—As Sir Wm. Thomson has referred in print to a conversation I had long ago with him on the subject, I take the opportunity of describing my recollection of the matter.

"I mentioned to him the perfect coincidence of bright and dark D, and a part at least of the reasons I had for attributing the latter to the vapour of sodium, using I think the dynamical illustration of the piano strings. I mentioned also, on the authority of Sir David Brewster, another case of coincidence (as was then supposed, though it has since been shown to be only a casual near agreement) of a series of bright lines in an artificial source of light with dark lines in the solar spectrum, from which it appeared to follow that potassium was present in the sun's atmosphere. On hearing this Thomson said something to this effect: 'Oh then, the way to find what substances are present in the sun and stars is to find what substances give bright lines coincident with the dark lines of those bodies.' I thought he was generalising too fast; for though *some* dark lines might thus be accounted for, I was disposed to think that the greater part of the non-terrestrial lines of the solar spectrum were due to the vapours of *compound* bodies existing in the higher and comparatively cool regions of the sun's atmosphere, and having (as we know is the case with peroxide of nitrogen and other coloured gases) the power of selective absorption changing rapidly and apparently capriciously with the refrangibility of the light.

"If (as I take for granted) Sir William Thomson is right as to the date [1852] when he began to introduce the subject into his lectures at Glasgow (Address at the Edinburgh Meeting of the British Association [1871], page xcv.†), he must be mistaken as to

[* For another statement, see the author's Burnett Lectures on Light, second series, 1885, pp. 42–50.]

[† This Presidential Address is reprinted in Lord Kelvin's *Popular Lectures and Addresses*, Vol. II, see pp. 169–175.]

the time when I talked with him about Foucault's discovery, for I feel sure I did not know it till 1855. Besides, when I heard it from Foucault's mouth, it fell in completely with my previous thoughts*.

"I have never attempted to claim for myself any part of Kirchhoff's admirable discovery, and cannot help thinking that some of my friends have been over zealous in my cause. As, however, my name has frequently appeared in print in connexion with it, I have been induced to put on paper a statement of the views I entertained and talked about, though without publishing.

"In ascribing to Stewart the discovery of the extension of Prevost's law of exchanges†, I do not forget that it was re-discovered by Kirchhoff, who, indeed, was the first to *publish* it in relation to light, though the transition from radiant heat to light is so obvious that it could hardly fail to have been made, as in fact it was made, by Stewart himself (see *Proceedings of the Royal Society*, Vol. x, p. 385‡). Nor do I forget that it is to Kirchhoff that we owe the admirable application of this extended law to the lines of the solar spectrum."

[* This is borne out by four existing letters, Feb. 24—Mar. 28, 1854, of Stokes to Thomson, with replies: also letters of Nov. 26 and Dec. 6, 1855, relating to the Foucault experiment. (Printed in Appendix to this volume.)]

[† See Kirchhoff's criticism on this and other matters in *Pogg. Ann.* CXVIII, 1862; *Ges. Abhandl.* pp. 625, 641; translated in *Phil. Mag.* XXV, 1863, pp. 250–262, as "Contributions towards the History of Spectrum Analysis and of the Analysis of the Solar Atmosphere"; and Stewart's rejoinder, *ibid.* pp. 354–360. Also Rayleigh, *loc. cit. supra*, and a rejoinder by Kayser, *Handbuch der Spectroscopie*, Vol. II, 1902, p. 10.]

[‡ "On the Light radiated by Heated Bodies," Feb. 7, 1860. It is convenient to record here that in a supplementary paper, received by the Royal Society on May 22, 1860, "On the Nature of the Light emitted from Heated Tourmaline," Balfour Stewart expresses his obligations to Prof. Stokes for suggesting "an apparatus" by which an investigation of the polarization of the light was successfully made; namely "a thick spherical cast-iron bomb, about 5 inches in external and 3 inches in internal diameter—the thickness of the shell being therefore 1 inch. It has a cover removable at pleasure. There is a small stand in the inside, upon which the substance under examination is placed, and when so placed it is precisely at the centre of the bomb. Two small round holes, opposite to one another, viz. at the two extremities of a diameter, are bored in the substance of the shell....Let the bomb with the substance on the stand be heated to a good red heat, and then withdrawn from the fire and allowed to cool...." See *Proc. Roy. Soc.* x, 1860, p. 503; *Phil. Mag.* XXI, 1861, p. 391. Kirchhoff had already made similar observations on a tourmaline heated in a Bunsen flame: cf. *Phil. Mag.* XX, 1860, p. 18.]

NOTE ON INTERNAL RADIATION.

[From the *Proceedings of the Royal Society*, XI, 1861, pp. 537–45.
Received *Dec.* 28, 1861.]

IN the eleventh volume of the *Proceedings of the Royal Society*, p. 193, is the abstract of a paper by Mr Balfour Stewart, in which he deduces an expression for the internal radiation in any direction within a uniaxal crystal from an equation between the radiations incident upon and emerging from a unit of area of a plane surface, having an arbitrary direction, by which the crystal is supposed to be bounded. With reference to this determination he remarks (p. 196), "But the internal radiation, if the law of exchanges be true, is clearly independent of the position of this surface, which is indeed merely employed as an expedient. This is equivalent to saying that the constants which define the position of the bounding surface must ultimately disappear from the expression for the internal radiation." This anticipation he shows is verified in the case of the expression deduced, according to his principles, for the internal radiation within a uniaxal crystal, on the assumption that the wave-surface* is the sphere and spheroid of Huygens.

In the case of an uncrystallized medium, the following is the equation obtained by Mr Stewart in the first instance.

* To prevent possible misapprehension, it may be well to state that I use this term to denote the surface, whatever it may be, which is the locus of the points reached in a given time by a disturbance propagated in all directions from a given point; I do not use it as a name for the surface defined analytically by the equation

$$(x^2+y^2+z^2)(a^2x^2+b^2y^2+c^2z^2)-a^2(b^2+c^2)x^2-b^2(c^2+a^2)y^2-c^2(a^2+b^2)z^2+a^2b^2c^2=0.$$

As the term *wave-surface* in its physical signification is much wanted in optics, the surface defined by the above equation should, I think, be called *Fresnel's surface*, or *the wave-surface of Fresnel*.

Let R, R' be the external and internal radiations in direetions OP, OP', which are connected as being those of an incident and refracted ray, the medium being supposed to be bounded by a plane surface passing through O. Let OP describe an elementary conical circuit enclosing the solid angle $\delta\phi$, and let $\delta\phi'$ be the elementary solid angle enclosed by the circuit described by OP'. Let i, i' be the angles of incidence and refraction. Of a radiation proceeding along PO, let the fraction A be reflected and the rest transmitted; and of a radiation proceeding internally along $P'O$ let the fraction A' be reflected and the rest transmitted. Then by equating the radiation incident externally on a unit of surface, in the directions of lines lying within the conical circuit described by OP, with the radiation proceeding in a contrary direction, and made up partly of a refracted and partly of an externally reflected radiation, we obtain

$$R \cos i\, \delta\phi = (1 - A') R' \cos i'\, \delta\phi' + A R \cos i\, \delta\phi,$$

or

$$(1 - A) R \cos i\, \delta\phi = (1 - A') R' \cos i'\, \delta\phi' \text{.........(1)}.$$

In the case of a crystal there are *two* internal directions of refraction, OP_1, OP_2, corresponding to a given direction PO of incidence, the rays along OP_1, OP_2 being each polarized in a particular manner. Conversely, there are two directions, P_1O, P_2O, in which a ray may be incident internally so as to furnish a ray refracted along OP, and in each case no second refracted ray will be produced, provided the incident ray be polarized in the same manner as the refracted ray OP_1 or OP_2. In the case of a crystal, then, equation (1) must be replaced by

$$(1 - A) R \cos i\, \delta\phi = (1 - A_1) R_1 \cos i_1\, \delta\phi_1 + (1 - A_2) R_2 \cos i_2\, \delta\phi_2 \text{...(2)}.$$

In the most general case it does not appear in what manner, if at all, equation (2) would split into two equations, involving respectively R_1 and R_2. For if an incident ray PO were so polarized as to furnish only one refracted ray, say OP_1, a ray incident along P_1O and polarized in the same manner as OP_1 would furnish indeed only one refracted ray, in the direction OP, but that would be polarized differently from PO; so that the two systems are mixed up together.

But if the plane of incidence be a principal plane, and if we may assume that such a plane is a plane of symmetry as regards the optical properties of the medium*, the system of rays polarized in and the system polarized perpendicularly to the plane of incidence will be quite independent of each other, and the equality between the radiation incident externally and that proceeding in the contrary direction, and made up partly of a refracted and partly of an externally reflected radiation, must hold good for each system separately. In this case, then, (2) will split into two equations, each of the form (1), R now standing for half the whole radiation, and R', A', etc. standing for R_1, A_1, etc., or R_2, A_2, etc., as the case may be. It need hardly be remarked that the value of A is different in the two cases, and that R' has a valuc which is no longer, as in the case of an isotropic medium, alike in all directions. In determining according to Mr Stewart's principles the internal radiation in any given direction within a uniaxal crystal, no limitation is introduced by the restriction of equation (1) to a principal plane, since we are at liberty to imagine the crystal bounded by a plane perpendicular to that containing the direction in question and the axis of the crystal.

Mr Stewart further reduces equation (1) by remarking that in an isotropic medium, as we have reason to believe, $A' = A$, and that the same law probably holds good in a crystal also, so that the equal factors $1 - A$, $1 - A'$ may be struck out. Arago long ago showed experimentally that light is reflected in the same proportion externally and internally from a plate of glass bounded by parallel surfaces; and the formulae which Fresnel has given to express, for the case of an isotropic medium, the intensity of reflected light, whether polarized in a plane parallel or perpendicular to the plane of incidence, are consistent with this law. In a paper published in the fourth volume of the *Cambridge and*

* According to Sir David Brewster (*Report of the British Association* for 1836, Part II, p. 13, and for 1842, Part II, p. 13), when light is incident on a plane surface of Iceland spar in a plane parallel to the axis, the plane of incidence, which is a principal plane, is not in general a plane of optical, any more than of crystalline symmetry as regards the phenomena of reflexion, although, as is well known, all planes passing through the axis are alike as regards internal propagation and the polarization of the refracted rays. Hence, strictly speaking, the statement as to the independence of the two systems of rays should be confined to the case in which the principal plane is also a plane of crystalline symmetry. As, however, the unsymmetrical phenomena were only brought out when the ordinary reflexion was weakened, almost annihilated, by the use of oil of cassia, we may conclude that under common circumstances they would be insensible.

Dublin Mathematical Journal (p. 1)*, I have given a very simple demonstration of Arago's law, based on the sole hypothesis that the forces acting depend only on the positions of the particles. This demonstration, I may here remark, applies without change to the case of a crystal whenever the plane of incidence is a plane of optical symmetry. It may be rendered still more general by supposing that the forces acting depend, not solely on the positions of the particles, but also on any differential coefficients of the coordinates which are of an even order with respect to the time,—a generalization which appears not unimportant, as it is applicable to that view of the mutual relation of the ether and ponderable matter, according to which the ether is compared to a fluid in which a number of solids are immersed, and which in moving as a whole is obliged to undergo local dislocations to make way for the solids.

On striking out the factors $1-A$ and $1-A'$, equation (1) is reduced to

$$\frac{R'}{R}=\frac{\cos i\,\delta\phi}{\cos i'\,\delta\phi'}\ \ldots\ldots\ldots\ldots\ldots\ldots\ldots\ldots(3).$$

In the case of an isotropic medium, R and R' are alike in all directions, and therefore the ratio of $\cos i\,\delta\phi$ to $\cos i'\,\delta\phi'$ ought to be independent of i, as it is very easily proved to be†. The same applies to a uniaxal crystal, so far as regards the ordinary ray. But as regards the extraordinary, it is by no means obvious that the ratio should be expressible in the form indicated—as a quantity depending only on the direction OP'. Mr Stewart has, however, proved that this is the case, independently of any restriction as to the plane of incidence being a principal plane, on the assumption that the wave-surface has the form assigned to it by Huygens.

It might seem at first sight that this verification was fairly adducible in confirmation of the truth of the whole theory, including the assumed form of the wave-surface. But a little consideration will show that such a view cannot be maintained. Huygens's construction links together the law of refraction and the form of the wave-surface, in a manner depending for its validity only on the most fundamental principles of the theory of undulations. The construction which Huygens applied to the

[* *Ante*, Vol. II, p. 91.]

[† It is that of μ'^2 to μ^2, $\delta\phi$ being a conical angle.]

ellipsoid is equally applicable to any other surface; it was a mere guess on his part that the extraordinary wave-surface in Iceland spar was an ellipsoid; and although the ellipsoidal form results from the imperfect dynamical theory of Fresnel, it is certain that rigorous dynamical theories lead to different forms of the wave-surface, according to the suppositions made as to the existing state of things. For every such possible form the ratio expressed by the right-hand member of equation (3) ought to come out in the form indicated by the left-hand member, and not to involve explicitly the direction of the refracting plane: and as it seemed evident that it could not be possible, merely by such general considerations as those adduced by Mr Stewart, to distinguish between those surfaces which were and those which were not dynamically possible forms of the wave-surface, I was led to anticipate that the possibility of expressing the ratio in question under the form indicated was a general property of surfaces. The object of the present Note is to give a demonstration of the truth of this anticipation, and thereby remove from the verification the really irrelevant consideration of a particular form of wave-surface; but it was necessary in the first instance to supply some steps of Mr Stewart's investigation which are omitted in the published abstract.

The proposition to be proved may be somewhat generalized, in a manner suggested by the consideration of internal reflexion within a crystal, or refraction out of one crystallized medium into another in optical contact with it. Thus generalized it stands as follows:—

Imagine any two surfaces whatsoever, and also a fixed point O; imagine likewise a plane Π passing through O. Let two points P, P', situated on the two surfaces respectively, and so related that the tangent planes at those points intersect each other in the plane Π, be called *corresponding points with respect to the plane* Π. Let P describe, on the surface on which it lies, an infinitesimal closed circuit, and P' the "corresponding" circuit; let $\delta\phi$, $\delta\phi'$ be the solid angles subtended at O by these circuits respectively, and i, i' the inclinations of OP, OP' to the normal to Π. Then shall the ratio of $\cos i\,\delta\phi$ to $\cos i'\,\delta\phi'$ be of the form $[P]:[P']$, where P depends only on the first surface and the position of P, and $[P']$ only on the second surface and the position of P'. Moreover, if either surface be a sphere having its centre at O, the corresponding quantity $[P]$ or $[P']$ shall be constant.

It may be remarked that the two surfaces may be merely two sheets of the same surface, or even two different parts of the same sheet.

Instead of comparing the surfaces directly with each other, it will be sufficient to compare them both with the same third surface; for it is evident that if the points P, P' correspond to the same point P_1, on the third surface, they will also correspond to each other. For the third surface it will be convenient to take a sphere described round O as centre with an arbitrary radius, which we may take for the unit of length. The letters P_1, i_1, ϕ_1 will be used with reference to the sphere.

Let the surface and sphere be referred to rectangular coordinates, O being the origin, and Π the plane of xy. Let x, y, z be the coordinates of P; ξ, η, ζ those of P_1. Then x, y, z will be connected by the equation of the surface, and ξ, η, ζ by the equation

$$\xi^2 + \eta^2 + \zeta^2 = 1.$$

According to the usual notation, let

$$\frac{dz}{dx} = p, \quad \frac{dz}{dy} = q, \quad \frac{d^2z}{dx^2} = r, \quad \frac{d^2z}{dx\,dy} = s, \quad \frac{d^2z}{dy^2} = t.$$

The equations of the tangent planes at P, P_1, X, Y, Z being the current coordinates, are

$$Z - z = p\,(X - x) + q\,(Y - y),$$
$$\xi X + \eta Y + \zeta Z = 1;$$

and those of their traces on the plane of xy are

$$pX + qY = px + qy - z,$$
$$\xi X + \eta Y = 1;$$

and in order that these may represent the same line, we must have

$$\xi = \frac{p}{px + qy - z}, \quad \eta = \frac{q}{px + qy - z} \quad \text{............(4).}$$

To the element $dx\,dy$ of the projection on the plane of xy of a superficial element at P, belongs the superficial element $dS = \sqrt{1 + p^2 + q^2}\,dx\,dy$, and to this again belongs the elementary solid angle $\frac{\cos\nu\,dS}{\rho^2}$, where $\rho = OP$, and ν is the angle between the normal at P and the radius vector. Hence the total solid angle within a small contour is $\frac{\cos\nu}{\rho^2}\sqrt{1+p^2+q^2}\iint dx\,dy$, the double integral being taken within the projection of that small contour. Also $\cos i = z/\rho$. Hence

$$\cos i\,\delta\phi = \frac{z\cos\nu}{\rho^3}\sqrt{1+p^2+q^2}\iint dx\,dy\,;$$

and applying this formula to the sphere by replacing $z\sqrt{1+p^2+q^2}$ by 1, ν by 0, and ρ by 1, we have

$$\cos i_1\,\delta\phi_1 = \iint d\xi\,d\eta,$$

the double integral being taken over the projection of the corresponding small area of the sphere.

Now by the well-known formula for the transformation of multiple integrals we have

$$\iint d\xi\,d\eta = \iint\left(\frac{d\xi}{dx}\frac{d\eta}{dy} - \frac{d\xi}{dy}\frac{d\eta}{dx}\right)dx\,dy\,;$$

and therefore

$$\frac{\cos i\,\delta\phi}{\cos i_1\,\delta\phi_1} = \frac{z\cos\nu\sqrt{1+p^2+q^2}}{\rho^3\left(\dfrac{d\xi}{dx}\dfrac{d\eta}{dy} - \dfrac{d\xi}{dy}\dfrac{d\eta}{dx}\right)}.$$

But the first of equations (4) gives

$$d\xi = \frac{(px+qy-z)\,dp - p\,(x\,dp+y\,dq)}{(px+qy-z)^2}$$

$$= \frac{\{(qy-z)\,r - pys\}\,dx + \{(qy-z)\,s - pyt\}\,dy}{(px+qy-z)^2}.$$

Similarly,

$$d\eta = \frac{\{(px-z)\,t - qxs\}\,dy + \{(px-z)\,s - qxr\}\,dx}{(px+qy-z)^2}.$$

Hence

$$\frac{d\xi}{dx}\frac{d\eta}{dy} - \frac{d\xi}{dy}\frac{d\eta}{dx} = \frac{V}{(px+qy-z)^4},$$

where

$$\begin{aligned} V &= \{(qy-z)\,r - pys\}\,\{(px-z)\,t - qxs\} \\ &\qquad - \{(qy-z)\,s - pyt\}\,\{(px-z)\,s - qxr\} \\ &= \{(qy-z)(px-z) - pqxy\}\,(rt-s^2) \\ &= z\,(z-px-qy)(rt-s^2). \end{aligned}$$

Hence

$$\frac{\cos i\,\delta\phi}{\cos i_1\,\delta\phi_1} = \frac{z\cos\nu\sqrt{1+p^2+q^2}}{\rho^3}\,\frac{(z-px-qy)^3}{z(rt-s^2)}.$$

But if ϖ be the perpendicular let fall from O on the tangent plane at P,

$$z-px-qy=\sqrt{1+p^2+q^2}\,.\,\varpi,$$

and therefore

$$\frac{\cos i\,\delta\phi}{\cos i_1\,\delta\phi_1} = \frac{\cos\nu\,.\,\varpi^3}{\rho^3}\,\frac{(1+p^2+q^2)^2}{rt-s^2}.$$

But $\varpi=\rho\cos\nu$. Also the quadratic determining the principal radii of curvature at P is

$$(rt-s^2)v^2+(\&c.)\,v+(1+p^2+q^2)^2=0;$$

and therefore if v_1, v_2 denote the principal radii of curvature,

$$v_1v_2=\frac{(1+p^2+q^2)^2}{rt-s^2}.$$

Hence

$$\frac{\cos i\,\delta\phi}{\cos i_1\,\delta\phi_1}=\cos^4\nu\,.\,v_1v_2 \quad\text{.................................}(5),$$

and

$$\frac{\cos i\,\delta\phi}{\cos i'\,\delta\phi'}=\frac{\cos i\,\delta\phi}{\cos i_1\,\delta\phi_1}\cdot\frac{\cos i_1\,\delta\phi_1}{\cos i'\,\delta\phi'}=\frac{\cos^4\nu\,.\,v_1v_2}{\cos^4\nu'\,.\,v_1'v_2'} \quad\text{.......}(6),$$

which proves the proposition enunciated.

In the particular case of an ellipsoid of revolution of which n is the axial and m the equatorial semi-axis, compared with a sphere of radius unity, both having their centres at O', one of the principal radii of curvature is the normal of the elliptic section, which by the properties of the ellipse is equal to $\frac{m}{n}m'$, m' denoting the semi-conjugate diameter; and the other is the radius of curvature of the elliptic section, or $\frac{m'^3}{mn}$. Also ϖ is the perpendicular let fall from the centre on the tangent line of the section. Hence from (5) or (6)

$$\frac{\cos i\,\delta\phi}{\cos i'\,\delta\phi'}=\frac{\varpi^4}{\rho^4}\cdot\frac{mm'}{n}\cdot\frac{m'^3}{mn}=\frac{\varpi^4m'^4}{n^2\rho^4}=\frac{m^4n^2}{\rho^4},$$

since $\varpi m'=mn$. This agrees with Mr Stewart's result (p. 197), since the R_e and $\frac{1}{2}R$ of Mr Stewart are the same as the R' and R of equation (3).

On the Intensity of the Light reflected from or transmitted through a Pile of Plates.

[From the *Proceedings of the Royal Society, January* 23, 1862.]

The frequent employment of a pile of plates in experiments relating to polarization suggests, as a mathematical problem of some interest, the determination of the mode in which the intensity of the reflected light, and the intensity and degree of polarization of the transmitted light, are related to the number of the plates, and, in case they be not perfectly transparent, to their defect of transparency.

The plates are supposed to be bounded by parallel surfaces, and to be placed parallel to one another. They will also be supposed to be formed of the same material, and to be of equal thickness, except in the case of perfect transparency, in which case the thickness does not come into account. The plates themselves and the interposed plates of air will be supposed, as is usually the case, to be sufficiently thick to prevent the occurrence of the colours of thin plates, so that we shall have to deal with intensities only.

On account of the different proportions in which light is reflected at a single surface according as the light is polarized in or perpendicularly to the plane of incidence, we must take account separately of light polarized in these two ways. Also, since the rate at which light is absorbed varies with its refrangibility, we must take account separately of the different constituents of white light. If, however, the plates be perfectly transparent, we may treat white light as a whole, neglecting as insignificant the chromatic variations of reflecting power. Let ρ be the fraction of the incident light reflected at the first surface of a plate. Then

$1-\rho$ may be taken as the intensity of the transmitted light*. Also, since we know that light is reflected in the same proportion externally and internally at the two surfaces of a plate bounded by parallel surfaces, the same expressions ρ and $1-\rho$ will serve to denote the fractions reflected and transmitted at the second surface. We may calculate ρ in accordance with Fresnel's formulæ from the expressions

$$\sin i' = \frac{\sin i}{\mu} \quad \text{.........................(1)},$$

$$\rho = \frac{\sin^2 (i-i')}{\sin^2 (i+i')}, \quad \text{or} = \frac{\tan^2 (i-i')}{\tan^2 (i+i')} \quad \text{...........(2)},$$

according as the light is polarized in or perpendicularly to the plane of incidence.

In the case of perfect transparency, we may in imagination make abstraction of the substance of the plates, and state the problem as follows:—There are $2m$ parallel surfaces (m being the number of plates) on which light is incident, and at each of which a given fraction ρ of the light incident upon it is reflected, the remainder being transmitted; it is required to determine the intensity of the light reflected from or transmitted through the system, taking account of the reflexions, infinite in number, which can occur in all possible ways.

This problem, the solution of which is of a simpler form than that of the general case of imperfect transparency, might be solved by a particular method. As, however, the solution is comprised in that of the problem which arises when the light is supposed to be partially absorbed, I shall at once pass on to the latter.

In consequence of absorption, let the intensity of light traversing a plate be reduced in the proportion of 1 to $1-q dx$ in passing over the elementary distance dx within the plate. Let T be the thickness of a plate, and therefore $T \sec i'$ the length of the path of the light within it. Then, putting for shortness

$$e^{-qT\sec i'} = g \quad \text{.........................(3)},$$

* In order that the intensity may be measured in this simple way, which saves rouble in the problem before us, we must define the intensity of the light transmitted across the first surface to mean what would be the intensity if the light were to emerge again into air across the second surface without suffering loss by absorption, or by reflexion at that surface.

1 to g will be the proportion in which the intensity is reduced by absorption in a single transit. The light reflected by a plate will be made up of that which is reflected at the first surface, and that which suffers 1, 3, 5, etc. internal reflexions. If the intensity of the incident light be taken as unity, the intensities of these various portions will be

$$\rho,\ (1-\rho)^2\rho g^2,\ (1-\rho)^2\rho^3 g^4,\ \text{etc.}$$

and if r be the intensity of the reflected light, we have, by summing a geometric series,

$$r = \rho + \frac{(1-\rho)^2 \rho g^2}{1-\rho^2 g^2} \quad \text{......................(4).}$$

Similarly, if t be the intensity of the transmitted light,

$$t = \frac{(1-\rho)^2 g}{1-\rho^2 g^2} \quad \text{........................(5),}$$

and we easily find

$$r = \rho + g\rho t;\ r + t = 1 - \frac{(1-\rho)(1-g)}{1-\rho g},$$

which is in general less than 1, but becomes equal to 1 in the limiting case of perfect transparency, in which case $g = 1$.

The values of μ, i, and q in any case being supposed known, the formulæ (1), (2), (3), (4), (5) determine r and t, which may now therefore be supposed known. The problem therefore is reduced to the following:—There are m parallel plates of which each reflects and transmits given fractions r, t of the light incident upon it: light of intensity unity being incident on the system, it is required to find the intensities of the reflected and refracted light.

Let these be denoted by $\phi(m)$, $\psi(m)$. Consider a system of $m+n$ plates, and imagine these grouped into two systems, of m and n plates respectively. The incident light being represented by unity, the light $\phi(m)$ will be reflected from the first group, and $\psi(m)$ will be transmitted. Of the latter the fraction $\psi(n)$ will be transmitted by the second group, and $\phi(n)$ reflected. Of the latter the fraction $\psi(m)$ will be transmitted by the first group, and $\phi(m)$ reflected, and so on. Hence we get for the light reflected by the whole system,

$$\phi(m) + (\psi m)^2 \phi(n) + (\psi m)^2 \phi(m)(\phi n)^2 + \ldots,$$

and for the light transmitted,

$$\psi(m)\psi(n) + \psi(m)\phi(n)\phi(m)\psi(n) + \psi(m)(\phi n)^2(\phi m)^2\psi(n) + \ldots,$$

which gives, by summing the two geometric series,

$$\phi(m+n) = \phi(m) + \frac{(\psi m)^2 \phi(n)}{1-\phi(m)\phi(n)} \quad \text{..........(6);}$$

$$\psi(m+n) = \frac{\psi(m)\psi(n)}{1-\phi(m)\phi(n)} \quad \text{..................(7).}$$

We get from (6)

$$\phi(m+n)\{1-\phi(m)\phi(n)\} = \phi(m) + \phi(n)\{(\psi m)^2 - (\phi m)^2\};$$

and the first member of this equation being symmetrical with respect to m and n, we get, by interchanging m and n and equating the results,

$$\phi(m) + \phi(n)\{(\psi m)^2 - (\phi m)^2\} = \phi(n) + \phi(m)\{(\psi n)^2 - (\phi n)^2\};$$

or $$\frac{1}{\phi(m)}\{1 + (\phi m)^2 - (\psi m)^2\} = \frac{1}{\phi(n)}\{1 + (\phi n)^2 - (\psi n)^2\},$$

which is therefore constant. Denoting this constant for convenience by $2\cos\alpha$, we have

$$(\psi m)^2 = 1 - 2\cos\alpha \,.\, \phi(m) + (\phi m)^2 \quad \text{...........(8).}$$

Squaring (7), and eliminating the function ψ by means of (8), we find

$$\{1-\phi(m)\phi(n)\}^2\{1-2\cos\alpha\,.\,\phi(m+n) + [\phi(m+n)]^2\}$$
$$= \{1-2\cos\alpha\,.\,\phi(m) + (\phi m)^2\}\{1-2\cos\alpha\,.\,\phi(n) + (\phi n)^2\} \text{...(9).}$$

From the nature of the problem, m and n are positive integers, and it is only in that case that the functions ϕ, ψ, as hitherto defined, have any meaning. We may, however, contemplate functions ϕ, ψ of a continuously changing variable, which are defined by the equations (6) and (7); and it is evident that if we can find such functions, they will in the particular case of a positive integral value of the variable be the functions which we are seeking.

In order that equations (6), (7) may hold good for a value zero of one of the variables, suppose n, we must have $\phi(0) = 0$, $\psi(0) = 1$. The former of these equations reduces (9) for $n = 0$ to an identical equation. Differentiating (9) with respect to n, and after differentiation putting $n = 0$, we find

$$\phi'(0)\phi(m)\{1-2\cos\alpha\,.\,\phi(m) + (\phi m)^2\} + \cos\alpha\,.\,\phi'(m) - \phi(m)\phi'(m)$$
$$= \cos\alpha\,.\,\phi'(0)\{1-2\cos\alpha\,.\,\phi(m) + (\phi m)^2\},$$

or dividing out by $\phi(m) - \cos\alpha$ (for $\phi(m) = \cos\alpha$ would only lead to $\phi(m) = \cos\alpha = 0$, $\psi(m) = C$),

$$\phi'(m) = \phi'(0)\{1 - 2\cos\alpha . \phi(m) + (\phi m)^2\} \quad \text{.........(10).}$$

Integrating this equation, determining the arbitrary constant by the condition that $\phi(m) = 0$ when $m = 0$, and writing β for $\sin\alpha . \phi'(0)$, we have

$$\phi(m) = \frac{\sin m\beta}{\sin(\alpha + m\beta)} \quad \text{.................(11).}$$

Substituting in (8) and reducing, we find

$$(\psi m)^2 = \frac{\sin^2\alpha}{\sin^2(\alpha + m\beta)} \quad \text{.................(12).}$$

But (8) was derived, not from (7) directly, but from (7) squared; and on extracting the square root of both sides of (12), we must choose that sign which shall satisfy (7), and therefore we must take the sign +, as we see at once on putting $m = n = 0$. The equation (12) on taking the proper root, and (11), may be put under the form

$$\frac{\phi(m)}{\sin(m\beta)} = \frac{\psi(m)}{\sin\alpha} = \frac{1}{\sin(\alpha + m\beta)} \quad \text{...........(13);}$$

and to determine the arbitrary constants α, β we have, putting $m = 1$, and $\phi(m) = r$, $\psi(m) = t$,

$$\frac{r}{\sin\beta} = \frac{t}{\sin\alpha} = \frac{1}{\sin(\alpha + \beta)} \quad \text{...............(14).}$$

We readily get from equations (13),

$$1 - \phi(m)\phi(n) = \frac{\sin\alpha \sin\{\alpha + (m + n)\beta\}}{\sin(\alpha + m\beta)\sin(\alpha + n\beta)};$$

$$\phi(m + n) - \phi(m) = \frac{\sin\alpha \sin n\beta}{\sin\{\alpha + (m + n)\beta\}\sin(\alpha + m\beta)};$$

whence the equations (6), (7) are easily verified. This verification seems necessary in logical strictness, because we have no right to assume *à priori* that it is possible to satisfy (6) and (7) for general values of the variables; and in deriving the equation (10), the equations (6) and (7) were only assumed to hold good for general values of m and infinitely small values of n.

The equations (13), (14) give the following *quasi*-geometrical construction for solving the problem:—Construct a triangle of which the sides represent in magnitude the intensity of the incident, reflected, and refracted light in the case of a single plate, and then, leaving the first side and the angle opposite to the third unchanged, multiply the angle opposite to the second by the number of plates; the sides of the new triangle will represent the corresponding intensities in the case of the system of plates. I say *quasi*-geometrical, because the construction cannot actually be effected, inasmuch as the first side of our triangle is greater than the sum of the two others, and the angles are imaginary.

To adapt the formulæ (13), (14) to numerical calculation, it will be convenient to get rid of the imaginary quantities. Putting

$$\sqrt{\{(1+r+t)(1+r-t)(1+t-r)(1-r-t)\}} = \Delta \ldots (15),$$

we have by the common formulæ of trigonometry,

$$\cos\alpha = \frac{1+r^2-t^2}{2r}; \quad \sin\alpha = \frac{\pm\sqrt{-1}\Delta}{2r};$$

whence, putting

$$\frac{1}{2r}(1+r^2-t^2+\Delta) = a \ \ldots\ldots\ldots\ldots\ldots\ldots (16),$$

we have

$$e^{\sqrt{-1}\alpha} = \cos\alpha + \sqrt{-1}\sin\alpha = a^{\mp 1}.$$

It is a matter of indifference which sign be taken: choosing the under signs, we have

$$2r\sin\alpha = -\sqrt{-1}\Delta, \quad e^{\sqrt{-1}\alpha} = a.$$

We have also

$$\cos\beta = \frac{1+t^2-r^2}{2t}, \quad \sin\beta = \frac{r}{t}\sin\alpha = -\frac{\sqrt{-1}\Delta}{2t},$$

no fresh ambiguity of sign being introduced. Putting therefore

$$\frac{1}{2t}(1+t^2-r^2+\Delta) = b \ \ldots\ldots\ldots\ldots\ldots\ldots (17),$$

we have

$$e^{\sqrt{-1}\beta} = b;$$

and equations (13) now give

$$\frac{\phi(m)}{b^m - b^{-m}} = \frac{\psi(m)}{a - a^{-1}} = \frac{1}{ab^m - a^{-1}b^{-m}} \ldots\ldots\ldots (18).$$

In the case of perfect transparency these expressions take a simpler form. If $r+t$ differ indefinitely little from 1, α and β will be indefinitely small. Making α and β indefinitely small in (13) and (14), and putting $1-r$ for t, we find

$$\frac{\phi(m)}{mr}=\frac{\psi(m)}{1-r}=\frac{1}{1+(m-1)r} \quad \text{..............(19)}.$$

In this case it is evident that each of the $2m$ reflecting surfaces might be regarded as a separate plate reflecting light in the proportion of ρ to 1, and therefore we ought also to have, writing $2m$ for m and ρ for r in the denominators of the equations (19),

$$\frac{\phi(m)}{2m\rho}=\frac{\psi(m)}{1-\rho}=\frac{1}{1+(2m-1)\rho} \quad \text{...........(20)*}.$$

It is easy to verify that when $g=1$ (4) reduces (19) to (20).

The following Table gives the intensity of the light reflected from or transmitted through a pile of m plates for the values 1, 2, 4, 8, 16, 32, and ∞ of m, for three degrees of transparency, and for certain selected angles of incidence. The assumed refractive index μ is 1·52. $\delta=1-e^{-qT}$ is the loss by absorption in a single transit of a plate at a perpendicular incidence, so that $\delta=0$ corresponds to perfect transparency. The most interesting angles of incidence to select appeared to be zero and the polarizing angle $\varpi=\tan^{-1}\mu$; but in the case of perfect transparency the result has also been calculated for an angle of incidence a little (2°) greater than the polarizing angle. ϕ denotes the intensity of the reflected and ψ that of the transmitted light, the intensity of the incident light being taken at 1000. For oblique incidences it was necessary to distinguish between light polarized in and light polarized perpendicularly to the plane of incidence; the suffixes 1, 2 refer to these two kinds respectively. For oblique incidences a column is added giving the ratio of ψ_1 to ψ_2, which may be taken as a measure of the defect of polarization of the transmitted light. No such column was required for $\delta=0$ and $i=\varpi$, because in this case $\psi_2=1000$.

* From a paper by M. Wild in *Poggendorff's Annalen* [Vol. IX, 1856, p. 240], I find that the formulæ for the particular case of perfect transparency have already been given by M. Neumann. His demonstration does not appear to have been published. [The question of the efficiency of a pile of plates was also considered by Fresnel himself; his solution, which neglects opacity, was published posthumously, *Œuvres Complètes*, Vol. II, 1868, pp. 789–92.]

	$\delta=0$									$\delta=\frac{1}{50}$						$\delta=\frac{1}{10}$					
	$i=0$		$i=\varpi$		$i=\varpi+2°$					$i=0$		$i=\varpi$				$i=0$		$i=\varpi$			
m	ϕ	ψ	ϕ_1	ψ_1	ϕ_1	ψ_1	ϕ_2	ψ_2	$\frac{\psi_1}{\psi_2}$	ϕ	ψ	ϕ_1	ψ_1	ψ_2	$\frac{\psi_1}{\psi_2}$	ϕ	ψ	ϕ_1	ψ_1	ψ_2	$\frac{\psi_1}{\psi_2}$
1	82	918	271	729	300	700	1	999	·701	80	900	265	711	976	·728	74	826	245	639	881	·725
2	151	849	426	574	459	541	2	998	·542	145	815	410	544	953	·571	125	686	351	435	777	·559
4	262	738	598	402	628	372	4	996	·373	244	679	555	355	908	·391	185	479	427	215	604	·357
8	416	584	749	251	771	229	8	992	·231	364	490	656	182	824	·221	229	237	451	57	365	·156
16	587	413	856	144	870	130	16	984	·132	464	276	695	58	679	·086	243	59	453	4	133	·030
32	740	260	922	78	931	69	32	968	·071	509	97	699	7	461	·014	244	4	453	0	18	·001
∞	1000	0	1000	0	1000	0	1000	0	·000	516	0	699	0	0	·000	244	0	453	0	0	·000

The intensity of the light reflected from an infinite number of plates, as we see from (18), is a^{-1}; and since a is changed into a^{-1} by changing the sign of α or of Δ,

$$a^{-1} = \frac{1}{2r}(1 + r^2 - t^2 - \Delta) \quad\text{.................}(21),$$

which is equal to 1 in the case of perfect transparency. Accordingly a substance which is at the same time finely divided, so as to present numerous reflecting surfaces, and which is of such a nature as to be transparent in mass, is brilliantly white by reflected light, —for example snow, and colourless substances thrown down as precipitates in chemical processes.

The intensity of the light reflected from a pile consisting of an infinite number of similar plates falls off rapidly with the transparency of the material of which the plates are composed, especially at small incidence. Thus at a perpendicular incidence we see from the above Table that the reflected light is reduced to little more than one half when 2 per cent. is absorbed in a single transit, and to less than a quarter when 10 per cent. is absorbed.

With imperfectly transparent plates, little is gained by multiplying the plates beyond a very limited number, if the object be to obtain light, as bright as may be, polarized by reflexion. Thus the Table shows that 4 plates of the less defective kind reflect 79 per cent., and 4 plates of the more defective as much as 94 per cent., of the light that could be reflected by a greater number, whereas 4 plates of the perfectly transparent kind reflect only 60 per cent.

The Table shows that while the amount of light transmitted at the polarizing angle by a pile of a considerable number of plates is materially reduced by a defect of transparency, its state of polarization is somewhat improved. This result might be seen without calculation. For while no part of the transmitted light which is polarized perpendicularly to the plane of incidence underwent reflexion, a large part of the transmitted light polarized the other way was reflected an even number of times; and since the length of path of the light within the absorbing medium is necessarily increased by reflexion, it follows that a defect of transparency must operate more powerfully in reducing the intensity of light polarized in, than of light polarized perpendicularly to the plane of polarization. But the Table also shows that a far better result can be obtained, as to the perfection of the polarization of the transmitted light, without any greater loss of illumination, by employing a larger number of plates of a more transparent kind.

Let us now confine our attention to perfectly transparent plates, and consider the manner in which the degree of polarization of the transmitted light varies with the angle of incidence.

The degree of polarization is expressed by the ratio of ψ_1 to ψ_2, which for brevity will be denoted by χ. When $\chi = 1$ there is no polarization; when $\chi = 0$ the polarization is perfect, in a plane perpendicular to the plane of incidence. Now ψ (which is used to denote ψ_1 or ψ_2 as the case may be) is given in terms of ρ by one of the equations (20), and ρ is given in terms of $i - i'$ and $i + i'$ by Fresnel's formulæ (2). Put

$$i - i' = \theta, \qquad i + i' = \sigma;$$

then, from (1),

$$\frac{di}{\tan i} = \frac{di'}{\tan i'} = \frac{d\theta}{\tan i - \tan i'} = \frac{d\sigma}{\tan i + \tan i'} = \cos i \cos i' d\omega, \text{ suppose,}$$

whence

$$d\theta = \sin\theta d\omega, \qquad d\sigma = \sin\sigma d\omega \quad \ldots\ldots\ldots\ldots(22);$$

and we see that i and ω increase together from $i = 0$ to $i = \frac{1}{2}\pi$. We have also

$$\rho_1 = \frac{\sin^2\theta}{\sin^2\sigma}, \quad d\rho_1 = \frac{2\sin\theta}{\sin^3\sigma}(\sin\sigma\cos\theta d\theta - \sin\theta\cos\sigma d\sigma)$$

$$= \frac{2\sin^2\theta}{\sin^2\sigma}(\cos\theta - \cos\sigma)\,d\omega;$$

$$\rho_2 = \frac{\tan^2\theta}{\tan^2\sigma}, \quad d\rho_2 = \frac{2\tan\theta}{\tan^3\sigma}(\tan\sigma\sec^2\theta d\theta - \tan\theta\sec^2\sigma d\sigma)$$

$$= \frac{2\sin^2\theta\cos\sigma}{\cos^3\theta\sin^2\sigma}(\cos\sigma - \cos\theta)\,d\omega = -\frac{\cos\sigma}{\cos^3\theta}d\rho_1.$$

Now $\cos\theta - \cos\sigma$ or $2\sin i \sin i'$ is positive; and $\cos\sigma$ is positive from $i = 0$ to $i = \varpi$, and negative from $i = \varpi$ to $i = \frac{1}{2}\pi$. But (20) shows that ψ decreases as ρ increases. From $i = 0$ to $i = \varpi$, ρ_1 increases and ρ_2 decreases, and therefore ψ_1 decreases and ψ_2 increases, and therefore on both accounts χ decreases. When $i = \varpi$, $d\rho_1/di$ is still positive, and therefore $d\psi_1/di$ negative, but ψ_2 has its maximum value 1, so that on passing through the polarizing angle χ still decreases, or the polarization improves. When the plates are very numerous, $\psi_2 = 1$ at the polarizing angle, and on both sides of it decreases rapidly, whereas ψ_1, which is always small, suffers no particular change about the polarizing angle. Hence in this case χ must be a minimum a little beyond the polarizing angle. Let us then seek the angle of incidence which makes χ a minimum in the case of an arbitrary number of plates.

We have from (20) and (2),

$$\chi = \frac{\sin^2\sigma - \sin^2\theta}{\sin^2\sigma + (2m-1)\sin^2\theta} \cdot \frac{\sin^2\sigma\cos^2\theta + (2m-1)\sin^2\theta\cos^2\sigma}{\sin^2\sigma\cos^2\theta - \sin^2\theta\cos^2\sigma}$$

$$= \frac{\sin^2\sigma\cos^2\theta + (2m-1)\sin^2\theta\cos^2\sigma}{\sin^2\sigma + (2m-1)\sin^2\theta}$$

$$= 1 - \frac{2m}{\operatorname{cosec}^2\theta + (2m-1)\operatorname{cosec}^2\sigma} \quad \text{.........................(23).}$$

Hence χ is a minimum along with $\operatorname{cosec}^2\theta + (2m-1)\operatorname{cosec}^2\sigma$. Differentiating, and taking account of the formulæ (22), we find, to determine the angle of maximum polarization, the very simple equation

$$\cos\theta\sin^2\sigma + (2m-1)\cos\sigma\sin^2\theta = 0 \quad \text{........(24).}$$

For any assumed value of i from ϖ to $\frac{1}{2}\pi$, this equation gives at once the value of m, that is, the number of plates of which a pile must be composed in order that the assumed incidence may be that of maximum polarization of the transmitted light. The equation may be put under the form

$$2m - 1 = -\frac{\tan\sigma}{\tan\theta} \cdot \frac{\sin\sigma}{\sin\theta} = \frac{1}{\sqrt{\rho_1\rho_2}}.$$

Now we have seen that both ρ_1 and ρ_2 continually increase, and therefore m continually decreases, from $i = \varpi$ to $i = \frac{1}{2}\pi$. At the first of these limits $\rho_2 = 0$, and therefore $m = \infty$. At the second $\rho_1 = \rho_2 = 1$, and therefore $m = 1$. Hence with a single plate the polarization of the transmitted light continually improves up to a grazing incidence, but with a pile of plates the polarization attains a maximum at an angle of incidence which approaches indefinitely to the polarizing angle as the number of plates is indefinitely increased.

Eliminating m from (23) and (24), we find

$$\chi = -\cos\theta\cos\sigma \quad \text{.....................(25),}$$

which determines for any pile χ_1, the defect of maximum polarization of the transmitted light, in terms of the angle of incidence for which the polarization is a maximum. We have, from (25), (22), and (24),

$$d\chi_1 = (\sin^2\theta \cos\sigma + \sin^2\sigma \cos\theta)\, d\omega = -2(m-1)\cos\sigma \sin^2\theta\, d\omega,$$

and $\cos\sigma$ is negative. Hence χ_1 decreases as ω (and therefore i) decreases, or as m increases. For $m = 1$, $i = \frac{1}{2}\pi$ and $\chi_1 = \mu^{-2}$; for $m = \infty$, $\cos\sigma = 0$, and therefore $\chi_1 = 0$, or the maximum polarization tends indefinitely to become perfect as the number of plates is indefinitely increased.

For a given number of plates the angle of maximum polarization may be readily found from (24) by the method of trial and error. But for merely examining the progress of the functions, instead of tabulating i for assumed values of m, it will serve equally well to tabulate m for assumed values of i. The following Table gives for assumed angles of incidence, decreasing by 5° from 90°, the number of plates required to make these angles the angles of maximum polarization of the transmitted light, and the value of χ_1, which determines the defect of polarization.

$i =$	90°	85°	80°	75°	70°	65°	60°	56° 40′ ($=\varpi$)
$m =$	1	1·330	1·944	2·913	4·921	9·775	30·372	∞
$\chi_1 =$	·433	·422	·390	·337	·265	·177	·075	0

REPORT ON DOUBLE REFRACTION.

[From the *Report of the British Association for the Advancement of Science*, Cambridge, 1862, pp. 253—282.]

I REGRET to say that in consequence of other occupations the materials for a complete report on Physical Optics, which the British Association have requested me to prepare, are not yet collected and digested. Meanwhile, instead of requesting longer time for preparation, I have thought it would be well to take up a single branch of the subject, and offer a report on that alone. I have accordingly taken the subject of double refraction, having mainly in view a consideration of the various dynamical theories which have been advanced to account for the phenomenon on the principle of transversal vibrations, and an indication of the experimental measurements which seem to me most needed to advance this branch of optical science. As the greater part of what has been done towards placing the theory of double refraction on a rigorous dynamical basis is subsequent to the date of Dr Lloyd's admirable report on "Physical Optics," I have thought it best to take a review of the whole subject, though at the risk of repeating a little of what is already contained in that report.

The celebrated theory of Fresnel was defective in rigour in two respects, as Fresnel himself clearly perceived. The first is that the expression for the force of restitution is obtained on the supposition of the *absolute* displacement of a molecule, whereas in undulations of all kinds the forces of restitution with which we are concerned are those due to *relative* displacements. Fresnel endeavoured to show, by reasoning professedly only probable, that while the *magnitude* of the force of restitution is altered in passing from absolute to relative displacements, the *law* of the force as to its dependence on the direction of vibration remains the same. The other point relates to the neglect of the component of the force in a direction perpendicular to the front of a wave. In the state of things supposed in the calculation of the forces of restitution called into play by absolute displacements, there is no immediate recognition of a wave at all, and a molecule is supposed to be as free to move in one direction as in another. But a displacement in a direction perpendicular to the front of a wave would call into play new forces of restitution having a resultant not in general in the direction of displacement; so that even the component of the force of restitution in a direction parallel to the front of a wave would have an expression altogether different from that determined by the theory of Fresnel. But the absolute displacements are only considered for the sake of obtaining results to be afterwards applied to relative displacements; and Fresnel distinctly makes the supposition that the ether is incompressible, or at least is sensibly so under the action of forces comparable with those with which we are concerned in the propagation of light. This supposition removes the difficulty; and though it increases the number of hypotheses as to the existing state of things, it cannot be objected to in point of rigour, unless it be that a demonstration might be required that incompressibility is not inconsistent with the assumed constitution of the ether, according to which it is regarded as consisting of distinct material points, symmetrically arranged, and acting on one another with forces depending, for a given pair, only on the distance. Hence the neglect of the force perpendicular to the fronts of the waves is not so much a new defect of rigour, as the former defect appearing under a new aspect.

I have mentioned these points because sometimes they are slurred over, and Fresnel's theory spoken of as if it had been rigorous throughout, to the injury of students and the retardation of the real progress of science; and sometimes, on the other hand, the grand advance made by Fresnel is depreciated on account of his theory not being everywhere perfectly rigorous. If we reflect on the state of the subject as Fresnel found it, and as he left it, the wonder is, not that he failed to give a rigorous dynamical theory, but that a single mind was capable of effecting so much.

The first deduction of the laws of double refraction, or at least of an approximation to the true laws, from a rigorous theory is due to Cauchy*, though Neumann† independently, and almost simultaneously, arrived at the same results. In the theory of Cauchy

* *Mémoires de l'Académie*, tom. x, p. 293.

† *Poggendorff's Annalen*, Vol. xxv, p. 418 (1832).

and Neumann the ether is supposed to consist of distinct particles, regarded as material points, acting on one another by forces in the line joining them which vary as some function of the distances, and the arrangement of these particles is supposed to be different in different directions. The medium is further supposed to possess three rectangular planes of symmetry, the double refraction of crystals, so far as has been observed, being symmetrical with respect to three such planes. The equations of motion of the medium are deduced by a method similar to that employed by Navier in the case of an isotropic medium. The equations arrived at by Cauchy, the medium being referred to planes of symmetry, contain nine arbitrary constants, three of which express the pressures in the principal directions in the state of equilibrium. Those employed by Neumann contain only six such constants, the medium in its natural state being supposed free from pressure.

In the theory of double refraction, whatever be the particular dynamical conditions assumed, everything is reduced to the determination of the velocity of propagation of a plane wave propagated in any given direction, and the mode of vibration of the particles in such a wave which must exist in order that the wave may be propagated with a unique velocity. In the theory of Cauchy now under consideration, the direction of vibration and the reciprocal of the velocity of propagation are given in direction and magnitude respectively by the principal axes of a certain ellipsoid, the equation of which contains the nine arbitrary constants, and likewise the direction-cosines of the wave-normal. Cauchy adduces reasons for supposing that the three constants G, H, I, which express the pressures in the state of equilibrium, vanish, which leaves only six constants. For waves perpendicular to the principal axes, the squared velocities of propagation and the corresponding directions of vibration are given by the following Table:—

Wave-normal		x	y	z
Direction of vibration	x	L	R	Q
	y	R	M	P
	z	Q	P	N

For waves *in these directions*, then, the vibrations are either wholly normal or wholly transversal. The latter are those with which we have to deal in the theory of light. Now, according to observation, in any one of the principal planes of a doubly refracting crystal, that ray which is polarized in the principal plane obeys the ordinary law of refraction. In order therefore that the conclusions of this theory should at all agree with observation, we must suppose that in polarized light the vibrations are parallel, not perpendicular, to the plane of polarization.

Let l, m, n be the direction-cosines of the wave-normal. In the theory of Cauchy and Neumann, the square v^2 of the velocity of propagation is given by a cubic of the form

$$v^6 + \alpha_2 v^4 + \alpha_4 v^2 + \alpha_6 = 0,$$

where α_2, α_4, α_6 are homogeneous functions of the 1st order as regards L, M, N, P, Q, R, and homogeneous functions of the orders 2, 4, 6 as regards l, m, n, involving even powers only of these quantities. For a wave perpendicular to one of the principal planes, that of yz suppose, the cubic splits into two rational factors, of which that which is of the first degree in v^2, namely,

$$v^2 - m^2R - n^2Q,$$

corresponds to vibrations perpendicular to the principal plane. This is the same expression as results from Fresnel's theory, and accordingly the section, by the principal plane, of one sheet of the wave-surface, which in this theory is a surface of three sheets, is an ellipse, and the law of refraction of that ray which is polarized perpendicularly to the principal plane agrees exactly with that given by the theory of Fresnel.

For the two remaining waves, the squared velocities of propagation are given by the quadratic

$$(v^2 - m^2M - n^2P)(v^2 - m^2P - n^2N) - 4m^2n^2P^2 = 0 \quad \ldots(1);$$

but according to observation the ray polarized in the principal plane obeys the ordinary law of refraction. Hence (1) ought to be satisfied by $v^2 - (m^2 + n^2)P = 0$, which requires that

$$(M - P)(N - P) = 4P^2,$$

on which supposition the remaining factor must evidently be linear as regards m^2, n^2, and therefore must be

$$v^2 - m^2M - n^2N,$$

since it gives when equated to zero $v^2 = M$, or $v^2 = N$ for $m = 1$, or $n = 1$. And since the same must hold good for each of the principal planes, we must have the three following relations between the six constants,

$$(M-P)(N-P)=4P^2;\quad (N-Q)(L-Q)=4Q^2;$$
$$(L-R)(M-R)=4R^2 \ldots\ldots\ldots\ldots\ldots\ldots (2).$$

The existence of six constants, of which only three are wanted to satisfy the numerical values of the principal velocities of propagation in a biaxal crystal, permits of satisfying these equations; so that the law that the ray polarized in the plane of incidence, when that is a principal plane, obeys the ordinary law of refraction is *not inconsistent* with Cauchy's theory. This simple law is, however, not in the slightest degree predicted by the theory, nor even rendered probable, nor have any physical conditions been pointed out which would lead to the relations (2); and, indeed, from the form of these equations, it seems hard to conceive what physical relations they could express. Hence an important desideratum would be left, even if the theory were satisfactory in all other respects.

The equation for determining v^2 virtually contains the theoretical laws of double refraction, which are embodied in the form of the wave-surface. The wave-surface of Cauchy and Neumann does not agree with that of Fresnel, except as to the sections of two of its sheets by the principal planes, the third sheet being that which relates to nearly normal vibrations. Nevertheless the first two sheets, being forced to agree in their principal sections with Fresnel's surface, differ from it elsewhere extremely little. In Arragonite, for instance, in a direction equally inclined to the principal axes, assuming Rudberg's indices* for the line D, I find that the velocities of propagation of the two polarized waves, according to the theory of Cauchy and Neumann, differ from those resulting from the theory of Fresnel only in the *tenth* place of decimals, the velocity in air being taken as unity. Such a difference as this would of course be utterly insensible in experiment. In like manner the directions of the planes of polarization according to the two theories, though not rigorously, are extremely nearly the same, the plane of polarization of a wave in which the

* *Annales de Chimie*, tom. XLVIII, p. 254 (1831).

vibrations are nearly transversal being defined as that containing the direction of propagation and the direction of vibration, in harmony with the previously established definition for the case of strictly transversal vibrations.

Hence as far as regards the laws of double refraction of the two waves which *alone* are supposed to relate to the visible phenomenon, and of the accompanying polarization, this theory, *by the aid of the forced relations* (2), is very successful. I am not now discussing the generality, or, on the contrary, the artificially restricted nature, of the fundamental suppositions as to the state of things, but only the degree to which the results are in accordance with observed facts. But as regards the third wave the case is very different. That theory should point to the necessary existence of such a wave consisting of strictly normal vibrations, and yet to which no known phenomenon can be referred, is bad enough; but in the present theory the vibrations are not even strictly normal, except for waves in a direction perpendicular to any one of the principal axes. In Iceland spar, for instance, for waves propagated in a direction inclined 45° to the axis, it follows from the numerical values of the refractive indices for the fixed line D given by Rudberg that the two vibrations in the principal plane which can be propagated independently of each other are inclined at angles of 9° 50′ and 80° 10′, or say 10° and 80°, to the wave-normal. We can hardly suppose that a mere change of inclination in the direction of vibration of from 10° to 80° with the wave-front makes all the difference whether the wave belongs to a long-known and evident phenomenon, no other than the ordinary refraction in Iceland spar, or not to any visible phenomenon at all.

It is true that before there can be any question of the third wave's being perceived it must be supposed excited, and the means of exciting it consist in the incident vibrations in air, which by hypothesis are strictly transversal. Hence we have to inquire whether the intensity of the third wave is such as to lead us to expect a sensible phenomenon answering to it. This leads us to the still more uncertain subject of the intensity of light reflected or refracted at the surface of a crystal—more uncertain because it not only depends on the laws of internal propagation, and involves all the hypotheses on which these laws are theoretically deduced, but requires fresh hypotheses as to the state of things at the confines of two media, introducing thereby fresh elements of uncertainty. But for our present purpose no exact calculation of intensities is required; a rough estimate of the intensity of the nearly normal vibrations is quite sufficient.

In order to introduce as little as possible relating to the theory of the intensity of reflected and refracted light, suppose the incident light to fall perpendicularly on the surface of a crystal, and let this be a surface of Iceland spar cut at an inclination of 45° to the axis. For a cleavage plane the result would be nearly the same. Let the incident light be polarized, and the vibrations be in the principal plane, which therefore, according to the theory now under consideration, must be the plane of polarization. The incident vibrations are parallel to the surface, and accordingly inclined at angles of 9° 50′ and 80° 10′ to the directions of the nearly transversal and nearly normal vibrations, respectively, within the crystal. Hence it seems evident that the amplitude of the latter must be of the order of magnitude of sin 9° 50′, or about $\frac{1}{5 \cdot 9}$, the amplitude of vibration in the incident light being taken as unity. The velocity of propagation of the nearly normal vibrations being to that of the nearly transversal roughly as $\sqrt{3}$ to 1, as will immediately be shown, it follows that the *vis viva* of the nearly normal would be to that of the nearly transversal vibrations in a ratio comparable with that of $\sqrt{3} \times \sin^2 9° \, 50'$ to 1, or about $\frac{1}{26}$ to 1. Hence the intensity of the nearly normal vibrations is by no means insignificant, and therefore it is a very serious objection to the theory that no corresponding phenomenon should have been discovered. It has been suggested by some of the advocates of this theory that the normal vibrations may correspond to heat. But the fact of the polarization of heat at once negatives such a supposition, even without insisting on the accumulation of evidence in favour of the identity of radiant heat and light of the same refrangibility.

But the objections to the theory on the ground of the absence of some unknown phenomenon corresponding with the third ray, to which the theory necessarily conducts, are not the only ones which may be urged against it in connexion with that ray. The existence of normal or nearly normal vibrations entails consequences respecting the transversal which could hardly fail to have been detected by observation. In the first place, the *vis viva* belonging to the normal vibrations is so much abstracted from the transversal, which alone by hypothesis constitute light, so that there is a loss of light inherent in the very act of passage from air into the crystal, or conversely, from the crystal into air. About $\frac{1}{26}$th of the whole might thus be expected to be lost at a single surface of Iceland spar, the surface being inclined 45° to the axis, and the light being incident perpendicularly, and being polarized in the principal plane; and the loss would amount to somewhere about $\frac{1}{13}$th in passage across a plate bounded by parallel surfaces, by which amount the sum of the reflected and transmitted light ought to fall short of the incident. And it is evident that something of the same kind must take place at other inclinations to the axis and at other incidences. The loss thus occasioned in multiplied reflexions could hardly have escaped observation, though it is not quite so great as might at first sight appear, as the transversal vibrations produced back again by the normal would presently become sensible.

But the most fatal objection of all is that urged by Green* against the supposition that normal vibrations could be propagated with a velocity comparable with those of transversal. As transversal vibrations are capable (according to the suppositions here combated) of giving rise at incidence on a medium to normal or nearly normal vibrations within it, so conversely the latter on arriving at the second surface are capable of giving rise to emergent transversal vibrations; so that not only would normal vibrations entail a loss of light in the quarter in which light is looked for, but would give rise to light (of small intensity it is true, but by no means imperceptible) in a quarter in which otherwise there would have been none at all. Thus in the case supposed above, the intensity of the light produced by nearly normal vibrations giving rise on emergence to transversal vibrations would be somewhere about the $(\frac{1}{26})^2$ or the $\frac{1}{676}$ of the incident light. In the case of light transmitted through a plate, the rays thus produced would be parallel to the incident, or to the emergent rays of the kind usually considered; but if the plate were wedge-shaped the two would come out in different directions, and with sunlight the former could not fail to be perceived. The only way apparently of

* *Camb. Phil. Trans.*, Vol. VII, p. 2. [Green's *Math. Papers*, p. 243.]

getting over this difficulty, is by making the perfectly gratuitous assumption that the medium, though perfectly transparent for the more nearly transversal vibrations, is intensely opaque for those more nearly normal.

Lastly, Green's argument respecting the necessity of supposing the velocity of propagation of normal vibrations very great has here full force as an objection against this theory. The constants P, Q, R are the squared reciprocals of the three principal indices of refraction, which are given by observation, and L, M, N are determined in terms of P, Q, R by the equations (2), by the solution of a quadratic equation. In the case of a uniaxal crystal everything is symmetrical about one of the axes, suppose that of z, which requires, as Cauchy has shown, that $L = M = 3R$, and $P = Q$; and of the equations (2) one is now satisfied identically, and the two others are identical with each other, and give

$$N = P + \frac{4P^2}{3R - P}.$$

For an isotropic medium we must have $L = M = N = 3P = 3Q = 3R$, and the three equations (2) are satisfied identically. The velocity of propagation of normal must be to that of transversal vibrations as $\sqrt{3}$ to 1, and cannot therefore be assumed to be what may be convenient for explaining the law of intensity of reflected light.

The theory which has just been discussed is essentially bound up with the supposition that in polarized light the vibrations are parallel, not perpendicular, to the plane of polarization. In prosecuting the study of light, Cauchy saw reason to change his views in this respect, and was induced to examine whether his theory could not be modified so as to be in accordance with the latter alternative. The result, constituting what may be called Cauchy's second theory, is contained in a memoir read before the Academy, May 20, 1839*. In this he refers to his memoir on dispersion, in which the fundamental equations are obtained in a manner somewhat different from that given in his "Exercices," but based on the same suppositions as to the constitution of the ether. In the new theory Cauchy retains the three constants G, H, I, expressing the pressures in equilibrium, which formerly he made vanish, the

* "Sur la Polarisation rectiligne, et la double Réfraction," *Mém. de l'Académie*, tom. XVIII, p. 153.

medium being supposed as before to be symmetrical with respect to three rectangular planes. The squares of the velocities of propagation, and the corresponding directions of vibration for the three waves which can be propagated in the direction of each of the principal axes, are given by the following Table.

Wave-normal		x	y	z
Direction of vibration	x	$L+G$	$R+H$	$Q+I$
	y	$R+G$	$M+H$	$P+I$
	z	$Q+G$	$P+H$	$N+I$

According to observation, in each of the principal planes the ray polarized in that plane obeys the ordinary law of refraction, and therefore if we suppose that in polarized light the vibrations, at least when strictly transversal, are perpendicular to the plane of polarization, we must assume that $R+H=Q+I$, $P+I=R+G$, $Q+G=P+H$, which are equivalent to only two distinct relations, namely,

$$P-G=Q-H=R-I \quad \text{.....................(3).}$$

For a wave parallel to one of the principal axes, as that of x, the direction of that axis is one of the three rectangular directions of vibration of the waves which are propagated independently. For such vibrations the velocity (v) of propagation is given by the formula

$$v^2=m^2(R+H)+n^2(Q+I),$$

which by (3) is reduced to

$$v^2=R+H=Q+I,$$

so that on the assumption that the velocity of propagation is the same for a wave perpendicular to the axis of y as for one perpendicular to the axis of z when the vibrations are parallel to the axis of x, the law of ordinary refraction in the plane of yz follows from theory.

For the two remaining waves which can be propagated independently in a given direction perpendicular to the axis of x, the vibrations are only approximately normal and transversal respectively. In fact, for the three waves which can travel independently in any given direction, the directions of vibration are not affected by the introduction of the constants expressing equilibrium-pressures, but only the velocities of propagation. The squares of the velocities of propagation of the two waves above mentioned are given as before by a quadratic; and in order that the velocity of propagation of the nearly transversal vibrations may be expressed by the formula

$$v^2 = c^2m^2 + b^2n^2 \dots\dots\dots\dots\dots\dots\dots\dots (4),$$

in conformity with the ellipsoidal form of the extraordinary wave surface in a uniaxal crystal, and the assumed elliptic form of the section of one sheet of the wave-surface in a biaxal crystal by a principal plane, the quadratic in question must split into two rational factors, which leads to precisely the same condition as before, namely that expressed by the first of equations (2); and by equating to zero the corresponding factor, we get

$$v^2 = (P + H)\,m^2 + (P + I)\,n^2,$$

which is in fact of the form (4). Applying the same to each of the other principal axes, we find again the three relations (2).

Hence Cauchy's second theory, in which it is supposed that in polarized light the vibrations (in air or in an isotropic medium) are *perpendicular* to the plane of polarization, leads like the first to laws of double refraction, and of the accompanying polarization, differing from those of Fresnel only by quantities which may be deemed insensible. This result is, however, in the present case only attained by the aid of *two* sets of forced relations, namely (2) and (3), that is, relations which there is nothing *à priori* to indicate, and which are not the expression of any simple physical idea, but are obtained by *forcing* the theory, which in its original state is of a highly plastic nature from the number of arbitrary constants which it contains, to agree with observation in some particulars, which being done, theory by itself makes known the rest. As regards the third ray by which this theory like its predecessor is hampered, there is nearly as much to be urged against the present theory as the former. There is, however, this difference, that, as there are only five relations, (2) and (3), between nine arbitrary constants, there remains one arbitrary constant in the expressions for the velocities of propagation after satisfying the numerical values of the three principal indices of refraction, by a proper disposal of which the objections which have been mentioned may *to a certain extent* be lessened, but by no means wholly overcome.

I come now to Green's theory, contained in a very remarkable memoir "On the Propagation of Light in Crystallized Media," read before the Cambridge Philosophical Society, May 20, 1839 *, and accordingly, by a curious coincidence, the very day that Cauchy's second theory was presented to the French Academy. Besides the great interest of the memoir in relation to the theory of light, Green has in it, as I conceive, given for the first time the true equations of equilibrium and motion of a homogeneous elastic solid slightly disturbed from its position of equilibrium, which is one of constraint under a uniform pressure different in different directions. In a former memoir † he had given the equations for the case in which the undisturbed state is one free from pressure ‡. When I speak of the true equations, I mean the equations which belong to the problem when not restricted in generality by arbitrarily assumed hypotheses, and yet not containing constants which are incompatible with any well-ascertained physical principle. It is right to mention, however, that on this point mathematicians are not agreed; M. de Saint-Venant, for instance, maintains the justice of the more restricted equations given by Cauchy §, though even he would not conceive the latter equations applicable to such solids as caoutchouc or jelly.

In these papers Green introduced into the treatment of the subject, with the greatest advantage, the method of Lagrange, in which the partial differential equations of motion are obtained from the variation of a single force-function, on the discovery of the proper form of which everything turns. Green's principle is thus enunciated by him:—"In whatever manner the elements of any material system may act on each other, if all the internal forces be multiplied by the elements of their respective directions, the total sum for any assigned portion of the mass will always be the exact

* *Camb. Phil. Trans.*, Vol. VII, p. 120. [Green's *Math. Papers*, p. 291.]

† "On the Reflexion and Refraction of Light," *Camb. Phil. Trans.* Vol. VII, p. 1. Read Dec. 11, 1837. [Green's *Math. Papers*, pp. 243, 280.]

‡ They are virtually given, though not actually written down at length.

§ *Comptes Rendus*, tom. LIII, p. 1105 (1861).

differential of some function." In accordance with this principle, the general equation may be put under the form

$$\iiint \rho dxdydz \left(\frac{d^2u}{dt^2}\delta u + \frac{d^2v}{dt^2}\delta v + \frac{d^2w}{dt^2}\delta w\right) = \iiint dxdydz\delta\phi \ldots (5),$$

where x, y, z are the equilibrium coordinates of any particle, ρ the density in equilibrium, u, v, w the displacements parallel to x, y, z, and ϕ the function in question. ϕ is in fact the function the variation of which in passing from one state of the medium to another, when multiplied by $dxdydz$, expresses the work given out by the portion of the medium occupying in equilibrium the elementary parallelepiped $dxdydz$, in passing from the first state to the second. The portion of the medium which in the state of equilibrium occupied the elementary parallelepiped becomes in the changed state an oblique-angled parallelepiped, whose edges may be represented by $dx(1+s_1)$, $dy(1+s_2)$, $dz(1+s_3)$, and the cosines of the angles between the second and third, third and first, and first and second of these edges by α, β, γ, which in case the disturbance be small will be small quantities only. It is manifest that the function ϕ must be independent of any linear or angular displacement of the element $dxdydz$, and depend only on the change of form of the element, and therefore on the six quantities s_1, s_2, s_3, α, β, γ, which may be expressed by means of the nine differential coefficients of u, v, w with respect to x, y, z, of which therefore ϕ is a function, but not any function, since it involves not nine, but only six independent variables. If the disturbance be small, the six quantities s_1, s_2, s_3, α, β, γ will be small likewise, and ϕ may be expressed in a convergent series of the form

$$\phi = \phi_0 + \phi_1 + \phi_2 + \phi_3 + \ldots,$$

where ϕ_0, ϕ_1, ϕ_2, ϕ_3, etc. are homogeneous functions of the six quantities, of the orders 0, 1, 2, 3, etc.; and if the motion be regarded as indefinitely small, the functions ϕ_3, $\phi_4 \ldots$ will be insensible, the left-hand member of equation (5) being of the second order as regards u, v, w. ϕ_0, being a constant, will not appear in equation (5), and ϕ_1 will be equal to zero in case the medium in its undisturbed state be free from internal pressure, but not otherwise. The function ϕ_2, being a homogeneous function of six independent variables of the second order, contains in its most general shape twenty-one arbitrary constants, and ϕ_1 which is of the first order introduces six more, so that the most general expression for ϕ contains no less than twenty-seven arbitrary constants, all which appear in the expressions for the internal pressures and in the partial differential equations of motion*.

The general expressions for the internal tensions in an elastic medium and the general equations of equilibrium or motion which were given by Cauchy, and which are written at length in the 4th volume of the *Exercices de Mathématiques*, contain twenty-one arbitrary constants when the undisturbed state of the medium is one of uniform constraint, and fifteen when it is one of freedom from pressure. In the latter case, Green's twenty-one constants are reduced to two, and Cauchy's fifteen to only one, when the medium is isotropic. Green's equations comprise Cauchy's as a particular case, as will be shown more at length further on. It becomes an important question to inquire whether Cauchy's equations involve some restrictive hypothesis as to the constitution of the medium, so as to be in fact of insufficient generality, or whether, on the other hand, Green's equations are reducible to Cauchy's by the introduction of some well-ascertained physical principle, and therefore contain redundant constants.

In the formation of Cauchy's equations, not only is the medium supposed to consist of material points acting on one another by forces which depend on the distance only (a supposition which, at least when coupled with the next, excludes the idea of molecular polarity), but it is assumed that the displacements of the *individual molecules* vary from molecule to molecule according to the variation of some continuous function of the coordinates; and accordingly the displacements u', v', w' of the molecule whose coordinates in equilibrium are $x + \Delta x$, $y + \Delta y$, $z + \Delta z$ are expanded by Taylor's theorem in powers of Δx, Δy, Δz, and the differential coefficients du/dx, etc. are put outside the sign of summation. The motion, varying from

* The twenty-seven arbitrary constants enter the equations of motion in such a manner as to be there equivalent to only twenty-six distinct constants, the physical interpretation of which analytical result will be found to be that a uniform pressure alike in all directions, in the undisturbed state of the medium, produces the same effect on the internal movements when the medium is disturbed as a certain internal elasticity, alike in all directions, and of a very simple kind, which is possible in a medium unconstrained in its natural state. The twenty-one arbitrary constants belonging to a medium unconstrained in its natural state are not reducible in the equations of motion, any more than in the expressions for the internal tensions, to a smaller number.

point to point, of the medium taken as a whole, or in other words the *mean* motion, in any direction, of the molecules in the neighbourhood of a given point, must not be confounded with the motion of the molecules *taken individually*. The medium being continuous, so far as anything relating to observation is concerned, the former will vary continuously from point to point. But it by no means follows that the motion of the molecules considered individually should vary from one to another according to some function of the coordinates. The motion of the individual molecules is only considered for the sake of deducing results from hypotheses as to the molecular constitution and molecular forces of the medium, and in it we are concerned only with the *relative* motion of molecules situated so close as to act sensibly on each other. It would seem to be very probable, *à priori*, that a portion by no means negligible of the relative displacement of a pair of neighbouring molecules should vary in an irregular manner from pair to pair; and indeed if the medium tends to relieve itself from a state of constrained distortion, this must necessarily be the case; and such a rearrangement must assuredly take place in fluids. The insufficient generality of Cauchy's equations is further shown by their being *absolutely incompatible* with the idea of incompressibility. We may evidently conceive a solid which resists compression of volume by a force incomparably greater than that by which it resists distortion of figure, and such a conception is actually realized in such a solid as caoutchouc or jelly.

I have not mentioned the hypothesis of what may be called, from the analogy of surfaces of the second order, a *central* arrangement of the molecules, that is, an arrangement such that each molecule is a centre with respect to which the others are arranged in pairs at equal distances in opposite directions, because the hypothesis was merely casually introduced as one mode of making certain terms vanish which are of a form that clearly ought not to appear in the expressions relating to the mean motion, with which alone we are ultimately concerned.

The arguments in favour of the existence of ultimate molecules in the case of ponderable matter appear to rest chiefly on the chemical law of definite proportions, and on the laws of crystallography, neither of which of course can be assumed to apply to the mysterious ether, of the very existence of which we have no direct evidence. If, for aught we know to the contrary, the very supposition of the existence of ultimate molecules as applied to the ether may entail consequences at variance with its real constitution, much more must the accessory hypotheses be deemed precarious which Cauchy found necessary in order to be able to deduce any results at all in proceeding by his method. There appears, therefore, no sufficient reason *à priori* for preferring the more limited equations of Cauchy to the more general equations of Green.

Green, on the other hand, takes his stand on the impossibility of perpetual motion, or in other words, on the principle of the conservation of work, which we have the strongest reasons for believing to be a general physical principle*. The number of arbitrary constants thus furnished in the case in which the undisturbed state of the medium is one of freedom from pressure is, as has been stated, twenty-one. Professor Thomson has recently put this result in a form which indicates more clearly the signification of the constants†, and at the end of his memoir promises to show how an elastic solid, which as a whole should possess this number of arbitrary constants, could be built up of isotropic matter.

Green supposes, in the first instance, that the medium is symmetrical with respect to planes in three rectangular directions, which simplifies the investigation and reduces the twenty-seven or twenty-one arbitrary constants to twelve (entering the partial differential equations of motion in such a manner as to be there equivalent to only eleven) or nine. It may be useful to give a Table of the constants employed by Green, with their equivalents in the theories of Cauchy and Neumann, the density of the medium at rest being taken equal to unity for the sake of simplicity. The Table is as follows:—

Green	ABC	GHI	LMN	PQR
Cauchy	GHI	LMN	PQR	PQR
Neumann	000	DCB	$A_{\prime}AA_{\prime\prime}$	$A_{\prime}AA_{\prime\prime}$;

so that Green's equations are reduced to Cauchy's by making

$$L = P, \qquad M = Q, \qquad N = R \text{ ....(6).}$$

* Whether vital phenomena are subject to this law is a question which we are not here called upon to discuss.

† "Elements of a Mathematical Theory of Elasticity," *Phil. Trans.* for 1856, p. 481. Read April 24, 1856. [Lord Kelvin's *Math. and Phys. Papers*, Vol. III, p. 84.]

For a plane wave propagated in any given direction there are three velocities of propagation, and three corresponding directions of vibration, which are determined by the directions of the principal axes of a certain ellipsoid $U=1$, which he proposes to call the ellipsoid of elasticity, the semiaxes at the same time representing in magnitude the squared reciprocals of the corresponding velocities of propagation; and Green has shown that U may be at once obtained from the function -2ϕ by taking that part only which is of the second order in u, v, w, and replacing u, v, w by x, y, z, and the symbols of differentiation $\frac{d}{dx}, \frac{d}{dy}, \frac{d}{dz}$ by the cosines of the angles which the wave-normal makes with the axes. This applies whether the medium be symmetrical or not with respect to the coordinate planes. Green then examines the consequences of supposing that for two of the three waves the vibrations are *strictly* in the front of the wave, as was supposed by Fresnel, and consequently that the vibrations belonging to the third wave are strictly normal. This hypothesis leads to five relations between the twelve constants, namely

$$G=H=I=\mu \text{ suppose}, \quad P=\mu-2L, \quad Q=\mu-2M, \quad R=\mu-2N \ldots(7);$$

and gives for the form of the fundamental function

$$-2\phi = 2A\frac{du}{dx} + 2B\frac{dv}{dy} + 2C\frac{dw}{dz}$$

$$+A\left\{\left(\frac{du}{dx}\right)^2 + \left(\frac{dv}{dx}\right)^2 + \left(\frac{dw}{dx}\right)^2\right\} + B\left\{\left(\frac{du}{dy}\right)^2 + \left(\frac{dv}{dy}\right)^2 + \left(\frac{dw}{dy}\right)^2\right\}$$

$$+C\left\{\left(\frac{du}{dz}\right)^2 + \left(\frac{dv}{dz}\right)^2 + \left(\frac{dw}{dz}\right)^2\right\} + \mu\left(\frac{du}{dx} + \frac{dv}{dy} + \frac{dw}{dz}\right)^2$$

$$+L\left\{\left(\frac{dv}{dz} + \frac{dw}{dy}\right)^2 - 4\frac{dv}{dy}\frac{dw}{dz}\right\} + M\left\{\left(\frac{dw}{dx} + \frac{du}{dz}\right)^2 - 4\frac{dw}{dz}\frac{du}{dx}\right\}$$

$$+N\left\{\left(\frac{du}{dy} + \frac{dv}{dx}\right)^2 - 4\frac{du}{dx}\frac{dv}{dy}\right\} \ldots\ldots\ldots\ldots\ldots\ldots\ldots\ldots(8),$$

from which the equations of motion, the expressions for the internal pressures, and the equation of the ellipsoid of elasticity may be at once written down.

The simpler case in which the medium in its natural state is supposed free from pressure is first considered*. Green shows

* The results obtained for this case remain the same if we suppose the medium in its undisturbed state to be subject to a pressure alike in all directions.

that the ellipse which is the section of the ellipsoid of elasticity by a diametral plane, parallel to the wave's front, if turned 90° in its own plane, belongs to a fixed ellipsoid, which gives at once Fresnel's elegant construction for the velocity of propagation and direction of the plane of polarization; but it is necessary to suppose that in polarized light the vibrations are parallel, not perpendicular, to the plane of polarization.

The general case in which the medium is not assumed to be symmetrical with respect to three rectangular planes, and in which therefore ϕ contains twenty-one arbitrary constants, is afterwards considered; and it is shown that the hypothesis of strict transversality leads to fourteen relations between them, leaving only seven constants arbitrary. But the function obtained on the assumption of planes of symmetry contains no fewer, for the four constants relating to these planes would be increased by three when the medium was referred to general axes. Hence therefore the existence of planes of symmetry is not an independent assumption, as in Cauchy's theory, but follows as a result.

In this beautiful theory, therefore, we are presented with no forced relations like Cauchy's equations; the result follows from the hypothesis of strictly transversal vibrations, to which Fresnel was led by physical considerations. The constant μ remains arbitrary, and it is easy to see that this constant expresses the square of the velocity of propagation of normal vibrations. Were this velocity comparable with the velocity of propagation of transversal vibrations, theory would lead us still to expect normal vibrations to be produced by light incident obliquely, though not by light incident perpendicularly, on the surface of a crystal, and the theory would still be exposed to many of the objections which have been already brought forward. But nothing hinders us from supposing, in accordance with the argument contained in Green's former paper, that μ is very great or sensibly infinite, which removes all the difficulty, since the motion corresponding to this term in the expression for -2ϕ would not be sensible except at a distance from the surface comparable with the length of a wave of light. Hence, although it might be said, so long as μ was supposed arbitrary, that the supposition of rigorous transversality had still something in it of the nature of a forced relation between constants, we see that the *single* supposition of incompressibility (under the action of forces at least comparable with those acting in the propagation of light)—the original supposition of Fresnel—introduced into the general equations, suffices to lead to the complete laws of double refraction as given by Fresnel. Were it not that other phenomena of light lead us rather to the conclusion that the vibrations are perpendicular, than that they are parallel to the plane of polarization, this theory would seem to leave us nothing to desire, except to prove that we had a right to neglect the *direct* action of the ponderable molecules, and to treat the ether within a crystal as a single elastic medium, of which the elasticity was different in different directions.

In his paper on Reflexion, Green had adopted the supposition of Fresnel, that the vibrations are perpendicular to the plane of polarization. He was naturally led to examine whether the laws of double refraction could be explained on this hypothesis. When the medium in its undisturbed state is exposed to pressure differing in different directions, six additional constants are introduced into the function ϕ, or three in case of the existence of planes of symmetry to which the medium is referred. For waves perpendicular to the principal axes, the directions of vibration and squared velocities of propagation are as follows:—

Wave-normal		x	y	z
Direction of vibration	x	$G+A$	$N+B$	$M+C$
	y	$N+A$	$H+B$	$L+C$
	z	$M+A$	$L+B$	$I+C$

Green assumes, in accordance with Fresnel's theory, and with observation if the vibrations in polarized light are supposed perpendicular to the plane of polarization, that for waves perpendicular to any two of the principal axes, and propagated by vibrations in the direction of the third axis, the velocity of propagation is the same. This gives three, equivalent to two, relations among the constants, namely,

$$A-L=B-M=C-N=\nu \text{ suppose} \quad\ldots\ldots\ldots\ldots(9),$$

which are equivalent to Cauchy's equations (3). The conditions that the vibrations are strictly transversal and normal respectively do not involve the six constants expressing the pressures in equilibrium, and therefore remain the same as before, namely (7). Adopting the relations (7) and (9), Green proves that for the two transversal waves the velocities of propagation and the azimuths of the planes of polarization are precisely those given by the theory of Fresnel, the vibrations in polarized light being now supposed *perpendicular* to the plane of polarization.

As to the wave propagated by normal vibrations, the square of its velocity of propagation is easily shown to be equal to

$$\mu + Al^2 + Bm^2 + Cn^2;$$

and as the constant μ does not enter into the expression for the velocity of propagation of transversal vibrations, the same supposition as before, namely that the medium is rigorously or sensibly incompressible, removes all difficulty arising from the absence of any observed phenomenon answering to this wave.

The existence of planes of symmetry is here *in part* assumed. I say *in part*, because Green shows that the six constants, expressing the pressures in equilibrium, enter the equation of the ellipsoid of elasticity under the form $K(x^2+y^2+z^2)$, where K is a homogeneous function of the six constants of the first order, and involves likewise the cosines l, m, n. Hence the directions of vibration are the same as when the six constants vanish; the velocities of propagation alone are changed; and as the existence of planes of symmetry for the case in which the six constants vanish was demonstrated, it is only requisite to make the very natural supposition that the planes of symmetry which must exist as regards the directions of vibration, are also planes of symmetry as regards the pressure in equilibrium.

We see then that this theory, which may be called Green's second theory, is in most respects as satisfactory (assuming for the present that Fresnel's construction does represent the laws of double refraction) as the former. I say *in most respects*, because, although the theory is perfectly rigorous, like the former, the equations (9) are of the nature of forced relations between the constants, not expressing anything which could have been foreseen, or even conveying when pointed out the expression of any simple physical relation.

The year 1839 was fertile in theories of double refraction, and on the 9th of December Prof. MacCullagh presented his theory to the Royal Irish Academy. It is contained in "An Essay towards a Dynamical Theory of Crystalline Reflexion and Refraction*." As indicated by the title, the determination of the intensities of the light reflected and refracted at the surface of a crystal is what the author had chiefly in view, but his previous researches had led him to observe that this determination was intimately connected with the laws of double refraction, and to seek to link together these laws as parts of the same system. He was led to apply to the problem the general equation of dynamics under the form (5), to seek to determine the form of the function ϕ (V in his notation), and then to form the partial differential equations of motion, and the conditions to be satisfied at the boundaries of the medium, by the method of Lagrange. He does not appear to have been aware at the time that this method had previously been adopted by Green. Like his predecessors, he treats the ether within a crystallized body as a single medium unequally elastic in different directions, thus ignoring any *direct* influence of the ponderable molecules in the vibrations. He assumes that the density of the ether is a constant quantity, that is, both unchanged during vibration, and the same within all bodies as in free space. We are not concerned with the latter of these suppositions in deducing the laws of internal vibrations, but only in investigating those which regulate the intensity of reflected and refracted light. He assumes further that the vibrations in plane waves, propagated within a crystal, are rectilinear, and that while the plane of the wave moves parallel to itself the vibrations continue parallel to a fixed right line, the direction of this right line and the direction of a normal to the wave being functions of each other,—a supposition which doubtless applies to all crystals except quartz, and those which possess a similar property.

In this method everything depends on the correct determination of the form of the function V. From the assumption that the density of the ether is unchanged by vibration, it is readily shown that the vibrations are entirely transversal. Imagine a system of plane waves, in which the vibrations are parallel to a fixed line in the plane of a wave, to be propagated in the crystal,

* *Memoirs of the Royal Irish Academy*, Vol. XXI, p. 17. [MacCullagh's *Collected Works*, pp. 145—193.]

and refer the crystal for a moment to the rectangular axes of x', y', z', the plane of $x'y'$ being parallel to the planes of the waves, and the axis of y' to the direction of vibration; and let κ be the angle whose tangent is $d\eta'/dz'$. With respect to the form of V, MacCullagh reasons thus:—"The function V can only depend upon the directions of the axes of x', y', z' with respect to fixed lines in the crystal, and upon the angle which measures the change of form produced in the parallelepiped by vibration. This is the most general supposition which can be made concerning it. Since, however, by our second supposition, any one of these directions, suppose that of x', determines the other two, we may regard V as depending on the angle κ and the direction of the axis of x' alone," from whence he shows that V must be a function of the quantities X, Y, Z, defined by the equations

$$X=\frac{d\eta}{dz}-\frac{d\zeta}{dy},\quad Y=\frac{d\zeta}{dx}-\frac{d\xi}{dz},\quad Z=\frac{d\xi}{dy}-\frac{d\eta}{dx}.$$

This reasoning, which is somewhat obscure, seems to me to involve a fallacy. If the form of V were known, the rectilinearity of vibration and the constancy in the direction of vibration for a system of plane waves travelling in any given direction would follow as a *result* of the solution of the problem. But in using equation (5) we are not at liberty to substitute for V (or ϕ) an expression which represents that function *only on the condition that the motion be what it actually is*, for we have occasion to take the variation δV of V, and this variation must be the most general that is geometrically possible though it be dynamically impossible. That the form of V, arrived at by MacCullagh, is inadmissible, is, I conceive, proved by its incompatibility with the form deduced by Green from the very same supposition of the *perfect* transversality of the transversal vibrations; for Green's reasoning is perfectly straightforward and irreproachable. Besides, MacCullagh's form leads to consequences absolutely at variance with dynamical principles*.

But waiving for the present the objection to the conclusion that V is a function of the quantities X, Y, Z, let us follow the consequences of the theory. The disturbance being supposed small, the quantities X, Y, Z will also be small, and V may be expanded in a series according to powers of these quantities;

* See Appendix.

and, as before, we need only proceed to the second order if we regard the disturbance as indefinitely small. The first term, being merely a constant, may be omitted. The terms of the first order MacCullagh concludes must vanish. This, however, it must be observed, is only true on the supposition that the medium in its undisturbed state is free from pressure. The terms of the second order are six in number, involving squares and products of X, Y, Z. The terms involving YZ, ZX, XY may be got rid of by a transformation of coordinates, when V will be reduced to the form

$$V = -\tfrac{1}{2}(a^2X^2 + b^2Y^2 + c^2Z^2) \text{ ....(10)},$$

the constant term being omitted, and the arbitrary constants being denoted by $-\frac{1}{2}a^2, -\frac{1}{2}b^2, -\frac{1}{2}c^2$. Thus on this theory the existence of principal axes is proved, not assumed. If MacCullagh's expression for V (10) be compared with Green's expression for ϕ (8) for the case of no pressure in equilibrium, so that $A = 0$, $B = 0$, $C = 0$, it will be seen that the two will become identical, provided first we omit the term $\mu\left(\frac{du}{dx} + \frac{dv}{dy} + \frac{dw}{dz}\right)^2$ in Green's expression, and secondly, we treat the symbols of differentiation as literal coefficients, so as to confound, for instance, $\frac{dv}{dy}\frac{dw}{dz}$ and $\frac{dv}{dz}\frac{dw}{dy}$. The term involving μ does not appear in the expressions for transversal vibrations, since for these $\frac{du}{dx} + \frac{dv}{dy} + \frac{dw}{dz} = 0$, and therefore does not affect the laws of the propagation of such vibrations, although it would appear in the problem of calculating the intensity of reflected and refracted light; and be that as it may, it follows from Green's rule for forming the equation of the ellipsoid of elasticity, that the laws of the propagation of transversal vibrations will be precisely the same whether we adopt his form of ϕ or V (for the case of no pressure in equilibrium) or MacCullagh's. Indeed, if we omit the term $\mu\left(\frac{du}{dx} + \frac{dv}{dy} + \frac{dw}{dz}\right)^2$, the partial differential equations of motion, on which alone depend the laws of internal propagation, would be just the same in the two theories*. Accordingly MacCullagh obtained, though inde-

* See Appendix. MacCullagh's reasoning appears to be so far correct as to have led to correct equations, although through a form of V which may, I conceive,

pendently of, and in a different manner from Green, precisely Fresnel's laws of double refraction and the accompanying polarization, on the condition, however, that in polarized light the vibrations are *parallel to* the plane of polarization.

It is remarkable that in the previous year MacCullagh, in a letter to Sir David Brewster*, published expressions for the internal pressures identical with those which result from Green's first theory, provided that in the latter the terms be omitted which arise from that term in ϕ which contains μ, a term which vanishes in the case of transversal vibrations propagated within a crystal. It does not appear how these expressions were obtained by MacCullagh; it was probably by a tentative process.

The various theories which have just been reviewed have this one feature in common, that in all, the direct action of the ponderable molecules is neglected, and the ether treated as a single vibrating medium. It was, doubtless, the extreme difficulty of determining the motion of one of two mutually penetrating media that led mathematicians to adopt this, at first sight, unnatural supposition; but the conviction seems by some to have been entertained from the first, and to have forced itself upon the minds of others, that the ponderable molecules must be taken into account in a far more direct manner. Some investigations were made in this direction by Dr Lloyd as long as twenty-five years ago†. Cauchy's later papers show that he was dissatisfied with the method, adopted in his earlier ones, of treating the ether within a ponderable body as a single vibrating medium‡; but he

be shown to be inadmissible. [The objection, however, is based on the hypothesis that the potential energy $(-V)$ is due entirely to elastic *deformation* of the medium. If the medium were a complex one containing kinetic molecules of permanent gyrostatic type, simple *rotation* would excite dynamical reaction, so that the objection would be evaded. The form of V adopted by MacCullagh is in fact analytically identical with the one appropriate to Clerk Maxwell's electric theory of light, of which, accordingly, MacCullagh's analysis constituted a development in advance. It was afterwards recognised (cf. p. 197 *infra*) by Sir George Stokes, that an explicit dynamical basis might possibly be found for this theory by passing beyond a hypothesis of elasticity of simple deformation. In his exposition of the theory MacCullagh very fully admitted that a dynamical groundwork for it had yet to be found. Cf. Larmor, *Phil. Trans.*, 1894.]

* *Phil. Mag.* for 1836, Vol. VIII, p. 103. [MacCullagh's *Collected Works*, p. 75.]

† *Proceedings of the Royal Irish Academy*, Vol. I, p. 10.

‡ See his optical memoirs published in the 22nd Volume of the *Mémoires de l'Académie.*

does not seem to have advanced beyond a few barren generalities, towards a theory of double refraction founded on a calculation of the vibrations of one of two mutually penetrating media. In the theory of double refraction advanced by Professor Challis*, the ether is assimilated to an ordinary elastic fluid, the vibrations of which are modified by resisting masses; and his theory leads him at once to Fresnel's elegant construction of the wave-surface by points. The theory, however, rests upon principles which have not received the general assent of mathematicians. In a work entitled "Light explained on the Hypothesis of the Ethereal Medium being a Viscous Fluid"†, Mr Moon has put in a clear form some of the more serious objections which may be raised against Fresnel's theory; but that which he has substituted is itself open to formidable objections, some of which the author himself seems to have perceived.

In concluding this part of the subject, I may perhaps be permitted to express my own belief that the true dynamical theory of double refraction has yet to be found.

In the present state of the theory of double refraction, it appears to be of especial importance to attend to a rigorous comparison of its laws with actual observation. I have not now in view the two great laws giving the planes of polarization, and the difference of the squared velocities of propagation, of the two waves which can be propagated independently of each other in any given direction within a crystal. These laws, or at least laws differing from them only by quantities which may be deemed negligible in observation, had previously been discovered by experiment; and the deduction of these laws by Fresnel from his theory, combined with the verification of the law, which his theory, correcting in this respect previous notions, first pointed out, that in each principal plane of a biaxal crystal the ray polarized in that plane obeys the ordinary law of refraction, leaves no reasonable doubt that Fresnel's construction contains the true laws of double refraction, at least in their broad features. But regarding this point as established, I have rather in view a verification of those laws which admit of being put to the test of experiment with extreme precision; for such verifications might often enable the

* *Cambridge Philosophical Transactions*, Vol. VIII, p. 524.

† Macmillan & Co., Cambridge, 1853.

mathematician, in groping after the true theory, to discard at once, as not agreeing with observation, theories which might present themselves to his mind, and on which otherwise he might have spent much fruitless labour.

To make my meaning clearer, I will refer to Fresnel's construction, in which the laws of polarization and wave-velocity are determined by the sections, by a diametral plane parallel to the wave-front, of the ellipsoid*

$$a^2x^2 + b^2y^2 + c^2z^2 = 1 \quad\text{.....................(11)},$$

where a, b, c denote the principal wave-velocities. The principal semiaxes of the section determine by their direction the normals to the two planes of polarization, and by their magnitude the reciprocals of the corresponding wave-velocities. Now a certain other physical theory which might be proposed† leads to a construction differing from Fresnel's only in this, that the planes of polarization and wave-velocities are determined by the section, by a diametral plane parallel to the wave-front, of the ellipsoid

$$\frac{x^2}{a^2} + \frac{y^2}{b^2} + \frac{z^2}{c^2} = 1 \quad\text{.........................(12)},$$

the principal semiaxes of the section determining by their direction the normals to the two planes of polarization, and by their magnitudes the corresponding wave-velocities. The law that the planes of polarization of the two waves propagated in a given direction bisect respectively the two supplemental dihedral angles made by planes passing through the wave-normal and the two optic axes, remains the same as before, but the positions of the optic axes themselves, as determined by the principal indices of refraction, are somewhat different; the difference, however, is but small if the differences between a^2, b^2, c^2 are a good deal smaller than the quantities themselves. Each principal section of the wave-surface, instead of being a circle and an ellipse, is a circle and an oval, to which an ellipse is a near approximation‡. The difference between

* It would seem to be just as well to omit the surface of elasticity altogether, and refer the construction directly to the ellipsoid (11).

[† This theory suggested itself independently to Lord Rayleigh, and is discussed by him in *Phil. Mag.* XLI, 1871, p. 519; *Scientific Papers*, I, p. 111.]

‡ The equation of the surface of wave-slowness in this and similar cases may be readily obtained by the method given by Professor Haughton in a paper "On the Equilibrium and Motion of Solid and Fluid Bodies," *Transactions of the Royal Irish Academy*, Vol. XXI, p. 172.

the inclinations of the optic axes, and between the amounts of extraordinary refraction in the principal planes, on the two theories, though small, are quite sensible in observation, but only on condition that the observations are made with great precision. We see from this example of what great advantage for the advancement of theory observations of this character may be*.

One law which admits of receiving, and which has received, this searching comparison with observation, is that according to which, in each principal plane of a biaxal crystal, the ray which is polarized in that plane obeys the ordinary law of refraction, and accordingly in a uniaxal crystal, in which every plane parallel to the axis is a principal plane, the so-called ordinary ray follows rigorously the law of ordinary refraction. This law was carefully verified by Fresnel himself in the case of topaz, by the method of cutting plates parallel to the same principal axis, or axis of elasticity, carefully working them to the same thickness, and then interposing them in the paths of two streams of light proceeding to interfere, as well as by the method of prismatic refraction; and he states as the result of his observations that he can affirm the law to be, at least in the case of topaz, mathematically exact. The same result follows from the observations by which Rudberg so accurately determined the principal indices of Arragonite and topaz†, for the principal fixed lines of the spectrum. Professor MacCullagh having been led by theoretical considerations to doubt whether, in Iceland spar for instance, the so-called ordinary ray rigorously obeyed the ordinary law of refraction, whether the refractive indices in the axial and equatorial directions were *strictly* the same, Sir David Brewster was induced to put the question to the test of a crucial experiment, by forming a compound prism consisting of two pieces of spar cemented together in the direction of the length of the prism, and so cut from the crystal that at a minimum deviation one piece was traversed axially and the other equatorially‡. The prism having been polished after cementing, so as to ensure the perfect equality of angle of the two parts, on viewing a slit through it the bright line D was seen unbroken in passing from one half to the other.

[* For Sir George Stokes' own observations on Iceland spar which rendered this theory inadmissible see *Roy. Soc. Proc.* xx, 1872, p. 443, reprinted *infra*.]

† *Annales de Chimie*, tom. XLVIII, p. 225 (1831).

‡ *Report of the British Association* for 1843, Trans. of Sect. p. 7.

More recently Professor Swan has made a very precise examination of the ordinary refraction in various directions in Iceland spar by the method of prismatic refraction*, from whence it results that for homogeneous light of any refrangibility the ordinary ray follows strictly the ordinary law of refraction.

It is remarkable that this simple law, which ought, one would expect, to lie on the very surface as it were of the true theory of double refraction, is not indicated *à priori* by most of the rigorous theories which have been advanced to account for the phenomenon. Neither of the two theories of Cauchy, nor the second theory of Green, lead us to expect such a result, though they furnish arbitrary constants which may be so determined as to bring it about.

The curious and unexpected phenomenon of conical refraction has justly been regarded as one of the most striking proofs of the general correctness of the conclusions resulting from the theory of Fresnel. But I wish to point out that the phenomenon is not competent to decide between several theories leading to Fresnel's construction as a near approximation. Let us take first internal conical refraction. The existence of this phenomenon depends upon the existence of a tangent plane touching the wave surface along a plane curve. At first sight this might seem to be a speciality of the wave-surface of Fresnel; but a little consideration will show that it must be a property of the wave-surface resulting from any reasonable theory. For, if possible, let the nearest approach to a plane curve of contact be a curve of double curvature. Let a plane be drawn touching the rim (as it may be called) of the surface, that is, the part where the surface turns over, in two points, on opposite sides of the rim; and then, after having been slightly tilted by turning about one of the points of contact, let it move parallel to itself towards the centre. The successive sections of the wave-surface by this plane will evidently be of the general character represented in the annexed figures, and in *four* positions

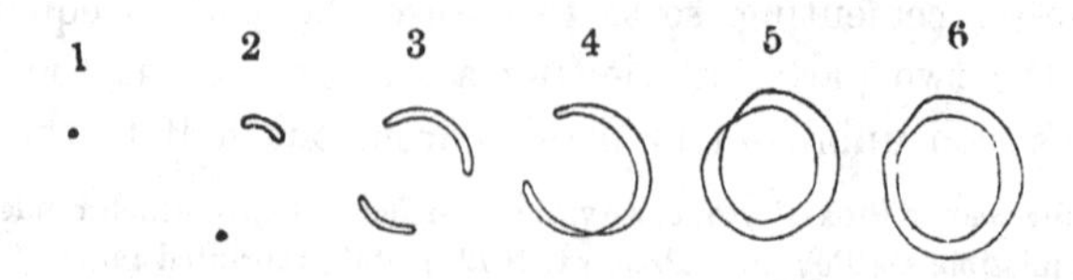

* *Transactions of the Royal Society of Edinburgh*, Vol. XVI, p. 375.

the plane will touch the surface in one point, as represented in Figs. 1, 2, 4, 5. Should the contacts represented in Figs. 4 and 5 take place simultaneously, they may be rendered successive by slightly altering the inclination of the plane. Hence in certain directions there would be *four* possible wave-velocities. Now the general principle of the superposition of small motions makes the laws of double refraction depend on those of the propagation of plane waves. But all theories respecting the propagation of a series of plane waves having a given direction, and in which the disturbance of the particles is arbitrary, but the same all over the front of a wave, agree in this, that they lead us to decompose the disturbance into three disturbances in three particular directions, to each of which corresponds a series of plane waves which are propagated with a determinate velocity. If the medium be incompressible, one of the wave-velocities becomes infinite, and one sheet of the wave-surface moves off to infinity. The most general disturbance, subject to the condition of incompressibility, which requires that there be no displacements perpendicular to the fronts of the waves, may now be expressed as the resultant of *two* disturbances, corresponding to displacements in particular directions lying in planes parallel to that of the waves, to each of which corresponds a determinate velocity of propagation. We see, therefore, that the limitation of the number of tangent planes to the wave-surface, which can be drawn in a given direction on one side of the centre, to two, or at the most three, is intimately bound up with the number of dimensions of space; so that the existence of the phenomenon of internal conical refraction is no proof of the truth of the particular form of wave-surface assigned by Fresnel rather than that to which some other theory would conduct. Were the law of wave-velocity expressed, for example, by the construction already mentioned having reference to the ellipsoid (12), the wave-surface (in this case a surface of the 16th degree) would still have plane curves of contact with the tangent plane, which in this case also, as in the wave-surface of Fresnel, are, as I find, circles, though that they should be circles could not have been foreseen.

The existence of external conical refraction depends upon the existence of a conical point in the wave-surface, by which the interior sheet passes to the exterior. The existence of a conical point is not, like that of a plane curve of contact, a necessary property of a wave-surface. Still it will readily be conceived that if Fresnel's wave-surface be, as it undoubtedly is, at least a near approximation to the true wave-surface, and if the latter have, moreover, plane curves of contact with the tangent plane, the mode by which the exterior sheet passes within one of these plane curves into the interior will be very approximately by a conical point; so that in the impossibility of operating experimentally on mere rays the phenomena will not be sensibly different from what they would have been had the transition been made rigorously by a conical point.

There is one direction within a biaxal crystal marked by a visible phenomenon of such a nature as to permit of observing the direction with precision, while it can also be calculated, on any particular theory of double refraction, in terms of the principal indices of refraction; I refer to the direction of either optic axis. Rudberg himself measured the inclination of the optic axes of Arragonite, probably with a piece of the same crystal from which his prisms were cut, and found it a little more than 32° as observed in air, but he speaks of the difficulty of measuring the angle with precision. The inclination within the crystal thence deduced is really a little greater than that given by Fresnel's theory; but in making the comparison Rudberg used the formula for the ray-axes instead of that for the wave-axes, which made the theoretical inclination in air appear about 2° greater than the observed*. A very exact measure of the angle between the optic axes of Arragonite for homogeneous light corresponding to the principal fixed lines of the spectrum has recently been executed by Professor Kirchhoff†, by a method which has the advantage of not making any supposition as to the direction in which the crystal is cut. The angle observed in air was reduced by calculation to the angle within the crystal, by means of Rudberg's indices for the principal axis of mean elasticity; and the result was compared with the angle calculated from the formula of Fresnel, on substituting for the constants therein contained the numerical values determined by Rudberg for all the three principal axes. The angle reduced from that observed in air proved to be from 13′ to 20′ greater than that calculated from Fresnel's formula. This small difference seems

* *Annales de Chimie*, tome XLVIII, p. 258 (1831).

† *Poggendorff's Annalen*, Vol. CVIII, p. 567 (1859).

to be fairly attributable to errors in the indices, arising from errors in the direction of cutting of the prisms employed by Rudberg. The angle measured by Kirchhoff would seem to have been trustworthy to within a minute or less.

It is doubtful, however, how far we may trust to the identity of the principal refractive indices in different specimens of the mineral. Chemical analysis shows that Arragonite is not pure carbonate of lime, but contains a variable though small proportion of other ingredients. To these variations doubtless correspond variations in the refractive indices; and De Senarmont has shown how the inclination of the optic axes of minerals is liable to be changed by the substitution one for another of isomorphous elements*. Moreover, M. Des Cloizeaux has recently shown that in felspar and some other minerals, which bear a high temperature without apparent change, the inclination of the optic axes is changed in a permanent manner by heat †; so that even perfect identity of chemical composition is not an absolute guarântee of optical identity in two specimens of a mineral of a given kind.

The exactness of the spheroidal form assigned by Huygens to the sheet of the wave-surface within Iceland spar corresponding to the extraordinary ray, does not seem to have been tested to the same degree of rigour as the ordinary refraction of the ordinary ray; for the methods employed by Wollaston ‡ and Malus § for observing the extraordinary refraction can hardly bear comparison for exactness with the method of prismatic refraction which has been applied to the ordinary ray; and observations on the *absolute* velocities of propagation in different directions within biaxal crystals are still almost wholly wanting. This has long been recognised as a desideratum, and it has been suggested to employ for the purpose the displacement of fringes of interference. It seems to me that a slight modification of the ordinary method of prismatic refraction would be more convenient and exact ||.

Let the crystal to be examined be cut, unless natural faces or cleavage planes answer the purpose, so as to have two planes inclined at an angle suitable for the measure of refractions; there

* *Annales de Chimie*, tome XXXIII, p. 391 (1851).

† *Annales de Mines*, tome II, p. 327 (1862).

‡ "On the Oblique Refraction of Iceland Spar," *Phil. Trans.* for 1802, p. 381.

§ *Mémoires de l'Institut; Sav. Étrangers*, tome II, p. 303 (1811).

[|| Cf. *ante*, Vol. I, p. 149 (1846).]

being at least two natural faces or cleavage-planes left undestroyed, so as to permit of an exact measure of the directions of any artificial faces. The prism thus formed having been mounted as usual, and placed in any azimuth, let the angle of incidence or emergence (according as the prism remains fixed or turns round with the telescope) be measured, by observing the light reflected from the surface, and likewise the deviation for several standard fixed lines in the spectrum of each refracted pencil. Let the prism be now turned into a different azimuth, and the deviations again observed, and so on. Each observation furnishes accurately an angle of incidence and the corresponding angle of emergence; for if ϕ be the angle of incidence, i the angle of the prism, D the deviation, and ψ the angle of emergence, $D=\phi+\psi-i$. But without making any supposition as to the law of double refraction, *or assuming anything beyond the truth of Huygens's principle*, which, following directly from the general principle of the superposition of small motions, lies at the very foundation of the whole theory of undulations, we may at once deduce from the angles of incidence and emergence the direction and velocity of propagation of the wave within the prism. For if a plane wave be incident on a plane surface bounding a medium of any kind, either ordinary or doubly refracting, it follows directly from Huygens's principle that the refracted wave or waves will be plane, and that if ϕ be the angle of incidence, ϕ' the inclination of a refracted wave to the surface, V the velocity of propagation in air, v the wave-velocity within the medium,

$$\frac{\sin\phi}{V}=\frac{\sin\phi'}{v}.$$

Hence if ϕ', ψ' be the inclinations of the refracted wave to the faces of our prism, we shall have the equations

$$v\sin\phi = V\sin\phi' \quad\text{......................(13)},$$

$$v\sin\psi = V\sin\psi' \quad\text{......................(14)},$$

$$\phi'+\psi'=i \quad\text{........................(15)}.$$

The equations (13) and (14) give, on taking account of (15),

$$v\sin\frac{\phi+\psi}{2}\cos\frac{\phi-\psi}{2}=V\sin\frac{i}{2}\cos\frac{\phi'-\psi'}{2} \quad\text{......(16)},$$

$$v\cos\frac{\phi+\psi}{2}\sin\frac{\phi-\psi}{2}=V\cos\frac{i}{2}\sin\frac{\phi'-\psi'}{2} \quad\text{......(17)},$$

whence by division

$$\tan\frac{\phi'-\psi'}{2}=\tan\frac{i}{2}\tan\frac{\phi-\psi}{2}\cot\frac{\phi+\psi}{2}\quad\text{.........(18).}$$

The equations (15) and (18) determine ϕ' and ψ', and then (16) gives v. Hence we know accurately the velocity of propagation of a wave, the normal to which lies in a plane perpendicular to the faces of the prism, and makes known angles with the faces, and is therefore known in direction with reference to the crystallographic axes. A single prism would enable the observer to explore the crystal in a series of directions lying in a plane perpendicular to its edge; but as these directions are practically confined to limits making no very great angles with a normal to the plane bisecting the dihedral angle of the prism, more than one prism would be required to enable him to explore the crystal in the most important directions; and it would be necessary for him to assure himself that the specimens of crystal, of which the different prisms are made, were strictly comparable with each other. It would be best, as far as practicable, to cut them from the same block.

The existence of principal planes, or planes of optical symmetry, for light of any given refrangibility, in those cases in which they are not determined by being at the same time planes of crystallographic symmetry, is a matter needing experimental verification. However, as no anomaly, so far as I am aware, has been discovered in the systems of rings seen with homogeneous light around the optic axes of crystals of the oblique or anorthic system, there is no reason for supposing that such planes do not exist.

APPENDIX.

Further Comparison of the Theories of Green, MacCullagh, *and* Cauchy.

In a paper "On a Classification of Elastic Media and the Laws of Plane Waves propagated through them," read before the Royal Irish Academy on the 8th of January, 1849*, Professor Haughton has made a comparative examination of different theories which have been advanced for determining the motion of elastic

* *Transactions of the Royal Irish Academy*, Vol. XXII, p. 97.

media, more especially those which have been applied to the explanation of the phenomena of light. Some of the results contained in this Appendix have already been given by Professor Haughton; in other instances I have arrived at different conclusions. In such cases I have been careful to give my reasons in detail.

Consider a homogeneous elastic medium, the parts of which act on one another only with forces which are insensible at sensible distances, and which in its undisturbed state is either free from pressure, or else subject to a pressure or tension which is the same at all points, though varying with the direction of the plane surface with reference to which it is estimated. Let x, y, z be the coordinates of any particle in the undisturbed state, $x+u$, $y+v$, $z+w$ the coordinates in the disturbed state, and for simplicity take the density in the undisturbed state as the unit of density. Then, according to the method followed both by Green and MacCullagh, the motion of the medium will be determined by the equation

$$\iiint\left(\frac{d^2u}{dt^2}\delta u+\frac{d^2v}{dt^2}\delta v+\frac{d^2w}{dt^2}\delta w\right)dx\,dy\,dz=\iiint\delta\phi\,dx\,dy\,dz\ldots(19),$$

where ϕ is the function due to the elastic forces. To this equation must be added, in case the medium be not unlimited, the terms relative to its boundaries.

The function ϕ multiplied by $dxdydz$ expresses the work given out by the element $dxdydz$ in passing from the initial to the actual state if we assume, as we may, the initial state for that in which $\phi=0$. According to the supposition with which we started, that the internal forces are insensible at sensible distances, the value of ϕ at any point must depend on the relative displacements in the immediate neighbourhood of that point, as expressed by the differential coefficients of u, v, w with respect to x, y, z. For the present let us make no other supposition concerning ϕ than this, that it is some function $(-f)$ of those nine differential coefficients; and let us apply the equation (19) to a limited portion of the medium bounded initially by the closed surface S. We must previously add the terms due to the action of the surrounding portion of the medium, which will evidently be of the form of a double integral having reference to the surface S, an element of which we may denote by dS. Hence we must add to the right-hand side of equation (19)

$$\iint E dS,$$

the expression for E having yet to be found.

Denoting for shortness the partial differential coefficients of $-\phi$ with respect to $\frac{du}{dx}$, $\frac{du}{dy}$, &c. by $f'\left(\frac{du}{dx}\right)$, $f'\left(\frac{du}{dy}\right)$, &c., we have

$$-\delta\phi = f'\left(\frac{du}{dx}\right)\delta\frac{du}{dx} + f'\left(\frac{du}{dy}\right)\delta\frac{du}{dy} + \&c.$$

$$= f'\left(\frac{du}{dx}\right)\frac{d\delta u}{dx} + f'\left(\frac{du}{dy}\right)\frac{d\delta u}{dy} + \&c.,$$

whence

$$-\iiint \delta\phi\, dx\, dy\, dz = \iiint f'\left(\frac{du}{dx}\right)\frac{d\delta u}{dx}\, dx\, dy\, dz$$
$$+ \iiint f'\left(\frac{du}{dy}\right)\frac{d\delta u}{dy}\, dx\, dy\, dz + \&c.$$

$$= \iint f'\left(\frac{du}{dx}\right)\delta u\, dy\, dz + \iint f'\left(\frac{du}{dy}\right)\delta u\, dz\, dx + \iint f'\left(\frac{du}{dz}\right)\delta u\, dx\, dy$$
$$+ \iint f'\left(\frac{dv}{dx}\right)\delta v\, dy\, dz + \&c.$$

$$-\iiint\left\{\delta u\frac{d}{dx}f'\left(\frac{du}{dx}\right) + \delta u\frac{d}{dy}f'\left(\frac{du}{dy}\right) + \delta u\frac{d}{dz}f'\left(\frac{du}{dz}\right) + \delta v\frac{d}{dx}f'\left(\frac{dv}{dx}\right)\right.$$
$$\left. + \&c.\right\} dx\, dy\, dz.$$

We must now equate to zero separately the terms in our equation involving triple and those involving double integrals. The result obtained from the former further requires that the coefficient of each of the independent quantities δu, δv, δw under the sign $\iiint$ shall vanish separately, whence

$$\left.\begin{aligned}
\frac{d^2u}{dt^2} &= \frac{d}{dx}f'\left(\frac{du}{dx}\right) + \frac{d}{dy}f'\left(\frac{du}{dy}\right) + \frac{d}{dz}f'\left(\frac{du}{dz}\right)\\
\frac{d^2v}{dt^2} &= \frac{d}{dx}f'\left(\frac{dv}{dx}\right) + \frac{d}{dy}f'\left(\frac{dv}{dy}\right) + \frac{d}{dz}f'\left(\frac{dv}{dz}\right)\\
\frac{d^2w}{dt^2} &= \frac{d}{dx}f'\left(\frac{dw}{dx}\right) + \frac{d}{dy}f'\left(\frac{dw}{dy}\right) + \frac{d}{dz}f'\left(\frac{dw}{dz}\right)
\end{aligned}\right\} \ldots\ldots(20)^*,$$

* These agree with Professor Haughton's equations (5).

equations which may be written in an abbreviated form as follows:—

$$\frac{d^2u}{dt^2} = -\left[\frac{d\phi}{du}\right], \quad \frac{d^2v}{dt^2} = -\left[\frac{d\phi}{dv}\right], \quad \frac{d^2w}{dt^2} = -\left[\frac{d\phi}{dw}\right] \quad \ldots(21),$$

where the expressions within crotchets denote differential coefficients taken in a conventional sense, namely by treating in the differentiation the symbols $\frac{d}{dx}, \frac{d}{dy}, \frac{d}{dz}$ as if they were mere literal coefficients, and prefixing to the whole term, and now regarding as a real symbol of differentiation, whichever of these three symbols was attached to the u, v, or w that disappeared by differentiation.

The equating of the double integrals gives

$$\iint E dS = \iint f'\left(\frac{du}{dx}\right) \delta u\, dy\, dz + \iint f'\left(\frac{du}{dy}\right) \delta u\, dz\, dx + \&c.$$

$$= \iint \left\{ \left[l f'\left(\frac{du}{dx}\right) + m f'\left(\frac{du}{dy}\right) + n f'\left(\frac{du}{dz}\right) \right] \delta u + [\&c.]\, \delta v + [\&c.]\, \delta w \right\} dS,$$

where l, m, n are the direction-cosines of the element dS of the surface which bounded the portion of the medium under consideration when it was in its undisturbed state. This expression leads us to contemplate the action of the surrounding medium as a tension having a certain value referred to a unit of surface in the undisturbed state. If P, Q, R be the components of this tension parallel to the axes of x, y, z, they must be the coefficients of δu, δv, δw under the sign $\iint$, so that

$$\left.\begin{aligned} P &= l f'\left(\frac{du}{dx}\right) + m f'\left(\frac{du}{dy}\right) + n f'\left(\frac{du}{dz}\right) \\ Q &= l f'\left(\frac{dv}{dx}\right) + m f'\left(\frac{dv}{dy}\right) + n f'\left(\frac{dv}{dz}\right) \\ R &= l f'\left(\frac{dw}{dx}\right) + m f'\left(\frac{dw}{dy}\right) + n f'\left(\frac{dw}{dz}\right) \end{aligned}\right\} \quad \ldots\ldots\ldots(22).$$

These formulæ give, in terms of the function ϕ, the components of the tension on a small plane which in its original position had any arbitrary direction. If we wish for the expressions for the components of the tensions on planes originally perpendicular to the axes of x, y, z, we have only to put in succession $l=1$, $m=1$, $n=1$, the other two cosines each time being equal to zero. If then

P_x, T_{yx}, T_{zx} denote the components in the direction of the axis of x of the tension on planes originally perpendicular to the axes of x, y, z, with similar notation in the other cases, we shall have

$$\left.\begin{aligned} P_x &= f'\left(\frac{du}{dx}\right), & T_{yz} &= f'\left(\frac{dw}{dy}\right), & T_{zy} &= f'\left(\frac{dv}{dz}\right) \\ P_y &= f'\left(\frac{dv}{dy}\right), & T_{zx} &= f'\left(\frac{du}{dz}\right), & T_{xz} &= f'\left(\frac{dw}{dx}\right) \\ P_z &= f'\left(\frac{dw}{dz}\right), & T_{xy} &= f'\left(\frac{dv}{dx}\right), & T_{yx} &= f'\left(\frac{du}{dy}\right) \end{aligned}\right\} \ldots(23)^*.$$

The formulæ hitherto employed are just the same whether we suppose the disturbance small or not; and we might express in terms of P_x, T_{yz}, &c. (and therefore in terms of ϕ), and of the differential coefficients of u, v, w with respect to x, y, and z, the components of the tension referred to a surface given in the actual instead of the undisturbed state of the medium, without supposing the disturbance small. As, however, the investigation is meant to be applied only to small disturbances, it would only complicate the formulæ to no purpose to treat the disturbance as of arbitrary magnitude, and I shall therefore regard it henceforth as indefinitely small.

On this supposition we may expand ϕ according to powers of the small quantities du/dx, &c., proceeding as far as the second order, the left-hand member of (19) being of the second order as regards u, v, w. The formulæ (22) or (23) show that ϕ will or will not contain terms of the first order according as the undisturbed state of the medium is one of uniform constraint, or of freedom from pressure.

In Green's first theory, and in the theory of MacCullagh, ϕ is supposed not to contain terms of the first order. Accordingly in considering the point with respect to which these two theories are at issue, I shall suppose the medium in its undisturbed state to be free from pressure. The tensions P, Q, R, P_x, &c. will now be small quantities of the first order, so that in the formulæ (22) and (23) we may suppose the tensions referred to a unit of surface in the actual or the undisturbed state of the medium indifferently, and

* These agree with Professor Haughton's equations at p. 100, but are obtained in a different manner.

may moreover in these formulæ, and in the expression for ϕ, take x, y, z for the actual or the original coordinates of a particle.

Green assumes as self-evident that the value of ϕ for any element, suppose that which originally occupied the rectangular parallelepiped $dxdydz$, must depend only on the change of form of the element, and not on any mere change of position in space. Any displacement which varies continuously from point to point must change an elementary rectangular parallelepiped into one which is oblique-angled, and the change of form is expressed by the ratios of the lengths of the edges to the original lengths, and by the angles which the edges make with one another or by their cosines. If the medium were originally in a state of constraint, ϕ would contain terms of the first order, and the expressions for the extensions of the edges and the cosines of the angles would be wanted to the second order, but when ϕ is wholly of the second order, those quantities need only be found to the first order. It is easy to see that to this order the extensions are expressed by

$$\frac{du}{dx}, \quad \frac{dv}{dy}, \quad \frac{dw}{dz} \quad \ldots\ldots\ldots\ldots\ldots\ldots\ldots(24),$$

and the cosines of the inclinations of the edges two and two by

$$\frac{dv}{dz}+\frac{dw}{dy}, \quad \frac{dw}{dx}+\frac{du}{dz}, \quad \frac{du}{dy}+\frac{dv}{dx} \quad \ldots\ldots\ldots\ldots(25),$$

and ϕ being a function of these six quantities, we have from (23)

$$T_{yz}=T_{zy}, \quad T_{zx}=T_{xz}, \quad T_{xy}=T_{yx} \ldots\ldots\ldots\ldots(26).$$

These are the relations pointed out by Cauchy between the nine components of the three tensions in three rectangular directions, whereby they are reduced to six. The necessity of these relations is admitted by most mathematicians.

Conversely, if we start with Cauchy's three relations (26), we have from (23)

$$f'\left(\frac{dw}{dy}\right)=f'\left(\frac{dv}{dz}\right), \quad f'\left(\frac{du}{dz}\right)=f'\left(\frac{dw}{dx}\right), \quad f'\left(\frac{dv}{dx}\right)=f'\left(\frac{du}{dy}\right) \ldots(27).$$

The integration of the first of these partial differential equations gives $f=$ a function of $\frac{dw}{dy}+\frac{dv}{dz}$ and of the seven other differential coefficients.

Substituting in the second of equations (27) and integrating, and substituting the result in the third and integrating again, we readily find

$$f = \text{a function of the six quantities (24) and (25).}$$

We see then that Green's axiom that the function ϕ depends only on the change of form of the element, and Cauchy's relations (26), are but different ways of expressing the same condition; so that either follows if the truth of the other be admitted.

Cauchy's equations were proved by applying the statical equations of moments of a rigid body to an elementary parallelepiped of the medium, and taking the limit when the dimensions of the element vanish. The demonstration is just the same whether the medium be at rest or in motion, since in the latter case we have merely to apply d'Alembert's principle. It need hardly be remarked that the employment of equations of equilibrium of a rigid body in the demonstration by no means limits the truth of the theorem to rigid bodies; for the equations of equilibrium of a rigid body are true of any material system. In the latter case they are not *sufficient* for the equilibrium, but all that we are concerned with in the demonstration of equations (26) is that they should be *true*.

On the other hand, the form of V or ϕ to which MacCullagh was led is that of a homogeneous function, of the second order, of the three quantities

$$\frac{dw}{dy}-\frac{dv}{dz},\quad \frac{du}{dz}-\frac{dw}{dx},\quad \frac{dv}{dx}-\frac{du}{dy}\quad \text{............(28)},$$

which, as is well known, are linear functions of the similarly expressed quantities referring to any other system of rectangular axes. On substituting in (23), we see that the normal tensions on planes parallel to the coordinate planes, and therefore on any plane since the axes are arbitrary, vanish, while the tangential tensions satisfy the three relations

$$T_{yz}=-T_{zy},\quad T_{zx}=-T_{xz},\quad T_{xy}=-T_{yx}\ \text{........(29)};$$

so that the equations of moment of an element are violated. The relative motion in the neighbourhood of a given point may be resolved, as is known, into three extensions (positive or negative) in three rectangular directions and three rotations. The directions of the axes of extension, and the magnitudes of the extensions, are determined by the six quantities (24) and (25), while the rotations or angular displacements are expressed by the halves of the three quantities (28). In this theory, then, the work stored up in an element of the medium would depend, not upon the change of form of the element, but upon its angular displacement in space.

It may be shown without difficulty that, according to the form of ϕ assumed by MacCullagh, the equations of moments are violated for a finite portion of the mass, and not merely for an element. Supposing for simplicity that the medium in its undisturbed state is free from pressure or tension, let us leave the form of ϕ open for the present, except that it is supposed to be a function of the differential coefficients of the first order of u, v, w with respect to x, y, z, and let us form the equation of moments round one of the axes, as that of x, for the portion of the medium comprised within the closed surface S. This equation is

$$\iiint\left\{-\frac{d^2w}{dt^2}y+\frac{d^2v}{dt^2}z\right\}dx\,dy\,dz+\iint(Ry-Qz)\,dS=0,$$

the double integrals belonging to the surface. Since all the terms in this equation are small, we may take x, y, z for the actual or the equilibrium coordinates indifferently. Substituting from equations (20), and integrating by parts, we find

$$\iint\left\{f'\left(\frac{dv}{dx}\right)z-f'\left(\frac{dw}{dx}\right)y\right\}dy\,dz+\iint\left\{f'\left(\frac{dv}{dy}\right)z-f'\left(\frac{dw}{dy}\right)y\right\}dz\,dx$$
$$+\iint\left\{f'\left(\frac{dv}{dz}\right)z-f'\left(\frac{dw}{dz}\right)y\right\}dx\,dy+\iint(Ry-Qz)\,dS$$
$$+\iiint\left\{f'\left(\frac{dw}{dy}\right)-f'\left(\frac{dv}{dz}\right)\right\}dx\,dy\,dz=0.$$

The double integrals in this equation destroy each other by virtue of (22), so that there remains

$$\iiint\left\{f'\left(\frac{dw}{dy}\right)-f'\left(\frac{dv}{dz}\right)\right\}dx\,dy\,dz=0 \quad \text{.........(30).}$$

But this equation cannot be satisfied, since the surface S within which the integration is to be performed is perfectly arbitrary, unless $f'\left(\frac{dw}{dy}\right)=f'\left(\frac{dv}{dz}\right)$ at all points. We are thus led back to the equations (27), which are violated in the theory of MacCullagh.

The form of the equations such as (30) is instructive, as pointing out the mode in which the condition of moments is violated. It is not that the resultant of the forces acting on an element of the medium does not produce its proper momentum in changing the motion of translation of the element; that is secured by the equations (20); but that a couple is supposed to act on each element to which there is no corresponding reacting couple.

The only way of escaping from these conclusions is by denying that the mutual action of two adjacent portions of the medium separated by a small ideal surface is capable of being represented by a pressure or tension, and saying that we must also take into account a couple; not, it is to be observed, a couple depending on variations of the tension (for that would be of a higher order and would vanish in the limit), but a couple ultimately proportional to the element of surface. But it would require a function ϕ of a totally different form to take into account the work of such couples; and indeed the method by which the expressions for the components of the tension have been here deduced seems to show that in the case of a function ϕ which depends only on the differential coefficients of the first order of u, v, w with respect to x, y, z, the mutual action of two contiguous portions of a medium *is* fully represented by a tension or pressure.

Indeed MacCullagh himself expressly disclaimed having given a mechanical theory of double refraction*. His methods have been characterized as a sort of mathematical induction, and led him to the discovery of the mathematical laws of certain highly important optical phenomena. The discovery of such laws can hardly fail to be a great assistance towards the future establishment of a complete mechanical theory.

I proceed now to form the function ϕ for Cauchy's most general equations.

If we have given the expressions for $\frac{d^2u}{dt^2}$, $\frac{d^2v}{dt^2}$, $\frac{d^2w}{dt^2}$ in terms of

* *Transactions of the Royal Irish Academy*, Vol. XXI, p. 50. It would seem, however, that he rather felt the want of a mechanical theory from which to deduce his form of the function ϕ or V, than doubted the correctness of that form itself. [In other words, his abstract theory fitted into the fundamental *dynamical* equation, that of least action; but concrete *mechanical* illustrations of it had not been found. The distinction thus indicated is now more widely recognised since its application by Kelvin and especially by Maxwell to electrical phenomena.]

the differential coefficients of u, v, w with respect to x, y, z, they do not suffice for the complete determination of the function ϕ, as appears from the equations (20) or (21); but if we have given the expressions for the tensions P_x, P_{yx}, &c., ϕ is completely determinate, as appears from equations (23). In using these equations, it must be remembered that the tensions are measured with reference to surfaces in the undisturbed state of the medium; and therefore, should the expressions be given with reference to surfaces in the actual state, they must undergo a preliminary transformation to make them refer to surfaces in the undisturbed state.

Supposing then the tensions expressed as required, in order to find ϕ we have only to integrate the total differential

$$-d\phi = P_x d\frac{du}{dx} + P_y d\frac{dv}{dy} + P_z d\frac{dw}{dz} + T_{yz} d\frac{dw}{dy} + T_{zx} d\frac{du}{dz}$$
$$+ T_{xy} d\frac{dv}{dx} + T_{zy} d\frac{dv}{dz} + T_{xz} d\frac{dw}{dx} + T_{yx} d\frac{du}{dy} \quad \text{...........(31)},$$

the nine differential coefficients, of which ϕ is a function, being regarded as independent variables. Should the three equations (27) be satisfied, the expression (31) will be simplified, becoming

$$-d\phi = P_x d\frac{du}{dx} + P_y d\frac{dv}{dy} + P_z d\frac{dw}{dz} + T_x d\left(\frac{dw}{dy} + \frac{dv}{dz}\right)$$
$$+ T_y d\left(\frac{du}{dz} + \frac{dw}{dx}\right) + T_z d\left(\frac{dv}{dx} + \frac{du}{dy}\right) \quad \text{...........(32)},$$

where T_x denotes T_{yz} or T_{zy}, and similarly for T_y, T_z.

The general expressions for the tensions resulting from Cauchy's method are written at length in the equations numbered 17 and 18, pp. 133, 134 of the 4th volume of his *Exercices de Mathématiques*, where the normal and tangential tensions, referred to surfaces in the actual state of the medium, are denoted by A, B, C, D, E, F. These expressions contain 21 arbitrary constants, of which six, $\mathfrak{A}, \mathfrak{B}, \mathfrak{C}, \mathfrak{D}, \mathfrak{E}, \mathfrak{F}$, denote the tensions in the state of equilibrium. If these be for the present omitted, the remaining terms will be wholly small quantities of the first order, and therefore the tensions may be supposed to be referred to a unit of surface in the actual, or in the undisturbed state of the medium indifferently. On substituting now for $P_x, P_y, P_z, T_x, T_y, T_z$ in

(32) the remaining parts of A, B, C, D, E, F (observing that the ξ, η, ζ in Cauchy's notation are the same as u, v, w), it will be seen that the right-hand member of the equation is a perfect differential, integrable at once by inspection, and giving

$$\left.\begin{aligned}
-2\phi = {} & L\left(\frac{du}{dx}\right)^2 + M\left(\frac{dv}{dy}\right)^2 + N\left(\frac{dw}{dz}\right)^2 + P\left\{\left(\frac{dv}{dz}+\frac{dw}{dy}\right)^2 + 2\frac{dv}{dy}\frac{dw}{dz}\right\} \\
& + Q\left\{\left(\frac{dw}{dx}+\frac{du}{dz}\right)^2 + 2\frac{dw}{dz}\frac{du}{dx}\right\} + R\left\{\left(\frac{du}{dy}+\frac{dv}{dx}\right)^2 + 2\frac{du}{dx}\frac{dv}{dy}\right\} \\
& + 2U\left\{\frac{du}{dx}\left(\frac{dv}{dz}+\frac{dw}{dy}\right) + \left(\frac{dw}{dx}+\frac{du}{dz}\right)\left(\frac{du}{dy}+\frac{dv}{dx}\right)\right\} \\
& + 2V'\left\{\frac{dv}{dy}\left(\frac{dw}{dx}+\frac{du}{dz}\right) + \left(\frac{du}{dy}+\frac{dv}{dx}\right)\left(\frac{dv}{dz}+\frac{dw}{dy}\right)\right\} \\
& + 2W''\left\{\frac{dw}{dz}\left(\frac{du}{dy}+\frac{dv}{dx}\right) + \left(\frac{dv}{dz}+\frac{dw}{dy}\right)\left(\frac{dw}{dx}+\frac{du}{dz}\right)\right\} \\
& + 2V\frac{du}{dx}\left(\frac{dw}{dx}+\frac{du}{dz}\right) + 2W'\frac{dv}{dy}\left(\frac{du}{dy}+\frac{dv}{dx}\right) + 2U''\frac{dw}{dz}\left(\frac{dv}{dz}+\frac{dw}{dy}\right) \\
& + 2W\frac{du}{dx}\left(\frac{du}{dy}+\frac{dv}{dx}\right) + 2U'\frac{dv}{dy}\left(\frac{dv}{dz}+\frac{dw}{dy}\right) + 2V''\frac{dw}{dz}\left(\frac{dw}{dx}+\frac{du}{dz}\right)
\end{aligned}\right\} \quad \text{..........(33)},$$

the arbitrary constant being omitted as unnecessary. We see that this is a homogeneous function of the second degree of the six quantities (24) and (25), but not the most general function of that nature, containing only 15 instead of 21 arbitrary constants.

Let us now form the part of the expression for ϕ involving the constants which express the pressures in the state of equilibrium. It will be convenient to effect the requisite transformation in the expressions for the tensions by two steps, first referring them to surfaces of the actual extent, but in the original position, and then to surfaces in the original state altogether.

Let P'_x, T'_{yz}, &c. denote the tensions estimated with reference to the actual extent but original direction of a surface, so that $P'_x\,dS$, for instance, denotes the component, in a direction parallel to the axis of x, of the tension on an elementary plane passing through the point (x, y, z) in such a direction that in the undisturbed state of the medium the same plane of particles was perpendicular to the axis of x, dS denoting the actual area of the element. Consider the equilibrium of an elementary tetrahedron of the medium, the sides of which are perpendicular to the axes of x, y, z, and the base in the direction of a plane which was perpendicular to the axis of x; and let l, m, n be the direction-cosines of the base; then

$$P'_x = lA + mF + nE,\quad T'_{xy} = lF + mB + nD,\quad T'_{xz} = lE + mD + nC \ldots (34);$$

but to the first order of small quantities

$$l = 1,\quad m = -\frac{du}{dy},\quad n = -\frac{du}{dz};$$

substituting in (34), and writing down the other corresponding equations, we have

$$\left.\begin{array}{ll} \multicolumn{2}{c}{P'_x = A - F\dfrac{du}{dy} - E\dfrac{du}{dz}} \\ \multicolumn{2}{c}{P'_y = B - D\dfrac{dv}{dz} - F\dfrac{dv}{dx}} \\ \multicolumn{2}{c}{P'_z = C - E\dfrac{dw}{dx} - D\dfrac{dw}{dy}} \\ T'_{yz} = D - C\dfrac{dv}{dz} - E\dfrac{dv}{dx} & T'_{zy} = D - F\dfrac{dw}{dx} - B\dfrac{dw}{dy} \\ T'_{zx} = E - A\dfrac{dw}{dx} - F\dfrac{dw}{dy} & T'_{xz} = E - D\dfrac{du}{dy} - C\dfrac{du}{dz} \\ T'_{xy} = F - B\dfrac{du}{dy} - D\dfrac{du}{dz} & T'_{yx} = F - E\dfrac{dv}{dz} - A\dfrac{dv}{dx} \end{array}\right\} \ldots (35).$$

Lastly, since an elementary area dS originally perpendicular to the axis of x becomes by extension $\left(1 + \dfrac{dv}{dy} + \dfrac{dw}{dz}\right) dS$, and similarly with regard to y and z, we have

$$\left.\begin{array}{l} P_x : P'_x = T_{xy} : T'_{xy} = T_{xz} : T'_{xz} = 1 + \dfrac{dv}{dy} + \dfrac{dw}{dz} \\ P_y : P'_y = T_{yz} : T'_{yz} = T_{yx} : T'_{yx} = 1 + \dfrac{dw}{dz} + \dfrac{du}{dx} \\ P_z : P'_z = T_{zx} : T'_{zx} = T_{zy} : T'_{zy} = 1 + \dfrac{du}{dx} + \dfrac{dv}{dy} \end{array}\right\} \ldots (36).$$

Expressing P_x, T_{xy}, &c. in terms of P'_x, T'_{xy}, &c. by (36), then P'_x, T'_{xy}, &c. in terms of A, B, C, D, E, F by (35), and lastly substituting for $A, B \ldots F$ the expressions given by Cauchy, we find

$$\left.\begin{aligned}
P_x &= \mathfrak{A}\left(1+\frac{du}{dx}\right)+\mathfrak{F}\frac{du}{dy}+\mathfrak{E}\frac{du}{dz}\\
P_y &= \mathfrak{B}\left(1+\frac{dv}{dy}\right)+\mathfrak{D}\frac{dv}{dz}+\mathfrak{F}\frac{dv}{dx}\\
P_z &= \mathfrak{C}\left(1+\frac{dw}{dz}\right)+\mathfrak{E}\frac{dw}{dx}+\mathfrak{D}\frac{dw}{dy}\\
T_{yz} &= \mathfrak{D}\left(1+\frac{dw}{dz}\right)+\mathfrak{F}\frac{dw}{dx}+\mathfrak{B}\frac{dw}{dy}\\
T_{zx} &= \mathfrak{E}\left(1+\frac{du}{dx}\right)+\mathfrak{D}\frac{du}{dy}+\mathfrak{C}\frac{du}{dz}\\
T_{xy} &= \mathfrak{F}\left(1+\frac{dv}{dy}\right)+\mathfrak{E}\frac{dv}{dz}+\mathfrak{A}\frac{dv}{dx}\\
T_{zy} &= \mathfrak{D}\left(1+\frac{dv}{dy}\right)+\mathfrak{E}\frac{dv}{dx}+\mathfrak{C}\frac{dv}{dz}\\
T_{xz} &= \mathfrak{E}\left(1+\frac{dw}{dz}\right)+\mathfrak{F}\frac{dw}{dy}+\mathfrak{A}\frac{dw}{dx}\\
T_{yx} &= \mathfrak{F}\left(1+\frac{du}{dx}\right)+\mathfrak{D}\frac{du}{dz}+\mathfrak{B}\frac{du}{dy}
\end{aligned}\right\}\quad\ldots\ldots\ldots(37).$$

Substituting now these expressions in (31) and integrating, we have

$$\left.\begin{aligned}
-2\phi = {}& \mathfrak{A}\left\{2\frac{du}{dx}+\left(\frac{du}{dx}\right)^2+\left(\frac{dv}{dx}\right)^2+\left(\frac{dw}{dx}\right)^2\right\}\\
&+\mathfrak{B}\left\{2\frac{dv}{dy}+\left(\frac{du}{dy}\right)^2+\left(\frac{dv}{dy}\right)^2+\left(\frac{dw}{dy}\right)^2\right\}\\
&+\mathfrak{C}\left\{2\frac{dw}{dz}+\left(\frac{du}{dz}\right)^2+\left(\frac{dv}{dz}\right)^2+\left(\frac{dw}{dz}\right)^2\right\}\\
&+2\mathfrak{D}\left\{\frac{dv}{dz}+\frac{dw}{dy}+\frac{du}{dy}\frac{du}{dz}+\frac{dv}{dy}\frac{dv}{dz}+\frac{dw}{dy}\frac{dw}{dz}\right\}\\
&+2\mathfrak{E}\left\{\frac{dw}{dx}+\frac{du}{dz}+\frac{du}{dz}\frac{du}{dx}+\frac{dv}{dz}\frac{dv}{dx}+\frac{dw}{dz}\frac{dw}{dx}\right\}\\
&+2\mathfrak{F}\left\{\frac{du}{dy}+\frac{dv}{dx}+\frac{du}{dx}\frac{du}{dy}+\frac{dv}{dx}\frac{dv}{dy}+\frac{dw}{dx}\frac{dw}{dy}\right\}
\end{aligned}\right\}\quad\ldots(38),$$

which is exactly Green's expression*, Green's constants $A, B \ldots F$ answering to Cauchy's $\mathfrak{A}$, $\mathfrak{B} \ldots \mathfrak{F}$. The sum of the right-hand members of equations (33) and (38) gives the complete expression for -2ϕ which belongs to Cauchy's formulæ. It contains, as we

* *Cambridge Philosophical Transactions*, Vol. VII, p. 127. [Green's *Math. Papers*, p. 298.]

see, 21 arbitrary constants, and is a particular case of the general form used by Green, which latter contains 27 arbitrary constants.

I have been thus particular in deducing the form of Green's function which belongs to Cauchy's expressions, partly because it has been erroneously asserted that Green's function does not apply to a system of attracting and repelling molecules, partly because, when once the function ϕ is formed, the short and elegant methods of Green may be applied to obtain the results of Cauchy's theory, and a comparison of the different theories of Green and Cauchy is greatly facilitated.

On the Long Spectrum of Electric Light.

[From the *Philosophical Transactions* for 1862. Received *June* 19, 1862.]

[Abstract, from the *Proceedings of the Royal Society*, XII, pp. 166–8.]

THE author's researches on fluorescence had led him to perceive that glass was opaque for the more refrangible invisible rays of the solar spectrum, and that electric light contained rays of still higher refrangibility, which were quite intercepted by glass, but that quartz transmitted these rays freely. Accordingly he was led to procure prisms and a lens of quartz, which, when applied to the examination of the voltaic arc, or of the discharge of a Leyden jar, by forming a pure spectrum and receiving it on a highly fluorescent substance, revealed the existence of rays forming a spectrum no less than six or eight times as long as the visible spectrum. This long spectrum, as formed by the voltaic arc with copper electrodes, was exhibited at a lecture given at the Royal Institution in 1853; but the author, for reasons he mentioned, did not then further pursue the subject. Having subsequently found that the spark of an induction-coil with a Leyden jar in connexion with the secondary terminals yielded a spectrum quite bright enough to work by, he resumed the investigation, and examined the spectra exhibited by a variety of metals as electrodes, as well as the mode of absorption of the rays of high refrangibility by various substances. The spectra of the metals may be viewed at pleasure by means of fluorescence, and the mode of absorption of the invisible rays by a given solution may be at once observed; but there are difficulties attending the preparation in this way of sufficiently accurate maps of the metallic lines; and the great liability of the rays of high refrangibility to be absorbed by impurities present in very minute quantity renders the certain determination of the optical character, in this respect, of substances which are only moderately opaque a matter of considerable difficulty. Having found that Dr Miller had been engaged independently at the same subject, working by photography, the author deemed it unnecessary to attempt a delineation of the metallic lines (for which, however, he has recently devised a practical method that was found to work satisfactorily), or to examine further the absorption of rays of high refrangibility by solutions of metallic salts, &c.

The present paper contains therefore mainly results obtained in other directions in the same wide field of research. Among the metals examined, the author had found aluminium the richest in invisible rays of extreme refrangibility; and accordingly aluminium electrodes were employed when the deportment of such rays had to be specially examined. As the bright aluminium lines of high refrangibility do not appear to have been taken by photography, a drawing of the aluminium spectrum is given, with zinc and cadmium for comparison.

The author has also described and figured the mode of absorption of the invisible rays by solutions of various alkaloids and glucosides. Bodies of these classes, he finds, are usually intensely opaque, acting on the invisible spectrum with an intensity comparable to that with which colouring matters act on the visible. This intensity of action causes the effect of minute impurities to disappear, and thereby increases the value of the characters observed. It very often happens that at some part or other of the long spectrum a band of absorption, or maximum of opacity, occurs; and the position of this band affords a highly distinctive character of the substance which produced it.

Among natural crystals, besides the previously known yellow uranite, the author found that in adularia, and felspar generally, a strong fluorescence is produced under the action of the rays of high refrangibility, referable not to impurities, but to the essential constituents of the crystal. A particular variety of fluor-spar shows also an interesting feature, though in this case referable to an impurity, exhibiting a well-marked reddish fluorescence under the exclusive influence of rays of the very highest refrangibility. This property renders such a crystal a useful instrument of research.

With some metals broad, slightly convex electrodes were found to have a great advantage over wires, exhibiting the invisible lines far more strongly, while with some metals the difference was not great.

The blue negative light formed when the jar is removed, and the electrodes are close together, was found to be exceedingly rich in invisible rays, especially invisible rays of moderate refrangibility. These exhibited lines independent of the electrodes, and therefore referable to the air. This blue light has a very appreciable duration, and is formed by what the author calls an arc discharge.

The paper concludes with some speculations as to the cause of the superiority of broad electrodes, and of the heating of the negative electrode.

Introduction.

THE experimental researches described in a former paper* led me indirectly to the conclusion that the electric spark, whether obtained directly from the prime conductor of an ordinary electrifying machine, or from the discharge of a Leyden jar, emits rays of very high refrangibility, surpassing in this respect any that reach us from the sun—and that these rays pass freely through quartz, while glass absorbs them, as it does also the most refrangible of the solar rays. I was induced in consequence to procure prisms and a lens of quartz, which were applied in the first instance to the examination of the solar spectrum, and which immediately revealed the existence of an invisible region extending as far beyond that previously known as the latter extends beyond the visible spectrum, and exhibiting a continuation of Fraunhofer's

* "On the Change of Refrangibility of Light," *Phil. Trans.* for 1852, p. 463. [*Ante*, Vol. III, p. 267 and Vol. IV, p. 1.]

lines*. A map of the new lines was exhibited at an evening lecture delivered before the British Association at their Meeting in Belfast in the autumn of the same year; and I then stated that I conceived we had obtained evidence that the limit of the solar spectrum in the more refrangible direction had been reached. In fact, the very same arrangement which revealed, by means of fluorescence, the existence of what were evidently rays of higher refrangibility coming from the electric spark failed to show anything of the kind when applied to the solar spectrum. At least, the only link in the chain of evidence which remained to be supplied by direct experiment related to the reflecting power, for rays of high refrangibility, of the metallic speculum of the heliostat which was employed to reflect the sun's rays into a convenient direction; and this was shortly afterwards tested by direct experiment, on rays from an electric discharge separated by prismatic refraction.

In making preparations for a lecture on the subject delivered at the Royal Institution in February 1853, in which I had the benefit of the kind assistance of Mr Faraday, recourse was naturally had to electric light, on account of the extraordinary richness which it had been found to possess in rays of high refrangibility. Although fully prepared to expect rays of much higher refrangibility than were found in the solar spectrum, I was perfectly astonished, on subjecting a powerful discharge from a Leyden jar to prismatic analysis with quartz apparatus, to find a spectrum extending no less than six or eight times the length of the visible spectrum, and could not help at first suspecting that it was a mistake arising from the reflexion of stray light. A similarly extensive spectrum was obtained from the voltaic arc, and this was sufficiently bright to be exhibited to the audience, the arc passing between copper electrodes, and the pure spectrum formed by quartz apparatus being received on a piece of uranium glass cut for the purpose. The spectrum thus formed was found to consist entirely of bright lines†, whereas the spectrum of the discharge of a Leyden jar had appeared (perhaps from not having been truly in focus) to be continuous, or at least not wholly discontinuous.

The mode of absorption of light by coloured solutions, as observed by the prism, affords in many cases most valuable characters of particular substances, which, strange to say, though so

* *Ibid.* p. 559.

† *Proceedings of the Royal Institution*, Vol. I, p. 264. [*Ante*, p. 28.]

easily observed, have till very lately been almost wholly neglected by chemists. Having obtained the long spectrum above mentioned, I could not fail to be interested with the manner in which substances, especially pure but otherwise imperfectly known organic substances, might behave as to their absorption of the rays of high refrangibility. But the difficulties attending the habitual use of a nitric-acid battery of 30 or 40 cells deterred me from entering on this investigation, and I determined to confine myself to the solar spectrum.

On account of some inconvenience attending the tarnishing of the speculum of my heliostat, I was induced to order a quartz plate, intended to be either silvered or coated with the usual amalgam of tin. On trying on a small scale the reflecting power of such plates with respect to the invisible rays, which may be done by means of fluorescence almost as easily as if those rays were visible*, I noticed a remarkable falling off in the reflecting power of the silvered plate for the most refrangible of the solar rays, which I readily found was due to a peculiarity of the metal silver. This metal is highly reflective for the invisible as it is for the visible rays up to about the fixed line S†, when its reflecting power falls off, with remarkable rapidity, and for the more refrangible rays of the solar spectrum is comparable with that of a vitreous substance rather than with that of a metal. Steel, gold, tin, &c. showed nothing of the kind, but copiously reflected the invisible rays.

A few years ago, as Dr Robinson was showing me some experiments with the induction coil, it seemed worth while to try whether the spark obtained when a Leyden jar has its coatings connected with the secondary terminals might not be sufficiently strong to exhibit by projection the long spectrum shown by electric light. On projecting a spectrum formed by a prism and lens of quartz on a piece of uranium glass, the long spectrum

* *Philosophical Transactions* for 1852, p. 537.

† According to the notation employed in the Map published in the *Philosophical Transactions* for 1859, Plate XLVII. In this Plate the group S should have been represented as three lines, of which the middle (specially named S) divides the interval between the 1st and 3rd in the proportion of 3 to 2 nearly, the spaces between the lines being a little darkened by shading. [This map was prepared by Prof. Stokes, and published by Bunsen and Roscoe as a base-line for their curve of photo-chemical intensity.

Kayser points out (*Handbuch der Spectroscopie*, I, p. 103) that the fundamental observation in the text is not quite accurate, the quasi-vitreous reflection extending only a limited distance beyond S.]

was in fact exhibited. It was not, indeed, so bright as when formed by means of a powerful voltaic battery, but nevertheless was quite bright enough to work by. It was discontinuous, consisting of bright lines. On changing the metals between which the spark passed, we found that the lines were changed, which showed clearly that they were due to the particular metals.

A wide field of research was thus thrown open to any one taking the very moderate trouble attending the use of an induction coil. It remained to study the lines given by different metals and gases, and the absorbing action of various substances with respect to the invisible rays of different refrangibilities.

Various observations were made from time to time in this subject. As regards the metallic lines, it is perfectly easy to view them at pleasure; but to obtain faithful delineations of them is another matter. Even an accomplished artist would find difficulty in obtaining by mere eye-sketching a faithful representation of an object which requires to be seen in the dark. I tried different methods without being able to satisfy myself as to the accuracy of the drawings which could be thus obtained, and frequently thought of resorting to photography.

Meanwhile the mode of absorption of the rays of high refrangibility by a good number of substances was observed. Nothing is easier, to a person provided with a cell with parallel faces of quartz, than to observe by means of fluorescence the mode of absorption of these rays by a given solution; but to draw safe conclusions as to the optical character in this respect of the substance deemed to be in solution is not so easy as it might appear; for the rays of high refrangibility are liable to be absorbed by an exceedingly small amount of an impurity which may chance to be present without the observer's knowledge. Thus I found that about a quarter of a square inch of clean filtering paper sufficiently contaminated the water contained in a small cell to interfere sensibly with its transparency. Should the solution be transparent there would be no difficulty, for the effect of an impurity would not be to render transparent a solution which otherwise would be opaque. Should it, on the other hand, absorb the invisible rays, or some of them, with great energy, or in a peculiar manner, we might again conclude that we had obtained the true character of the substance deemed to be observed. The most remarkable example of this kind which I met with among inorganic colourless solutions was in the case of nitric acid and its salts, such as nitrate of potash, soda, ammonia, baryta, which absorb the rays of high refrangibility with great energy and in a peculiar manner, exhibiting a maximum of opacity followed by a maximum of transparency, beyond which the absorption becomes still more energetic than before. But if the solution should be found to absorb the rays of high refrangibility with only moderate energy, it would be left doubtful whether the observed absorption might not be due to some impurity; and I did not see how this doubt could be solved otherwise than by a laborious system of recrystallizations.

After having obtained these results, I found by conversation with my friend Dr [W. A.] Miller that he also had been engaged at the same subject, working by photography, and had prepared a number of photographs of metallic spectra, and studied by the same means the absorption of the rays of high refrangibility by a great variety of substances, chiefly inorganic acids, bases, and salts, and the commoner organic bodies. Although a large part of the task which I had proposed to myself has thus been accomplished in another way, there are many results which I have met with which are not likely to have been obtained by one working by photography, and I have therefore thought it well to draw up a paper embodying these results, and thus forming, as it were, a supplement to the paper by Dr Miller.

Preparation of a Screen by means of a Salt of Uranium.

Few substances are more powerfully fluorescent than several of the salts of sesquioxide of uranium; and a piece of glass coloured by uranium and polished along at least two planes at right angles to each other is exceedingly convenient, from its powerful fluorescence and its permanence, for a screen on which to receive a spectrum. Nevertheless such a screen, which must be viewed in particular directions in order to get the strongest effect, is in many cases less convenient than a screen would be which was prepared by means of a highly fluorescent powder treated like a water colour, which could be viewed in all directions indifferently. This is especially the case in taking measures by a method which will be mentioned presently. Besides, I find an excellent piece of such glass defective in fluorescent power as regards the extreme lines shown by aluminium; and some specimens are defective to a much greater extent, which is doubtless due to impurities. Accordingly I have long regarded it as a desideratum to obtain by precipitation an insoluble or very sparingly soluble salt of sesquioxide of uranium which should be as fluorescent as the best salts of that base, and which might be treated like a water colour. I have now succeeded in preparing such a salt, though not by direct precipitation.

The ordinary phosphate obtained by precipitation, the composition of which, independently of water of hydration, is $PO_5\,(U_2O_3)_2HO$*, is only slightly fluorescent. If, however, this salt, with as much water as remains when it is washed by decantation, be put into a saucer, a little free phosphoric or sulphuric acid added, and then crystals of phosphate of soda, phosphate of ammonia, microcosmic salt, or borax be added in excess, the original salt is gradually changed into one which is powerfully fluorescent. The change seems to take place most rapidly with borax; but as an excess of this salt is liable slowly to decompose the fluorescent salt first formed, it is better to employ a phosphate. The quantity of acid should be sufficient to leave a decided acid reaction when the liquid is fully saturated by the alkaline phosphate employed. The change may be watched by observing from time to time the fluorescence of the salt by daylight, with the aid of absorbing media. It is complete in a few days at furthest, when the salt is ready to be collected.

This requires precaution, as the salt is quickly decomposed by dilute acids (and accordingly by its own mother-liquor if diluted), and even, though more slowly, by pure water, with the formation apparently of the original phosphate. It is also decomposed, at least in time, by alkaline carbonates, with the formation of a beautiful yellow non-fluorescent salt resembling the precipitate given by alkaline carbonates in salts of sesquioxide of uranium. The salt may be collected by adding at once, instead of water, a saturated solution of borax, in quantity at least sufficient to destroy the acid reaction. The salt is then poured off in suspension from any undissolved crystals of the alkaline phosphate employed, and collected on a filter. A pressed cake of this salt, or a porous tile on which the salt is spread, having been moistened with a solution of borax, forms an admirable screen, and is what I have chiefly employed of late. It shows, of course, the visible as

[* This formula was constructed on the supposition, then current, that the atomic weight of oxygen is 8, and of uranium 120. Cf. also p. 216 *infra*.]

well as the invisible rays—the former by ordinary scattering, the latter by fluorescence.

From the circumstances of its formation, the salt is probably (abstraction being made of the water of hydration) the original phosphate with the equivalent of constitutional water replaced by an equivalent of an alkali, which would make it analogous to the highly fluorescent natural yellow uranite. At any rate this hypothesis guides us to its successful preparation, the conditions of which it would not have been easy to make out by observation alone. Without the use of free acid the fluorescence is not fully developed, which is accounted for by the insolubility of the original phosphate and the fluorescent salt, which presents an obstacle to the complete conversion of the one into the other.

Metallic Lines.

These may be viewed, as already mentioned, by passing the spark of an induction coil between two electrodes formed of the metal to be examined (the secondary terminals being respectively in connexion with the coatings of a jar of suitable size), forming a pure spectrum by a prism and lens of quartz, the faces of the prism being equally inclined to the axis of the crystal, and the lens being cut perpendicular to the axis, and receiving the spectrum on a suitable screen, for which, if a fluorescent liquid be employed, it is to be placed in a quartz-faced vessel, in default of which a piece of filtering paper may be saturated with the liquid.

If the visible spectrum and the very beginning of the invisible be excepted, the lines thus seen vary from metal to metal, and therefore are to be referred to the metal and not to the air. They are further distinguished from air lines by being formed only at an almost insensible distance from the tips of the electrodes, whereas air lines would extend right across. The spectrum is far too extended to allow us to regard the whole at once as in the position of minimum deviation; and if the prism be placed at all near the electrodes, without which we should have comparatively little light to work with, the effect of the different divergency, converted by the lens into convergency, of the rays in the primary and secondary planes is very great. In order to obtain a pure spectrum, the screen must be in focus as regards the primary plane; and if a particular point P of the spectrum be at a minimum deviation, the lines immediately about P are reduced almost to points, which are the images, for light of that refrangibility, of the tips of the electrodes, or, to speak more exactly, of the part of the spark just outside the tips. But in the secondary plane the rays on one side of P have not yet reached their focus, and on the other side have passed it; so that the image of a point is a line, the primary focal line, of a length increasing on receding from P in either direction, and accordingly the spectral image of either tip, assumed to be a mere point, would be a pair of slender triangles vertically opposite, and having their common vertex at P, their lengths lying in the plane of refraction. The invisible spectrum is in fact made up of two such pairs of triangles corresponding to the two tips respectively, as may be readily seen when the electrodes are not too close. At a distance from P at which the length of the primary focal line becomes equal to that of the image of the spark, the two lines which are the images, for rays of the refrangibility answering to that distance, of the tips of the electrodes meet in the middle of the spectrum, and beyond that distance they overlap, so that a line appears to run across the spectrum, though it relates to rays which emanated only from the immediate neighbourhood of the tips of the electrodes, as may be seen by turning the prism till that part of the spectrum is at a minimum deviation, and focusing afresh.

Besides the bright lines, evidently due to metals, which have been mentioned, other weaker light is perceptible, too faint for precise observation. A portion of this is probably due to the air.

The chief part of the visible spectrum as seen by projection appears plainly to belong to the air; for the lines stretch across the interval separating the electrodes, while the lines belonging to the metals extend but a little way, even in the visible spectrum, and the former reappear when the electrodes are changed. With some metals, however, lines belonging to the metal appear in the visible spectrum which are comparable in strength with the invisible lines of high refrangibility; but in general it is rather remarkable how poor is the visible spectrum, and even the invisible region for a good distance beyond, compared with the part of the spectrum of still higher refrangibility, with respect to strong lines characteristic of the metal.

I have lately adopted a mode of laying down positions in the invisible spectrum which is extremely simple and convenient, and yields results agreeing well with one another. It might be applied to the formation of maps of the metallic lines; but this is unnecessary, as the subject has been worked out by Dr Miller. It is still useful, however, for laying down the positions of bands of absorption, being more convenient and exact than estimating their place with reference to the known metallic lines.

The method is as follows. The quartz prism is placed on a block, raising it to a convenient height above a long drawing-board, to which the block is screwed, and is fixed at pleasure by a screw pressing upon it from above. The lens is fixed in a blackened board screwed edgeways to the drawing-board near the prism, so as to be ready to receive the rays of all refrangibilities after refraction through the prism. The focal length of the lens actually used was about 12 inches, and its diameter 1¼ inch. A convenient distance of the spark from the prism having been selected (I chose 30 inches), the drawing-board was turned round till it attained such a position that, on placing the prism in the position of minimum deviation for the middle of the long spectrum, the rays belonging to that part fell perpendicularly, or nearly so, on the lens, which had previously been placed so that this should be a convenient position relatively to the drawing-board. The prism was then fixed by its screw, and to mark the angle of incidence a pin was placed at the edge of the shadow of one of the blocks. On account of the increasing refraction by the lens of rays of increasing refrangibility, the locus of the foci of the different rays formed an arc of a curve, or nearly a straight line, lying very obliquely to the axes of the pencils coming through the lens. The projection of this line on the board having been marked, a line was drawn bisecting this at right angles, and at a point in the latter line situated 11⅜ inches from the former*, the board was pierced for the insertion of a pivot, which carried two wooden rulers, which could be clamped together at any convenient angle. The shorter of these carried a vertical needle, which as the ruler was turned moved in front of the focus of the different rays at the distance of about a quarter of an inch. The longer ruler carried a pricker, destined to mark on a sheet of paper, temporarily fastened to the drawing-board, the position of any object observed. Thus the prism, the lens, the axis of motion of the needle and pricker, and the pin for fixing the angle of incidence retained an invariable relative position when the drawing-board was moved.

* A longer distance would have been better.

In observing, the electrodes were placed at the proper distance, and the board turned till the edge of the shadow fell on the pin. The rulers were then turned together till any bright line or other object was eclipsed by the needle, and its place was then pricked down. To obtain a fixed point of reference, I generally pricked down the position of the extreme red visible on a screen, such as a piece of paper; but if great accuracy were required, it might be better to employ a well-marked green air line.

The metals the spectra of which I have observed are Platinum, Palladium, Gold, Silver, Mercury, Antimony, Bismuth, Copper, Lead, Tin, Nickel, Cobalt, Iron, Cadmium, Zinc, Aluminium, Magnesium. Several of these show invisible lines of extraordinary strength, which is especially the case with zinc, cadmium, magnesium, aluminium, and lead, which last, in a spectrum not generally remarkable, contains one line surpassing perhaps all the other metals. Other metals exhibit lines which in certain parts of the spectrum are both bright and numerous; so that, in taking a rough view of the whole, certain parts of the spectrum are bright and tolerably continuous, while other parts are comparatively weak. This grouping of the lines is especially remarkable in copper, nickel, cobalt, iron, and tin. Of the metals mentioned, magnesium gives by far the shortest spectrum, ending in a very bright line, beyond which, however, excessively faint light may be perceived to a distance about as great as the extent of the longer spectra. Aluminium, on the other hand, stands at the head of the above metals for richness in rays of the very highest refrangibility; and it is to this part of the spectrum that the strong lines above mentioned belong. In calling these lines strong, it must be understood that some allowance is made for their very high refrangibility; for when observed as above described they do not appear *absolutely* quite so strong as the bold lines of zinc or cadmium. This is partly due to the defective transparency of quartz, which for this part of the spectrum shows itself by no means perfect; and indeed the highest aluminium line, which is a double line, can only be seen by rays which pass through the prism near its edge.

The following figure exhibits the principal lines of aluminium, with zinc and cadmium for comparison. In the first of the aluminium lines represented, I could not make out the division into two parts corresponding to the tips of the electrodes.

R denotes the extreme red visible on a screen; the lines in the visible spectrum are omitted, as this has been made the subject of elaborate researches by others. The horizontal distances are proportional to the distances of the several pricks from that belonging to the extreme red, and therefore vary as the chords of the arcs described by the pricker. This tends to correct to a certain extent the exaggeration of the more refrangible end of the spectrum arising from the mode adopted of laying down the positions of the lines. The lowest row of lines in the figure, which is placed here for the sake of comparison, will be referred to further on.

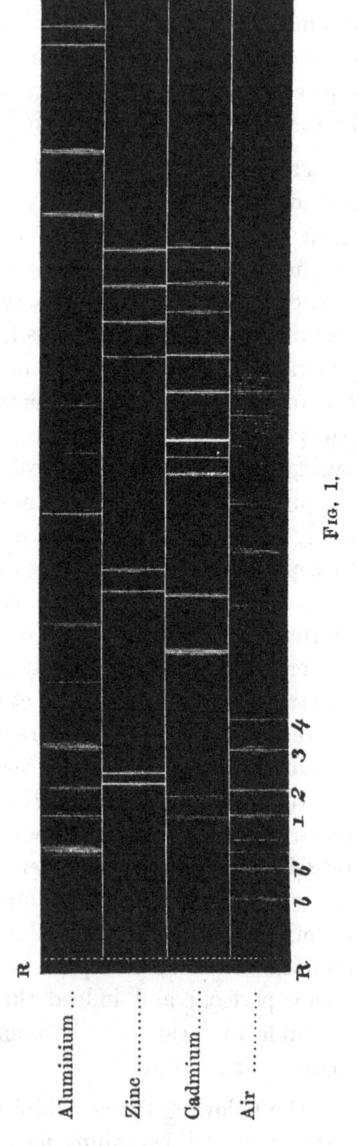

Fig. 1.

Besides the lens above mentioned, I sometimes employ in a different manner another of $\frac{1}{2}$-inch diameter and $2\frac{1}{2}$ inches focal length, and accordingly large for its focal length. This is used for forming an image of the spark, which is received on the substance that is to be examined, or that is used for examining the spark. The difference of focal length for the different rays is so enormous that, while one part of the spectrum is in focus, other parts are utterly out of focus, and thus we may judge in a general way of the refrangibility of the rays by which any particular effect is produced. In this way such concentration of the rays is obtained, that effects may be studied which would not bear examination by prismatic analysis. In speaking of this lens I shall call it the 2·5-inch lens, from its focal length.

Absorption of the invisible rays by Alkaloids, Glucosides, &c.

Before examining these substances it is requisite to dissolve them, and we must first inquire into the transparency of the solvent. Fortunately the most useful of all solvents, water, is transparent when pure; and as to reagents, we may employ sulphuric or hydrochloric acid for an acid, these acids being transparent, and ammonia, suppose, for an alkali. In speaking of a substance as transparent, I wish it only to be understood that it is of a transparency comparable with quartz. As to ammonia, although it absorbs the more refrangible rays when in quantity (unless the observed absorption were due to some impurity), it may be deemed transparent in the small quantity which alone it is requisite to employ. Even alcohol, which in the state in which it is to be had is defective in transparency, is sufficiently transparent to be employed as a solvent for such substances as those under consideration, provided it be used in small thickness only.

The alkaloids and glucosides which I have examined are almost without exception intensely opaque for a portion at least of the invisible rays, absorbing them with an energy comparable for the most part to that with which colouring matters (such as alizarine, &c.) absorb the visible rays. The mode of absorption also is frequently, I might almost say generally, highly characteristic; so that by this single property they might be distinguished one from another. It frequently happens too that the mode of absorption decidedly changes according as the solution is acid or alkaline, which assists still further in the discrimination.

In the examination I sometimes employ a small cell with parallel faces of quartz, sometimes a wedge-shaped vessel, having its inclined faces also of quartz, but more commonly the former. The cell being filled with the solvent, a minute quantity of the substance is introduced, and the progress of the absorption is watched as the substance gradually dissolves, the fluid meantime being of course stirred up. In this way it is easy to seize the most characteristic phase of the absorption, which may be then registered by the pricking instrument. When minima of opacity occur, it is best to seize that stage of the absorption at which they are well developed. When no minima occur, a greater or less part of the more refrangible region is quickly absorbed, after which the absorption creeps on towards the less refrangible side. When once it has become tolerably stationary, the limit of the rays transmitted may be marked. It seems desirable not to go beyond this point in the absorption, lest some possible impurity in the substance examined, which if it had formed the whole of the specimen would have absorbed rays of lower refrangibility, should begin to make itself perceived, and its mode of absorption should be mistaken for that of the substance professed to be examined.

All the metallic spectra are discontinuous, which prevents the mode of absorption of even a solid or liquid from being observed quite so well as in the solar spectrum, even independently of the greater intensity of the latter, and would greatly interfere with the observation of narrow bands like those shown by the absorption of certain gases in the visible spectrum, and of which chlorous acid gas (ClO_4) shows a splendid system in the invisible part of the solar spectrum. Should a general absorption take place in a part of the spectrum where previously a bright group of lines was seen, with weaker light for some distance on both sides, it is evident that at a certain stage of the absorption the bright group would be left isolated, and the effect might be mistaken for a maximum of transparency. In doubtful cases of this kind it is requisite to change the electrodes, so as to use the spectrum of some other metal; but practically the difficulty is not so great as might be supposed.

It is desirable to choose a metal which gives a spectrum that is bright and tolerably continuous in the region in which the distinctive features of the absorption are most likely to occur. For general use in the examination of substances such as here considered, I prefer tin—the electrodes (or one of them at least) being broad, for a reason which will be mentioned presently. Tin, indeed, is weak in the most refrangible region, though after a long interval of weakness it shows one pretty strong line between the second and third of the strong aluminium lines; but with these substances the distinctive features of the absorption hardly ever occur so late. For combined strength and continuity, copper answers well for the highly refrangible region in which tin is weak; while mercury, which may be employed in the form of amalgamated zinc, is the richest metal for the invisible region just beyond the visible spectrum; but I have employed tin almost exclusively.

The following figure gives the bands of absorption observed in

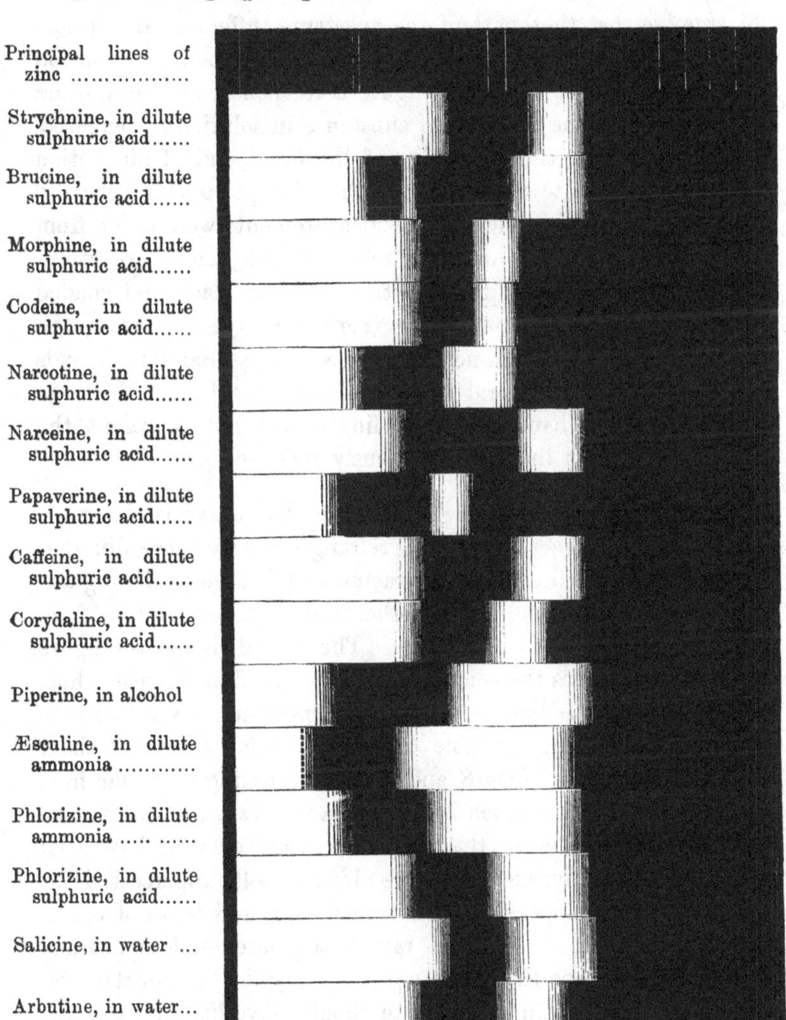

Fig. 2.

solutions of several alkaloids and glucosides. The bold lines of zinc are given as points of reference; but the observations were made with electrodes of tin. The border on the left is the limit of the red light visible on a screen.

Although the central part of the maxima of transparency in this figure is generally left white to save trouble, the reader must not suppose that that part of the spectrum suffers no absorption. On the contrary, it is more or less weakened when the solution has the strength to which the figure corresponds, and disappears altogether when the quantity of substance in solution is increased, while at the same time the edge of the first band of absorption creeps on *a little* towards the red, the absorption being usually pretty definite at this edge. The measurements were taken from the points where the light ceased to be sensible, which are represented in the figure by the junction of plain black and shaded white. The shading merely represents the general effect, the gradation of illumination not having been registered. It extends in the figure, as a general rule, too far to the left of the edge of the first black band, and accordingly does not represent the absorption at that limit as sufficiently definite.

A glance at the figure will show how distinctive is the mode of absorption of the rays of high refrangibility by these different substances. Indeed this one character would serve to distinguish all these substances one from another, unless it be morphine from codeine, and caffeine from salicine. The dotted line in the figure for æsculine denotes the commencement of the fluorescence, which is situated near the line G of the solar spectrum. A solution of brucine cuts off the invisible end of the solar spectrum about midway between the lines S and T, and accordingly not far from the end of the region which it requires a quartz prism and lens to see. Accordingly, when these substances are examined by solar light their distinctive characters are almost wholly unperceived, the solutions of some appearing quite transparent, and those of others merely cutting off the extreme rays to a greater or less distance. With æsculine alone the maximum of opacity lies within the solar spectrum; but even in this case we should have little idea of the great increase of transparency about to take place.

The effect of acids and alkalies on all the glucosides referred to in the figure presents one uniform feature. When a previously neutral solution is rendered alkaline, the absorption begins somewhat earlier, when rendered acid somewhat later. With salicine there is merely an indication of this change, falling within the limits of errors of observation; but in the other cases it is quite perceptible, and with phlorizine the shifting of the band of absorption produced by an acid is very large. Fraxine (or paviine) agrees remarkably with æsculine in all its optical characters; the maximum of absorption is merely situated a little nearer to the red, and the tint of the fluorescent light corresponds to a slightly lower mean refrangibility.

Quinine presents no decided maximum of transparency. With this and the other bases observed, with one exception, the absorption, if changed at all, is changed in an opposite manner to the glucosides when the base is set free by ammonia.

Bands of absorption occur also with neutral substances, for example coumarine and paranaphthaline, which last exhibits a system of such bands in the invisible part of the solar spectrum.

Aconitine, atropine, and solanine exhibit no bands of absorption, but merely a general opacity for the more refrangible rays. The last, indeed, when dissolved in dilute sulphuric acid, is, for this class of bodies, remarkably transparent; while when the base is set free the solution, contrary to what takes place with the other bases, becomes much more opaque, but the absorption is vague. I am not sure, however, how far the purity of the specimen examined may be trusted, though it was white, and regularly crystallized. It would be easy to examine more such substances; but what precedes is sufficient to show the value of the study of the absorption of the rays of high refrangibility, as affording distinctive characters of substances little known.

Minerals.

I have examined a large number of minerals by the rays from the induction spark, both as to their transparency and as to their fluorescence. The transparency of those crystals which were of such a form as to permit it, was examined by holding them in front of a pure spectrum formed on a fluorescent screen. The fluorescence was sought for by forming an image of the spark, for which aluminium electrodes were employed, by the 2·5-inch lens, holding the mineral first at the focus of the visible rays, and then moving it up towards the lens, and watching for any image which might be formed by the rays of higher refrangibility. Should such be observed, its nature was further demonstrated by interposing in the path of the rays a very thin piece of mica. This cut off the image by intercepting the invisible rays, with respect to which, except a small portion of the lowest refrangibility, mica is intensely opaque.

Carbonate of lime, the sulphates of lime, baryta, and strontia, and colourless fluor-spar, were found transparent (sulphate of strontia less so), at least in the qualified sense above mentioned, thus demonstrating the transparency of carbonic, sulphuric, and probably hydrofluoric acid, and of the bases, lime, baryta, strontia*. But this subject would be better followed out by salts artificially prepared, and has been investigated by Dr Miller. In two cases results of considerable interest were obtained with reference to fluorescence.

At the time of writing my first paper, on the change of refrangibility of light, I had found but one mineral, yellow uranite, to the essential constituents of which the property of fluorescence plainly belongs†. In many other cases, both before and since that time, I have observed with solar light fluorescence in minerals, but always apparently having reference to unknown impurities, and therefore to my mind of much inferior interest. By means of the induction spark, employed as above described, I have found one more fluorescent mineral‡.

On receiving the image on adularia, and focusing it for the rays of highest refrangibility, a pair of bluish dots were seen, which were the images of the tips of the electrodes exhibited by fluorescence. As the appearance was everywhere the same, on natural faces and cleavage planes alike, and the same was observed with colourless felspars generally from different localities, it is

[* This inference seems to be open to question, as also the ascription of the opacity of mica to peroxide of iron.]

† *Philosophical Transactions* for 1852, p. 524. [*Ante*, Vol. III, p. 352.]

‡ The method by which M. Edmond Becquerel has examined the fluorescence of minerals (*Annales de Chimie*, sér. III, tom. LVII, p. 43) does not permit of distinct vision of the specimen from the distance of a few inches, which seems to me necessary to allow the observer to judge whether the fluorescence which may be observed is due to the essential constituents of the crystal or to accidental impurities.

doubtless a property of the silicate of alumina and potash constituting the crystal. Some specimens, it is true, did not show the effect so strongly as adularia or moonstone; but this is easily explained by the greater purity of the latter varieties. For the fluorescence extended to a very sensible though small depth within the crystal, and yet the rays producing it were cut off by a film of mica much thinner than paper. The intense opacity of mica is doubtless due to peroxide of iron, which nevertheless forms no more than perhaps 5 per cent of the mineral. Hence a very small percentage of peroxide of iron, or any other impurity having a similar absorbing action, would suffice greatly to reduce the quantity of fluorescent light emitted.

In a concentrated solar beam passed through a suitable absorbing medium, adularia did not show the least sign of fluorescence, in which respect it notably differs from common glass, such as window-glass.

The other case of interest relates to a particular variety of fluor-spar found at Alston Moor in Cumberland. This variety is very pale by transmitted light, being in part of a brownish purple colour, shows a strong blue fluorescence, and is eminently phosphorescent on exposure to the electric spark. On presenting such a crystal to the spark passing between aluminium electrodes, besides the usual blue fluorescence there is seen another of a reddish colour, extending not near so far into the crystal. On receiving on the crystal the image of the spark, and moving the crystal from the focus of the invisible rays towards the lens, it was soon in best focus for the rays producing the blue fluorescence. It had to be moved much nearer to the lens before it came into focus for the rays producing the reddish fluorescence, and was then at the distance at which a well-defined image of the tips of the electrodes is formed on the uranium salt; which proves that the reddish fluorescence was produced by the rays belonging to the bright lines (considered as a whole) of aluminium of extreme refrangibility.

The crystal which showed this effect best was externally colourless for about the $\frac{1}{20}$th of an inch, which stratum showed no fluorescence when examined in this way. Then came one or two strata, parallel to the faces of the cube, showing the ruddy fluorescence, and exhausting apparently the rays capable of producing that effect. The blue fluorescence extended much deeper, and presented a stratified appearance, as Sir David Brewster long ago observed.

On admitting a pencil concentrated by a quartz lens parallel to and almost grazing a face of the cube, so that the rays traversed the colourless stratum, the reddish fluorescence was observed in the stratum which produced it to a long distance from the face by which the rays were admitted, which demonstrates the transparency of fluoride of calcium for the rays of very high refrangibility.

The property of exhibiting such a well-marked effect under the exclusive influence of rays of extreme refrangibility, renders such a crystal a useful instrument of research. Several other metals besides aluminium show the reddish fluorescence; but none of those examined showed it so well, partly because it is evidently produced more copiously by aluminium electrodes, and partly because it is less masked by the blue fluorescence, the spectrum of aluminium being rather wanting in brightness until the region of extreme refrangibility is reached.

If the crystal be held near the electrodes, and observed while their distance changes, it will be found that on passing from the greatest striking distance the reddish fluorescence decidedly improves. On still further diminishing the distance between the electrodes, the reddish fluorescence appears still to increase; though whether this is a real absolute increase or only an increasing preponderance over the blue, it is not easy in this way to say for certain. Hence the copiousness of rays of high refrangibility increases at first, and continues to increase *relatively* if not absolutely. It is supposed that the jar is sufficiently large to prevent the discharge from degenerating into what will be presently described as the arc discharge.

If the crystal be held close to the contact-breaker when the secondary terminals are separated, and the effect be compared with that of the secondary discharge (a jar being in connexion, as has been supposed all along), the electrodes being of platinum for fairness of comparison, it will be found that the proportion of rays of extremely high refrangibility is decidedly greater for the spark at the contact-breaker than for the secondary discharge.

On forming by the 2·5-inch lens an image of the spark from aluminium electrodes, and placing a crystal, such as that above mentioned, in the focus of the rays producing the reddish fluorescence, it is easy to determine the transparency or opacity of substances for those rays, the alteration of the focus by the introduction of a thick plate being of course borne in mind, and the crystal moved accordingly. The rays forming the image have had to pass only through air, and through a very small thickness of quartz, before reaching the crystal. In this way I have found that even quartz itself in very moderate thickness is opaque for these rays; but different specimens, or different parts of the same specimen, vary in this respect. I possess a large plate 0·42 inch thick, cut perpendicular to the axis of the crystal, which is generally transparent, but is slightly brownish on one side, to the distance of about half an inch from the face of the hexagonal prism. The *colourless* part of this plate, beyond a little distance from the brownish part, is opaque for the rays in question*, while the *brownish* part is nearly transparent. It may be inferred that the colourless part contains a minute quantity of some impurity capable of absorbing these rays, which does not exist, at least to the same extent, in the brownish part, although the latter is not perfectly pure silica, as is shown by its colour. On the whole, I am disposed to think that quartz, if it were *rigorously* pure, would be transparent. We see at any rate how difficult it is to draw certain conclusions respecting the transparency or opacity of a substance which, in the state of purity in which it may be obtained, shows only a slight defect of transparency.

I tried reflecting the rays from the spark by a fine Munich grating, but the light was far too faint to be of any use. Possibly a large and very closely ruled plane speculum, with a concave speculum instead of a lens, might give light which it would be possible to observe. But at present I have not found any sufficiently

* It should be mentioned that this part contains those delicate, definitely directed, elongated laminae or crystals, hardly visible except in a beam of sunlight, which are called by practical opticians 'blue shoots.' An examination of a number of cut pieces of quartz lent me by Mr Darker confirms me in the suspicion that such crystals are more defective in transparency than other colourless specimens for the rays of extreme refrangibility. July 1862.

marked effects referable to rays of still higher refrangibility to make it worth trying.

The same crystal which showed the reddish fluorescence was eminently phosphorescent, with a blue colour. The phosphorescence, like the fluorescence, was arranged in strata parallel to the faces of the cube, and, like the reddish but unlike the blue fluorescence, was not perceptible beyond a moderate distance from the surface at which the exciting rays had entered. On forming an image of the discharge by the 2·5-inch lens, focusing the crystal for the rays producing the reddish fluorescence, fixing it there, and breaking the circuit after the induction coil had worked for a little while, a dart of blue phosphorescent light was seen in the crystal at the focus of the lens. On focusing for the rays most efficient in producing the blue fluorescence, the reddish was diffused over a broad portion of the strata producing it; and on repeating the above experiment in this position of the crystal, the blue phosphorescence was seen similarly diffused. This shows that the rays of extremely high refrangibility are those most efficient in producing the blue phosphorescence.

[We may suppose that the blue fluorescence, the reddish fluorescence, and the blue phosphorescence are due to the action of the assemblage of heterogeneous exciting rays on the same substance (doubtless some impurity taken up during crystallization), or on two or three distinct substances. The blue fluorescence is produced abundantly at a depth within the crystal at which the two other effects are invisible; but this alone is no proof of a diversity in the nature of the substance acted on, because the rays producing the two latter effects would have been absorbed before arriving at such a depth. Hence it is among the early strata, in crossing which rays capable of producing each of the three effects are still vigorous, that evidence must be sought, in the coincidence or non-coincidence of the strata in which the three effects are respectively perceived, of the probable identity or certain diversity of nature of the substance acted on. At the time when this paper was read I fancied I had observed slight discrepancies as to coincidence in the strata. But a renewed examination, in which a larger number of specimens were observed, leads me to regard the fancied discrepancies as too doubtful to rely upon and to overpower the increasing weight of evidence on the other side.

The blue fluorescence may be observed in the early strata (which ordinarily, at least with electrodes of aluminium and several other metals, show a red) by absorbing the more refrangible of the exciting rays by a suitable plate of quartz, or else by substituting for aluminium some metal, such as magnesium, which is poor in rays of extreme refrangibility. On the other hand, the red fluorescence really existing in the early strata, when it is overpowered by the blue, may be seen by viewing the crystal through a solution of chromate of potash, which greatly enfeebles the blue fluorescence, while at the same time it transmits enough of the spectrum to allow the unabsorbed residue to be at once distinguishable by its colour (green) from the red fluorescence. In this way the red fluorescence may be readily perceived even with electrodes of magnesium. Again, a particular stratum which showed a blue fluorescence when acted on by rays which entered by a face of the cube, and before reaching it had to traverse some other strata showing fluorescence, exhibited a red fluorescence when acted on by rays which fell on it directly, having been admitted through an octahedral face.

It is more difficult to decide as to the identity or diversity of the strata showing respectively red fluorescence and blue phosphorescence, because the two effects are observed in a different way; but as far as I could decide, the strata appeared to correspond.

On the whole, then, I am disposed to think it probable that it is the same substance which, in consequence of the action of rays beginning with a part of the violet and extending from thence onwards, exhibits a blue fluorescence, which, in consequence of the action of rays of extreme refrangibility, exhibits a red fluorescence, and which, in consequence of the action of rays of a similar refrangibility, exhibits a powerful blue phosphorescence. At least, if the substances be different they would appear to have coexisted in solution, and so to have been taken up together in the crystallization of the mineral. I should mention, however, that it is contrary to all my experience that the fluorescence of a single substance (*i.e.* not a mixture) should thus, as it were, take a fresh start *with a totally different colour* on proceeding onwards in the spectrum; but then my experience is derived mainly from the examination of substances in the comparatively short solar spectrum.—July, 1862.]

I have said that the phosphorescence was produced in certain strata within the crystal. These strata were in some places sharply terminated, so as to be foreshortened into well-defined lines. On watching the phosphorescence, there was nothing to be seen at all like conduction; the strata remained sharply defined as long as the light was strong enough to enable one to judge. This is at variance with one of the two results which, on the authority of others, I formerly mentioned as indicating a distinction between phosphorescence and fluorescence*. On trying shortly afterwards along with Mr Faraday, I could not obtain either of these results. One of them, that relating to apparent conduction, which was obtained by MM. Biot and A. C. Becquerel, has since been explained by M. Edmond Becquerel as an illusion of observation†. The other, that relating to the production of phosphorescence in Canton's phosphorus by rays which had traversed a strong solution of bichromate of potash, I am, after a conversation with Dr Draper, still unable to explain.

Advantage of Broad Electrodes.

At first I employed by preference wires or sharp pieces of metal for electrodes, in consequence of the greater facility with which the discharge passed, and the larger quantity of light given out by the spark. Certain considerations, however, led me to try broad electrodes; and I accordingly procured electrodes of the common metals shaped like small watch-glasses, about an inch in diameter. These showed in some cases a most marked superiority over thin wires, exhibiting the invisible metallic lines in far greater strength, while with some metals there was not much difference. With copper, for example, the superiority was very great, with iron it was comparatively small.

Instead of electrodes of this shape, it is sufficient to take two pieces of thick foil, make them slightly cylindrical by means of a round ruler, or a pencil, and mount them with their convexities opposed and the axes of the cylinders crossed.

* *Philosophical Transactions* for 1852, p. 547. [*Ante*, Vol. III, p. 385.]

† *Annales de Chimie*, tom. LV, (1859), p. 112.

Besides copper, silver, tin, and aluminium show a great advantage of flat electrodes, and lead a moderate advantage, while with zinc, as with iron, sharp electrodes are nearly as good. Brass agrees in this respect with zinc, and not with copper, though it shows the copper lines very strongly.

With such electrodes, however, the spark dances about; and its unsteadiness is objectionable in some experiments. A good part of the advantage of flat electrodes is however retained if one only be flat, especially if this be negative, and the spark is now steadier. Instead of using the end of a wire to combine with a flat electrode, it seems rather better, according to a plan suggested to me by Dr Miller, to bend a wire to a gentle curve lying in a vertical plane passing through the prism; or the edge of a flat piece of metal may be similarly employed.

On forming an image of the spark between a sharp and a flat electrode of copper, and receiving it on a fluorescent screen, the flat electrode gave the brighter of the two images already mentioned, and that, whether the electrode were positive or negative.

On similarly forming an image of the spark between two very broad electrodes, and focusing for the rays of highest refrangibility, the image did not, as usual, consist of two separate dots; but, whether it was, that, from the shortness of the spark, the two ran into one, or that the rays belonging to the metallic lines of high refrangibility were emitted throughout the whole length of the spark, I am not quite certain; but I incline to the latter opinion, as a separation of the discharge into two portions, corresponding to the immediate neighbourhood of the two electrodes respectively, could hardly have escaped detection had it existed.

Arc Discharge, and Lines of Blue Negative Light.

On diminishing the distance between the electrodes, formed suppose of copper wires, the brightness of the metallic lines at first improves, and afterwards changes but little, or, if anything, rather falls off. On still further diminishing the distance, so that the electrodes almost touch, and the discharge passes with little noise, a new set of strong lines make their appearance in the invisible region of moderate refrangibility. In this mode of discharge, in which the negative electrode, if at all thin, quickly becomes red-hot and fuses, the jar has not much influence, and the lines in question are still better seen when it is suppressed altogether. To show them to perfection, it is best to take a flat negative electrode, so as to carry off the heat, and not to hide from the prism any part of the blue negative light, and a sharp positive electrode almost touching the former. In this way the visible discharge is reduced almost wholly to an insignificant-looking star of blue light; but it is wonderful how strong an effect it is capable of producing in the invisible region. The most striking part of the invisible spectrum consists of four bright lines, numbered 1, 2, 3, 4 in Fig. 1 [p. 214], situated not far from the visible spectrum. These are followed, after a nearly dark interval, by light arranged in masses resembling in its general aspect the groups of copper lines (from which, however, it differs), but not strong enough to be resolved or accurately measured. The figure represents also a couple of blue bands (b, b') seen by projection. These are not seen on looking at the blue light directly with a flint-glass prism of 60°, because everything is seen in too great detail. Most of the air-lines in the invisible spectrum, especially the bands beyond line 4, have an ill-defined look, and would probably be resolved did the intensity of the light permit.

The appearance just described is independent of the nature of the electrodes, and therefore is to be referred to the air, and not to the metal. On viewing in a moving mirror the star of light producing this effect, it is found to have a considerable duration.

On slightly separating the electrodes, forming an image of the discharge with the 2·5-inch lens, and receiving it on a cake of the uranium salt, a very strong fluorescence was seen over the image of the blue disk when the lens was focused for a point a little beyond the visible spectrum. On moving the lens onwards, the fluorescence produced by the rays belonging to this image spread out into a ring; and on moving still further, a tolerably well-defined image of the whole discharge was perceived. Of this the part belonging to the blue disk was the brightest, and was surrounded concentrically by the ring before mentioned, now still further widened. The image of the remainder of the discharge was brightest where it was most contracted at the positive electrode. The discharge generally was perhaps of slightly higher refrangibility than the blue disk, even excluding from the latter the rays belonging to the ring. It thus appears that the four bright lines figured were produced mainly by the blue negative light.

The mode of transition of the discharge may be studied by placing the electrodes at the greatest striking-distance and making them gradually approach. At first there passes a clean bright spark making a sharp report, and not resolved by a revolving mirror. The invisible spectrum which this shows is too faint for precise observation; the visible spectrum shows chiefly air-lines. As the electrodes approach, the spark becomes clothed by the well-known yellowish envelope capable of being blown aside, and the blue negative light begins to appear. A moving mirror, as M. Lissajous has already observed*, shows an instantaneous spark at the commencement, in point of time, of the envelope and blue negative light, both which are drawn out, indicating a very appreciable duration. On making the electrodes approach somewhat nearer, the spark diminishes, and the envelope is formed in perfection, especially with broad electrodes. The air-lines now begin to show themselves well, but are brightest on the side of the spectrum answering to the blue negative light.

It might be supposed at first sight that the permanence of the yellowish and of the blue light only indicated a glow of appreciable duration left by a sensibly instantaneous discharge; but several circumstances indicate that the discharge itself lasts, and that it is *under* its action that the glow takes place†. The action, I am persuaded, is this: a spark first passes; and this enables a continuous discharge to pass, which is due, in part at least, to the inductive action of the still falling magnetism, just as a voltaic arc may be started in a powerful battery by passing an electric spark between the slightly separated electrodes; and the glowing of the air *under the action* of this discharge produces the yellowish envelope and blue negative light. Thus, when the

* See Du Moncel, *Recherches sur la non-homogénéité de l'étincelle d'induction*, p. 107.

† Although this view may be considered already established (see the work by the Vicomte Du Moncel just quoted), the observations here mentioned will not, I hope, be altogether useless.

electrodes are nearly at the greatest distance at which this sort of discharge takes place, the blue negative light is seen pretty sharply terminated in a moving mirror. Were it a dying glow, it ought to fade away; but if produced under a discharge, it ought to cease almost abruptly, inasmuch as at this distance of the electrodes a continuous discharge is unable to pass when the tension has sunk much below that under which it was first produced.

The same conclusion may be drawn from an effect which I once obtained, the exact conditions for the production of which it is not easy to hit off. With a jar in connexion, each discharge due to a single breach of contact appeared in a moving mirror as a bright spark joined to a spark less bright by the blue negative light, and also by the yellowish or reddish light, brightest close to the positive electrode. Were the blue light due to a glow, it ought to be reinforced instead of being put out by the second spark, whereas the explanation of the result is easy on the supposition of a continuous discharge. The first spark started a continuous discharge, which emptied the jar less fast than it was filled by the secondary coil; so that presently another discharge took place, which emptied the jar so that a continuous discharge could no longer pass.

On viewing the broad discharge formed without a jar when the electrodes are at a moderate distance, through a revolving disk of black paper with a single hole near the circumference, while the envelope was being blown aside, so as to get a succession of momentary views of the discharge, the envelope was seen *extravagantly bent*, as a flexible conductor might have been—not *torn across*, as a column might have been which was heated by a *previous* spark. The central spark, of course, was usually missing, as it is sensibly instantaneous.

I have spoken of the arc, and especially the blue negative light, as exhibiting air-lines. The arc, however, is liable to be coloured not only by casual dust (as when it passes partly through the flame of a spirit-lamp with a salted wick, when it is coloured yellow by sodium), but also by matter torn from the positive electrode. This is well seen with electrodes of aluminium, when the arc or a portion of it is frequently coloured green. This green light has a very sensible duration, and a distinctive prismatic composition, and is brighter towards the positive than towards the negative electrode, but is not confined to the immediate neighbourhood of that electrode (extending indeed sometimes over almost the whole length of the arc), in which respect, and in its duration, it differs from the light of the spark proper*. With aluminium opposed to another metal, as copper or iron, the green light is seen only when the aluminium is positive. Even with aluminium this light may generally be got rid of by making the electrodes approach; and it is the arc in what may thus be deemed its normal state that was observed for the construction of the last line of Fig. 1, though I have not at present noticed variations in the invisible corresponding with those in the visible spectrum of the arc discharge.

On the Cause of the Advantage of Broad Electrodes; and on the Heating of the Negative Electrode.

Although the spark appears instantaneous when viewed in a moving mirror, it must yet occupy a certain time; so that we have in fact a brief electric current, to which we may apply Ohm's laws. The electromotive force is here the difference of tensions of the coatings of the jar. As to the resistance, the short metallic part of the circuit may be neglected, and we need only attend to the place of the discharge. The resistance here may be divided into that due to the air and that due to the parts of the electrodes close to the points of discharge. That the latter is by no means insignificant, may be inferred from the enormous temperature to which minute portions of the electrodes are raised, as indicated by the excessively high refrangibility of the rays emitted by the metals, in the state doubtless of vapour. By the use of flat electrodes the striking-distance is materially diminished, without any change in the difference of tension of the coatings of the jar. Hence the electricity which it contains passes at a higher velocity, and therefore produces a more powerful effect on the metals.

The injurious effect of the introduction of a small resistance was very strikingly shown with broad, slightly curved copper

* The *outer* part of the jar-spark between aluminium electrodes has the same green colour and prismatic composition, though in this case the green light is sensibly instantaneous. July, 1862.

electrodes, three inches in diameter, by leading wires from a coating of the jar into a tumbler of water, and from thence to the corresponding electrode, when the spark became quite insignificant in comparison to what it had been.

With one sharp and one flat electrode placed near together, bright sparks passed when the connexion was metallic, and the invisible spectrum then showed the copper lines, with one or two air-lines not conspicuous; but when water was interposed the spark was greatly reduced, and the invisible spectrum showed the air-lines. In both cases the spark was followed by an arc discharge, as might be seen in a moving mirror; and in the latter case the arc discharge was increased in consequence of the diminution of the spark, which, though necessary to start it, was formed at its expense; and as in the arc discharge the jar was idle, the increase of resistance in a circuit already comprising the secondary coil was unimportant.

The fact that the blue negative light which appears when the arc discharge is formed shows air-lines, points to the air as the seat of the intense action which there takes place; and the very high refrangibility of some of the rays emitted, and the copiousness of those rays, indicate how intense that action is. The heating of the negative electrode seems to be a secondary effect, not due to the direct passage of the electricity through the metal (for the section through which it passes is not by any means small), but to the heat communicated from the film of air investing it. Small as is the mass of the film compared with that of the portion of the electrode adjacent to it, the rate at which heat is communicated is enormous. Thus with a positive point nearly touching underneath a negative electrode of platinum foil containing water, the foil is kept red-hot under the water, though the mere passage of electricity through the metal would be quite inadequate to produce that effect. Corresponding to the heating of the electrode by the air is the cooling of the air by the electrode; and such a powerful abstraction of heat can hardly take place without altering the state of the film of air in relation to its power of conducting electricity. This would seem to be the reason why the film of air in contact with the negative electrode behaves so differently from any arbitrary section of the column along which the discharge takes place, and from offering greater resistance becomes the seat of a more intense emission of highly refrangible rays. At the positive electrode, at which, for whatever reason, the issue of electricity is confined almost to a point, nothing of this kind takes place; but, from the contraction of the section through which the electricity has to pass in the electrode, a minute portion of the metal of which it is composed is so highly acted on that matter belonging to the electrode is liable to appear in the arc.

These views lead to curious speculations respecting the negative light in highly exhausted tubes, and respecting the remarkable reversion of heating-effect which Mr Gassiot has obtained according as the discharge is intermittent or continuous*, but I forbear to speculate further.

* *Proceedings of the Royal Society*, Vol. XI, p. 329.

ON THE CHANGE OF FORM ASSUMED BY WROUGHT IRON AND OTHER METALS WHEN HEATED AND THEN COOLED BY PARTIAL IMMERSION IN WATER. By Lieut.-Col. H. CLARK, R.A., F.R.S. Note appended by Prof. STOKES.

[From the *Proceedings of the Royal Society*, XIII, pp. 471-2. Received Feb. 9, 1863.]

THE cause of the curious phenomenon described by Colonel Clark in the preceding paper seems to be indicated by some of the figures, especially those relating to hollow cylinders of wrought iron, which are very instructive.

Imagine such a cylinder divided into two parts by a horizontal plane at the water-line, and in this state immersed after heating. The under part, being in contact with water, would rapidly cool and contract, while the upper part would cool but slowly. Consequently by the time the under part had pretty well cooled, the upper part would be left jutting out; but when both parts had cooled, their diameters would again agree. Now in the actual experiment this independent motion of the two parts is impossible, on account of the continuity of the metal; the under part tends to pull in the upper, and the upper to pull out the under. In this contest the cooler metal, being the stronger, prevails, and so the upper part gets pulled in, a little above the water-line, while still hot. But it has still to contract on cooling; and this it will do to the full extent due to its temperature, except in so far as it may be prevented by its connexion with the rest. Hence, on the whole, the effect of this cause is to leave a permanent contraction a little above the water-line; and it is easy to see that the contraction must be so much nearer to the water-line as the thickness of the metal is less, the other dimensions of the hollow cylinder and the nature of the metal being given. When the hollow cylinder is very short, so as to be reduced to a mere hoop, the same cause operates; but there is not room for more than a general inclination of the surface, leaving the hoop bevelled.

But there is another cause of deformation at work, the operation of which is well seen in Figs. 2 and 3. Imagine a mass of metal heated so as to be slightly plastic, and then rapidly cooled over a large part of its surface. In cooling, the skin at the same time contracts and becomes stronger, and thereby tends to squeeze out its contents. This accounts for the bulging of the ends of the solid cylinders of wrought iron and the rents seen in their cylindrical surface. The skin at the bottom is of course as strong as at the sides in the part below the water-line; but a surface which resists extension far more than bending has far less power to resist pressure of the nature of a fluid pressure when plane than when convex. The effect of the cause first explained is also manifest in these cylinders, although it is less marked than in the case of the hollow cylinders, as might have been expected.

The tendency of the cooled skin of a heated metallic mass to squeeze out its contents appears to be what gives rise to the bulging seen near the water-line in the hollow cylinder of brass. Wrought iron, being highly tenacious even at a comparatively high temperature, resists with great force the sliding motion of the particles which must take place in order that the tendency of the cooled skin to squeeze out its contents may take effect; but brass, approaching in its hotter parts more nearly to the state of a molten mass, exhibits the effect more strongly. It seems probable that even in the case of brass a *very* thin hollow cylinder would exhibit a contraction just above the water-line. Should there be a metal or alloy which about the temperatures with which we have to deal was stronger hot than cold, the effect of the cause first referred to would be to produce an expansion a little below the water-line.

ON THE SUPPOSED IDENTITY OF BILIVERDIN WITH CHLOROPHYLL, WITH REMARKS ON THE CONSTITUTION OF CHLOROPHYLL.

[From the *Proceedings of the Royal Society*, XIII, pp. 144-5, Feb. 25, 1864.]

I HAVE lately been enabled to examine a specimen, prepared by Professor Harley, of the green substance obtained from the bile, which has been named biliverdin, and which was supposed by Berzelius to be identical with chlorophyll. The latter substance yields with alcohol, ether, chloroform, &c., solutions which are characterized by a peculiar and highly distinctive system of bands of absorption, and by a strong fluorescence of a blood-red colour. In solutions of biliverdin these characters are *wholly wanting*. There is, indeed, a vague minimum of transparency in the red; but it is totally unlike the intensely sharp absorption-band of chlorophyll, nor are the other bands of chlorophyll seen in biliverdin. In fact, no one who is in the habit of using a prism could suppose for a moment that the two were identical; for an observation which can be made in a few seconds, which requires no apparatus beyond a small prism, to be used with the naked eye, and which as a matter of course *would* be made by any chemist working at the subject, had the use of the prism made its way into the chemical world, is sufficient to show that chlorophyll and biliverdin are quite distinct.

I may take this opportunity of mentioning that I have been for a good while engaged at intervals with an optico-chemical examination of chlorophyll. I find the chlorophyll of land-plants to be a mixture of four substances, two green and two yellow, all possessing highly distinctive optical properties. The green substances yield solutions exhibiting a strong red fluorescence; the yellow substances do not. The four substances are soluble in the same solvents, and three of them are extremely easily decomposed by acids or even acid salts, such as binoxalate of potash; but by proper treatment each may be obtained in a state of very approximate isolation, so far at least as coloured substances are concerned. The *phyllocyanine* of Fremy* is mainly the product of decomposition by acids of one of the green bodies, and is naturally a substance of a nearly neutral tint, showing however extremely sharp bands of absorption in its neutral solutions, but dissolves in certain acids and acid solutions with a green or blue colour. Fremy's *phylloxanthine* differs according to the mode of preparation. When prepared by removing the green bodies by hydrate of alumina and a little water, it is mainly one of the yellow bodies; but when prepared by hydrochloric acid and ether, it is mainly a mixture of the same yellow body (partly, it may be, decomposed) with the product of decomposition by acids of the second green body. As the mode of preparation of *phylloxanthine* is rather hinted at than described, I can only conjecture what the substance is; but I suppose it to be a mixture of the second yellow substance with the products of decomposition of the other three bodies. Green sea-weeds (*Chlorospermeæ*) agree with land-plants, except as to the relative proportion of the substances present; but in olive-coloured sea-weeds (*Melanospermeæ*) the second green substance is replaced by a third green substance, and the first yellow substance by a third yellow substance, to the presence of which the dull colour of those plants is due. The red colouring-matter of the red sea-weeds (*Rhodospermeæ*), which the plants contain in addition to chlorophyll, is altogether different in its nature from chlorophyll, as is already known, and would appear to be an albuminous substance. I hope, before long, to present to the Royal Society the details of these researches.

* *Comptes Rendus*, tom. L, p. 405.

On the Discrimination of Organic Bodies by their Optical Properties.

[Discourse delivered at the Royal Institution of Great Britain, Mar. 4, 1864, *Roy. Inst. Proc.* IV, pp. 223—231: also *Phil. Mag.*, XXVII, 1864, pp. 388—95, *Ann. der Phys.*, CXXVI, 1865, pp. 619—23, *Journ. de Pharm.*, I, 1865, pp. 292—8.]

THE chemist who deals with the chemistry of inorganic substances has ordinarily under his hands bodies endowed with very definite reactions, and possessing great stability, so as to permit of the employment of energetic reagents. Accordingly he may afford to dispense with the aids supplied by the optical properties of bodies, though even to him they might be of material assistance. The properties alluded to are such as can be applied to the scrutiny of organic substances; and therefore the examination of the bright lines in flames and incandescent vapours is not considered. This application of optical observation, though not new in principle (for it was clearly enunciated by Mr Fox Talbot more than thirty years ago), was hardly followed out in relation to chemistry, and remained almost unknown to chemists until the publication of the researches of Professors Bunsen and Kirchhoff, in consequence of which it has now become universal.

But while the chemist who attends to inorganic compounds may confine himself without much loss to the generally recognized modes of research, it is to his cost that the organic chemist, especially one who occupies himself with proximate analysis, neglects the immense assistance which in many cases would be afforded him by optical examination of the substances under his hands. It is true that the method is of limited application, for a great number of substances possess no marked optical characters; but when such substances do present themselves, their optical characters afford facilities for their chemical study of which chemists generally have at present little conception.

Two distinct objects may be had in view in seeking for such information as optics can supply relative to the characters of a chemical substance. Among the vast number of substances which chemists have now succeeded in isolating or preparing, and which in many cases have been but little studied, it often becomes a question whether two substances, obtained in different ways, are or are not identical. In such cases an optical comparison of the bodies will either add to the evidence of their identity, the force of the additional evidence being greater or less according as their optical characters are more or less marked, or will establish a difference between substances which might otherwise erroneously have been supposed to be identical.

The second object is that of enabling us to follow a particular substance through mixtures containing it, and thereby to determine its principal reactions before it has been isolated, or even when there is small hope of being able to isolate it; and to demonstrate the existence of a common proximate element in mixtures obtained from two different sources. Under this head should be classed the detection of mixtures in what were supposed to be solutions of single substances*.

Setting aside the labour of quantitative determinations carried out by well-recognized methods, the second object is that the attainment of which is by far the more difficult. It involves the methods of examination required for the first object, and more besides; and it is that which is chiefly kept in view in the present discourse.

The optical properties of bodies, properly speaking, include every relation of the bodies to light; but it is by no means every such relation that is available for the object in view. Refractive power, for instance, though constituting, like specific gravity, &c., one of the characters of any particular pure substance, is useless for the purpose of following a substance in a mixture containing it. The same may be said of dispersive power. The properties which are of most use for our object are, first absorption, and secondly fluorescence.

* The detection of mixtures by the microscopic examination of intermingled crystals properly belongs to the first head, the question which the observer proposes to himself being, in fact, whether the pure substances forming the individual crystals are or are not identical.

Colour has long been employed as a distinctive character of bodies; as, for example, we say that the salts of oxide of copper are mostly blue. The colour, however, of a body gives but very imperfect information respecting that property on which the colour depends; for the same tint may be made up in an infinite number of ways from the constituents of white light. In order to observe what it is that the body does to each constituent, we must examine it in a pure spectrum. [The formation of a pure spectrum was then explained, and such a spectrum was formed on a screen by aid of the electric light. On holding a cell containing a salt of copper in front of the screen, and moving it from the red to the violet, it was shown to cast a shadow in the red as if the fluid had been ink, while in the blue rays it might have been supposed to have been water. Chromate of potash similarly treated gave the reverse effect, being transparent in the red, and opaque in the blue. Of course the transition from transparency to opacity was not abrupt; and for intermediate colours the fluids caused a partial darkening. Indeed, to speak with mathematical rigour, the darkening is not absolute even when it appears the greatest; but the light let through is so feeble that it eludes our senses. In this way the behaviour of the substance may be examined with reference to the various kinds of light one after another; but in order to see at one glance its behaviour with respect to all kinds, it is merely requisite to hold the body so as to intercept the whole beam which forms the spectrum—to place it, for instance, immediately in front of the slit.]

To judge from the two examples just given, it might be supposed that the observation of the colour would give almost as much information as analysis by the prism. [To show how far this is from being the case, two fluids very similar in colour, port-wine and a solution of blood, were next examined. The former merely caused a general absorption of the more refrangible rays; the latter exhibited two well-marked dark bands in the yellow and green.] These bands, first noticed by Hoppe, are eminently characteristic of blood, and afford a good example of the facilities which optical examination affords for following a substance which possesses distinctive characters of this nature. On adding to a solution of blood a particular salt of copper (any ordinary copper salt, with the addition of a tartrate to prevent precipitation, and then carbonate of soda), a fluid is obtained utterly unlike blood in colour, but showing the characteristic bands of blood, while at the same time a good deal of the red is absorbed, as it would be by the copper salt alone. On adding, on the other hand, acetic acid to a solution of blood, the colour is merely changed to a browner red, without any precipitate being produced. Nevertheless, in the spectrum of this fluid the bands of blood have wholly vanished, while another set of bands less intense, but still very characteristic, make their appearance. This alone, however, does not decide whether the colouring matter is decomposed or not by the acid; for as blood is an alkaline fluid, the change might be supposed to be merely analogous to the reddening of litmus. To decide the question, we must examine the spectrum when the fluid is again rendered alkaline, suppose by ammonia, which does not affect the absorption bands of blood. The direct addition of ammonia to the acid mixture causes a dense precipitate, which contains the colouring matter, which may, however, be separated by the use merely of acetic acid and ether, of which the former has been already used, and the latter does not affect the colouring matter of blood. This solution gives the same characteristic spectrum as blood to which acetic acid has been added; but now there is no difficulty in obtaining the colouring matter in an ammoniacal solution. In the spectrum of this solution, the sharp absorption bands of blood do not appear, but instead thereof there is a single band a little nearer to the red, and comparatively vague [this was shown on a screen]. This difference of spectra decides the question, and proves that hæmatine (the colouring matter prepared by acid, &c.) is, as Hoppe stated, a product of decomposition.

The spectrum of blood may be turned to account still further in relation to the chemical nature of that substance. The colouring matter contains, as is well known, a large quantity of iron; and it might be supposed that the colour was due to some salt of iron, more especially as some salts of peroxide of iron, sulphocyanide for instance, have a blood-red colour. But there is found a strong general resemblance between salts of the same metallic oxide as regards the character of their absorption. Thus the salts of sesquioxide of uranium show a remarkable system of bands of absorption in the more refrangible part of the spectrum. The number and position of the bands differ a little from one salt to another; but there is the strongest family likeness between the different salts. Salts of sesquioxide of iron in a similar manner have a family likeness in the vagueness of the absorption, which creeps on from one part of the spectrum to another without presenting any rapid transitions from comparative transparency to opacity and the converse. [The spectrum of sulphocyanide of peroxide of iron was shown, for the sake of contrasting with blood.] Hence the appearance of such a peculiar system of bands of absorption in blood would negative the supposition that its colour is due to a salt of iron as such, even had we no other means of deciding. The assemblage of the facts with which we are acquainted seems to show that the colouring matter is some complex compound of the five elements, oxygen, hydrogen, carbon, nitrogen, and iron, which, under the action of acids and otherwise, splits into hæmatine and an albuminous substance.

This example was dwelt on, not for its own sake, but because general methods are most readily apprehended in their application to particular examples. To show one example of the discrimination which may be effected by the prism, the spectra were exhibited of the two kinds of red glass which (not to mention certain inferior kinds) are in common use, and which are coloured, one by gold, and the other by suboxide of copper. Both kinds exhibit a single band of absorption near the yellow or green; but the band of the gold glass is situated very sensibly nearer to the blue end of the spectrum than that of the copper glass.

In the experiments actually shown, a battery of fifty cells and complex apparatus were employed, involving much trouble and expense. But this was only required for projecting the spectra on a screen, so as to be visible to a whole audience. To see them, nothing more is required than to place the fluid to be examined (contained, suppose, in a test-tube) behind a slit, and to view it through a small prism applied to the naked eye, different strengths of solution being tried in succession. In this way the bands may be seen by anyone in far greater perfection than when, for the purpose of a lecture, they are thrown on a screen.

In order to be able to examine the peculiarities which a substance may possess in the mode in which it absorbs light, it is not essential that the substance should be in solution, and viewed by transmission. Thus, for example, when a pure spectrum is thrown on a sheet of paper painted with blood, the same bands are seen in the yellow and green region as when the light is transmitted through a solution of blood and the spectrum thrown on a white screen. This indicates that the colour of such a paper is in fact due to absorption, although the paper is viewed by reflected light. Indeed, by far the greater number of coloured objects which are presented to us, such as green leaves, flowers, dyed cloths, though ordinarily seen by reflexion, owe their colour to absorption. The light by which they are seen is, it is true, reflected, but it is not *in reflexion* that the preferential selection of certain kinds of rays is made which causes the objects to appear coloured. Take, for example, red cloth. A small portion of the incident light is reflected at the outer surfaces of the fibres, and this portion, if it could be observed alone, would be found to be colourless. The greater part of the light penetrates into the fibres, when it immediately begins to suffer absorption on the part of the colouring matter. On arriving at the second surface of the fibre, a portion is reflected and a portion passes on, to be afterwards reflected from, or absorbed by, fibres lying more deeply. At each reflexion the various kinds of light are reflected in as nearly as possible the same proportion; but in passing across the fibres, in going and returning, they suffer very unequal absorption on the part of the colouring matter, so that in the aggregate of the light perceived the different components of white light are present in proportions widely different from those they bear to each other in white light itself, and the result is a vivid colouring.

There are, however, cases in which the different components of white light are reflected with different degrees of intensity, and the light becomes coloured by regular reflexion. Gold and copper may be referred to as examples. In ordinary language we speak of a soldier's coat as red, and gold as yellow. But these colours belong to the substances in two totally different senses. In the former case the colouring is due to absorption, in the latter case to reflexion. In the same sense, physically speaking, in which a soldier's coat is red, gold is not yellow but blue or green. Such is, in fact, the colour of gold by transmission, and therefore as the result of absorption, as is seen in the case of gold-leaf, which transmits a bluish-green light, or of a weak solution of chloride of gold after the addition of protosulphate of iron, when the precipitated metallic gold remains in suspension in a finely-divided state, and causes the mixture to have a blue appearance when seen by transmitted light. In this case we see that while the substance copiously reflects and intensely absorbs rays of all kinds, it more copiously reflects the less refrangible rays, with respect to which it is more intensely opaque.

All metals are, however, highly opaque with regard to rays of all colours. But certain non-metallic substances present themselves which are at the same time intensely opaque with regard to one part of the spectrum, and only moderately opaque or even pretty transparent with regard to another part. Carthamine, murexide, and platinocyanide of magnesium may be mentioned as examples. Such substances reflect copiously, like a metal, those rays with respect to which they are intensely opaque, but more feebly, like a vitreous substance, those rays for which they are tolerably transparent. Hence, when white light is incident upon them the regularly-reflected light is coloured, often vividly, those colours preponderating which the substance is capable of absorbing with intense avidity. But perhaps the most remarkable example known of the connexion between intense absorption and copious reflexion occurs in the case of crystals of permanganate of potash. These crystals have a metallic appearance, and reflect a greenish light. They are too dark to allow the transmitted light to be examined; and even when they are pulverized, the fine purple powder they yield is too dark for convenient analysis of the transmitted light. But the splendid purple solution which they yield may be diluted at pleasure, and the analysis of the light transmitted by it presents no difficulty. The solution absorbs principally the green part of the spectrum; and when it is not too strong, or used in too great thickness, five bands of absorption, indicating minima of transparency, make their appearance [these were shown on a screen]. Now, when the green light reflected from the crystals is analyzed by a prism, there are observed *bright* bands, indicating maxima of reflecting power, corresponding in position to the *dark* bands in the light transmitted by the solution. The fifth bright band, indeed, can hardly, if at all, be made out; but the corresponding dark band is both less strong than the others and occurs in a fainter part of the spectrum. When the light is reflected at a suitable angle, and is analyzed both by a Nicol's prism, placed with its principal section in the plane of incidence, and by an ordinary prism, the whole spectrum is reduced to the bands just mentioned. The Nicol's prism would under these circumstances extinguish the light reflected from a vitreous substance, and transmit a large part of the light reflected from a metal. Hence we see that as the refrangibility of the light gradually increases, the substance changes repeatedly, as regards the character of its reflecting power, from vitreous to metallic and back again, as the solution (and therefore it may be presumed the substance itself) changes from moderately to intensely opaque, and conversely.

These considerations leave little doubt as to the chemical state of the copper present in a certain glass which was exhibited. This glass was coloured only in a very thin stratum on one face. By transmission it cut off a great deal of light, and was bluish. By reflexion, especially when the colourless face was next the eye, it showed a reddish light visible in all directions, and having the appearance of coming from a fine precipitate, though it was not resolved by the microscope, at least with the power tried. It evidently came from a failure in an attempt to make one of the ordinary red glasses coloured by suboxide of copper, and the only question was as to the state in which the copper was present. It could not be oxide, for the quantity was too small to account for the blueness, and in fact the glass became sensibly colourless in the outer flame of a blowpipe. Analysis of the transmitted light by the prism showed a small band of absorption in the place of the band seen in those copper-red glasses which are not too deep; and therefore a small portion of copper was present in the state of suboxide, *i.e.* a silicate of that base. The rest was doubtless present as metallic copper, arising from over-reduction in the manufacture; and accordingly the blue colour, which would have been purer if the suboxide had been away, indicates the true colour of copper by transmitted light, quite in conformity with what we have seen in the case of gold. Hence, in both metals alike, the absorbing and the reflecting powers are, on the whole, greater for the less than for the more refrangible colours, the law of variation with refrangibility being of course somewhat different in the two cases.

Time would not permit of more than a very brief reference to the second property to which the speaker had referred as useful in tracing substances in impure solutions—that of fluorescence. The phenomenon of fluorescence consists in this, that certain substances, when placed in rays of one refrangibility, emit during the time of exposure compound light of lower refrangibility. When a pure fluorescent substance (as distinguished from a mixture) is examined in a pure spectrum, it is found that on passing from the extreme red to the violet and beyond, the fluorescence commences at a certain point of the spectrum, varying from one substance to another, and continues thence onwards, more or less strongly in one part or another according to the particular substance. The colour of the fluorescent light is found to be nearly constant throughout the spectrum. Hence, when in a solution presented to us, and examined in a pure spectrum, we notice the fluorescence taking, as it were, a fresh start, *with a different colour*, we may be pretty sure that we have to deal with a mixture of two fluorescent substances.

It might be inferred *à priori*, that fluorescence at any particular part of the spectrum would necessarily be accompanied by absorption, since otherwise there would be a creation of *vis viva*; and experience shows that rapid absorption (such as corresponds to a well-marked minimum of transparency indicated by a determinate band of absorption in the transmitted light) is accompanied by copious fluorescence. But experience has hitherto also shown, what could not have been predicted, and may not be universally true*, that conversely, absorption is accompanied, in the case of a fluorescent substance, by fluorescence.

* Fluorescent substances, like others, doubtless absorb the invisible heat-rays lying beyond the extreme red, in a manner varying from one substance to another. Hence, if we include such rays in the incident spectrum, we have an example of absorption not accompanied by fluorescence. But the invisible heat-rays differ from those of the visible spectrum (as there is every reason to believe) only in the way that the visible rays of one part of the spectrum differ from those of another, that is, by wave-length, and consequently by refrangibility, which depends on wave-length. Hence it is not improbable that substances may be discovered which absorb the visible rays in some parts of the spectrum less refrangible than that at which the fluorescence commences; and *mixtures* possessing this property may be made at pleasure. Nevertheless the speaker has not yet met with a pure fluorescent substance which exhibits this phenomenon. [See *ante*, Vol. III, p. 411, note added in 1901.]

From what precedes it follows that the colour of the fluorescent light of a solution, even when the incident light is white, or merely sifted by absorption, may be a useful character. To illustrate this, the electric light, after transmission through a deep-blue glass, was thrown on solutions in weak ammonia of two crystallized substances, æsculine and fraxine, obtained from the bark of the horse-chestnut; the latter of the two occurs also in the bark of the ash, in which, indeed, it was first discovered. Both solutions exhibited a lively fluorescence; but the colour was different, being blue in the case of æsculine, and bluish-green in the case of fraxine. A purified solution obtained from the bark exhibits a fluorescence of an intermediate colour, which would suffice to show that æsculine would not alone account for the fluorescence of the solution of the bark.

When a substance possesses well-marked optical properties, it is in general nearly as easy to follow it in a mixture as in a pure solution. But if the problem which the observer proposes to himself be, Given a solution of unknown substances which presents well-marked characters with reference to different parts of the spectrum, to determine what portion of these characters belongs to one substance, and what portion to another, it presents much greater difficulties. It was with reference to this subject that the second of the objects mentioned at the beginning of the discourse had been spoken of as that the attainment of which was by far the more difficult. The problem can, in general, be solved only by combining processes of chemical separation, especially fractional separation, with optical observation. When a solution has thus been sufficiently tested, those characters which are found always to accompany one another, in, as nearly as can be judged, a constant proportion, may, with the highest probability, be regarded as belonging to one and the same substance. But while a combination of chemistry and optics is in general required, important information may sometimes be obtained from optics alone. This is especially the case when one at least of the substances present is at the same time fluorescent and peculiar in its mode of absorption.

To illustrate this the case of chlorophyll was referred to. An eminent French chemist, M. Frémy, proposed to himself to examine whether the green colour were due to a single substance, or to a mixture of a yellow and a blue substance. By the use of merely neutral bodies, he succeeded in separating chlorophyll into a yellow substance, and another which was green, but inclining a little to blue; but he could not in this way get further in the direction of blue. He conceived, however, that he had attained his object by dissolving chlorophyll in a mechanical mixture of ether and hydrochloric acid, the acid on separation showing a fine blue colour, while the ether was yellow. Now solutions of chlorophyll in neutral solvents, such as alcohol, ether, &c., show a lively fluorescence of a blood-red colour; and when the solution is examined in a pure spectrum, the red fluorescence, very copious in parts of the red, comparatively feeble in most of the green, is found to be very lively again in the blue and violet. Now a substance of a pure yellow colour, and exercising its absorption therefore, as such substances do, on the more refrangible rays, would not show a pure red fluorescence. Either it would be non-fluorescent, or the fluorescence of its solution would contain (as experience shows) rays of refrangibilities reaching, or nearly so, to the part of the spectrum at which the fluorescence, and therefore the absorption, commences; and therefore the fluorescent light could not be pure red, as that of chlorophyll is found to be even in the blue and violet. The yellow substance separated by M. Frémy, by the aid of neutral reagents, is, in fact, non-fluorescent. Hence the powerful red fluorescence in the blue and violet can only be attributed to the substance exercising the well-known powerful absorption in the red, which substance must therefore powerfully absorb the blue and violet. We can affirm, therefore, *à priori*, that if this substance were isolated it would *not* be blue, but only a somewhat bluer green. The blue solution obtained by M. Frémy owes in fact its colour to a product of decomposition, which when dissolved in neutral solvents is not blue at all, but of a nearly neutral tint, showing, however, in its spectrum extremely sharp bands of absorption.

ON THE APPLICATION OF THE OPTICAL PROPERTIES OF BODIES TO THE DETECTION AND DISCRIMINATION OF ORGANIC SUBSTANCES.

[A Discourse delivered before the Fellows of the Chemical Society, *June* 2, 1864 : *Chem. Soc. Journal*, II, 1864, pp. 303—18.]

THE optical properties of bodies, properly speaking, include every phenomenon in which ponderable matter is related to light by virtue of its molecular constitution, and not merely of its external form. Many of these, however, are of no use in helping us to follow a particular substance through mixtures or solutions containing it, though they may be useful as additional characters of substances which have been obtained in a state of isolation. Take for example refractive power. The refractive power of a pure substance, like its specific gravity, is one of the characters, the assemblage of which serves to distinguish it from other bodies; but as all bodies in nature refract light, solvents and body dissolved alike, though not to the same extent, the observation of the refractive power of a mixture would help us in disentangling its constituents. The same may be said of dispersive power. Circular polarization again belongs to the same class of properties; though from the fact that all inorganic, and a number of organic solvents are destitute of it, it is sometimes employed to trace a substance, but of course only in those cases in which we know or may presume that there is but one substance present which possesses the property in any marked degree.

If we exclude the emission of definite rays by flames and electric discharges, which is made known by spectral analysis, and which, though most valuable for the detection of elementary bodies, is of little or no avail for even the simplest and most stable compounds, and cannot of course be applied to organic analysis, there remain three phenomena in which bodies are related to light in a manner varying from one ray to another, not in a gradual, regular way like the refractive index, but in a way depending on something in the molecular constitution of the particular body in question, and changing sometimes in an apparently capricious way from one ray to another. These are (1) absorption, (2) fluorescence, (3) coloured reflexion.

I. The colour of substances has long been used as an important character; thus for example it is a character of the salts of oxide of copper, to yield in general blue solutions. In all cases in which colour is presented to us, we must, in considering the physical cause of the phenomenon, revert, in the first instance, to Newton's discovery of the compound nature of white light, and inquire how it comes to pass, that the homogeneous constituents of white light are presented to us in different proportions from those in which they occur in white light itself. Now, if a coloured solution be examined in homogeneous light of any kind, *i.e.*, light of definite refrangibility, it is found that the transmitted light, retaining all the properties of the incident light*, becomes feebler and feebler, as the thickness of the stratum through which it passes is increased. A portion of the incident light continually disappears as the rays traverse the solution, and is said to be *absorbed*. The incident light being by hypothesis homogeneous, and the quantity of light which is absorbed in passing across a given stratum being, as experiment shows, proportional to the quantity which falls upon it, it readily follows, that the intensity of the light which escapes absorption decreases in geometrical progression, as the length of path of the rays within the solution increases in arithmetical. The rate of absorption changes in general from one set of homogeneous rays to another. For one part of the spectrum, the absorption produced by the solution may be very powerful, for another part

* Sir David Brewster indeed conceived that he had succeeded in analysing by absorption light which was homogeneous as regards refrangibility. But though the direct judgment of the senses, *when the experiment is made in Sir David Brewster's manner*, is in accordance with his statement, as might be expected from his well known accuracy, the inference that a real analysis has taken place may be considered to have been disproved by subsequent researches.

very weak, while for another part again the solution may be sensibly transparent like water.

To determine the absolute rate of absorption for homogeneous light of a given kind would be useless, unless the body to be examined were isolated; for not only would foreign substances present contribute to the observed absorption, unless, indeed, they happened to be perfectly transparent with respect to the part of the spectrum selected, but even in an otherwise colourless solution, it would be necessary to estimate the quantity of substance present, since the rate of absorption would depend on the strength of the solution. Hence, *absolute* absorbing power is of no more avail for our purpose than refractive power.

But the *relative* absorption of different parts of the spectrum is what may be observed at a glance (of course, qualitatively only, not quantitatively); it is in general independent of the degree of dilution of the solution, the solvents being supposed colourless, and *when the substance to be observed has in this respect well marked characters,* may be observed to a very great extent independently of coloured impurities, even though they may be sufficient to change the colour very greatly.

For this purpose, nothing more is required than to form in any manner a pure spectrum, and interpose the coloured solution anywhere in the path of the rays forming it. The simplest mode of obtaining a pure spectrum, when we have no occasion to place objects in it, is to view a slit held against a luminous background, through a prism applied to the eye. Hence the following simple arrangement may be adopted.

A small prism is to be chosen, which may be made of rather dense flint glass ground to an angle of about 60°, and need not be larger than is sufficient just to cover the eye comfortably. The top and bottom should be flat, for convenience of holding the prism between the thumb and fore-finger, and of laying it down on an end, so as not to scratch or dirty the faces. This forms the only apparatus required beyond what the observer may readily make for himself. The slit may conveniently be made by taking a board 6 inches square, or a little longer in a horizontal direction, making an oblong aperture in it in a vertical direction, and adapting to the aperture two pieces of thin metal to form clean cheeks to the slit. One of the metal pieces should be moveable, to allow of altering the breadth of the slit. About the fiftieth of an inch is a suitable breadth for ordinary purposes. The board and metal pieces should be well blackened.

On holding the board at arm's length against the sky or a luminous flame, the slit being, we will suppose, in a vertical direction, and viewing the line of light thus formed through the prism held close to the eye, with its edge vertical, a pure spectrum is obtained at a proper azimuth of the prism. Turning the prism round its axis alters the focus, and the proper focus is got by trial. The whole of the spectrum is not, indeed, in perfect focus at once, so that in scrutinizing one part after another it is requisite to turn the prism a little. When daylight is used, the spectrum is known to be pure by its showing the principal fixed lines; in other cases the focus is got by the condition of seeing distinctly the other objects, whatever they may be, which are presented in the spectrum. The use of a prism in this way is at the first moment a little puzzling, but soon becomes perfectly easy. To observe the absorption-spectrum of a liquid, an elastic band is put round the board near the top, and a test-tube containing the liquid is slipped under the band, which holds it in its place behind the slit. The spectrum is then observed just as before, the test-tube being turned from the eye.

To observe the whole progress of the absorption, different degrees of strength must be used in succession, beginning with a strength which does not render any part of the spectrum absolutely black, unless it be one or more very narrow bands, as otherwise the most distinctive features of the absorption might be missed. If the solution be contained in a wedge-shaped vessel instead of a test-tube, the progress of the absorption may be watched in a continuous manner by sliding the vessel before the eye; but for actual work this is an unnecessary luxury. Some observers prefer using a wedge-shaped vessel in combination with the slit, the slit being perpendicular to the edge of the wedge. In this case each element of the slit forms an elementary spectrum corresponding to a thickness of the solution, which increases in a continuous manner from the edge of the wedge, where it vanishes.

In many cases nothing is observed, beyond a general absorption of one or other end of the spectrum, or of its middle part, and the prism gives little information beyond what is got by the eye, by observing the *succession of colours* produced by different thicknesses of the liquid. And here it may be remarked in passing, with reference to the description of pure substances, that in specifying only one colour, that corresponding to a considerable thickness, as is commonly done by chemists, the peculiar features of the absorption are left almost wholly undescribed. Thus of two solutions, one might be pink when dilute, passing on to red with increase of strength or thickness, another yellow, passing through orange to red. These would commonly be described as red, yet the series of tints indicates an utter difference in the mode of absorption, the middle of the spectrum in the one case, and the most refrangible end in the other, being the most powerfully attacked.

But in some cases, especially with substances of intense colorific power, the mode of absorption is eminently characteristic. Two or more dark bands are seen in the spectrum, indicating maxima of absorption; and the positions of these bands, their relative intensity, and their other features, form altogether a series of characters the distinctive nature of which is such as those who have neglected the use of the prism have little conception of. They render it perfectly easy in many cases to follow a particular substance among a host of impurities. For each coloured substance produces its own absorption, independently of the others (supposing the substances do not chemically react on each other), so that, unless the part of the spectrum in which the distinctive bands, or most of them, occur, is wholly absorbed by the impurities, the presence of the substance can still be recognised. Such a complete obliteration is the less likely to occur, for this reason, that when the characters of the solution are so strongly marked, it almost always happens that a comparatively small quantity of the substance suffices to produce the effect, and the solution must consequently be so much diluted that the effect of the impurities comparatively disappears.

Nor is this all. When a substance exhibits marked characters of one kind in one solvent, it often happens that it shows different and no less marked characters in a solvent of a different nature. Not only does this furnish additional characters by which the substance can be distinguished from others, but it is valuable for following the substance when involved in impurities; for the nature of the impurities may be such as to mask the substance in one solvent and not in another. This is especially the case where one solvent is alkaline and the other acid; but differences are sometimes observed even with two neutral solvents.

To illustrate these principles, we may refer to the colouring matters of madder. Alizarin and purpurin both yield highly distinctive spectra, the former, however, only in the case of solutions containing caustic alkali*, whereas most solutions of the latter are highly distinctive. Madder itself contains, either directly or as the result of decomposition, a number of substances which in alkaline solution absorb that part of the spectrum in which the distinctive bands of purpurin occur. Hence, in a mixture obtained from madder, and containing, we will suppose, purpurin in comparatively small quantity, the presence of purpurin would be masked by the other substances *in an alkaline solution.* But in ether or acidulated alcohol the other substances yield spectra showing nothing particular, and interfering comparatively little with the distinctive bands of purpurin; while in an alum-liquor solution made by boiling, not only are the purpurin-bands, which in this solvent occur at a lower refrangibility than with ether, more effectually separated from the absorption produced by the associated substances, but those substances themselves are also in good measure excluded.

For an example of the necessity of attending to the nature of the solvent, even in the case of different neutral solvents, we may refer to a yellow substance which is one of the constituents of the green colouring matter of leaves. The alcoholic solution of this substance exhibits two characteristic bands of absorption, the first of which is situated immediately adjacent to the line F on the more refrangible side. The solution in bisulphide of carbon exhibits two similar bands, but much less refrangible, the line F now nearly bisecting the bright interval between the first and second dark bands. The substance is very easily decomposed by acids, and even by acid salts, yielding a product of decomposition which, in alcoholic solution, exhibits two bands of absorption like the parent substance, but a good deal more refrangible. There is

* On boiling with an alkaline carbonate the same spectrum is obtained, but not perfectly developed. An alcoholic solution with a little caustic potash introduced gives it in perfection.

the same change of position as in the former case, in passing from alcohol to bisulphide of carbon, so that the solution of the product of decomposition in bisulphide of carbon agrees almost exactly, in colour and spectrum, with that of the parent substance in alcohol.

Not only is an examination of the absorption-spectrum of a substance useful for enabling us to follow the substance through mixed solutions, but it sometimes reveals relationships in cases in which they might not be suspected, if the origin of the substances were unknown. Thus, the purpureïn of Dr Stenhouse dissolves in ether or acidulated alcohol with a red colour, while that of the same solutions of purpurin is yellow. But the prism reveals in both cases alike the existence of three bands of absorption, of similar breadth, while the purpureïn bands are situated nearer the red than those of purpurin, by about one interval. This example shows how deeply seated in the molecular constitution of a body may in some cases be the cause which produces the bands.

Hitherto it has been supposed that the peculiarities of absorption of a substance were known, and applied to the detection of that substance in a mixture. But the question may arise:—Given a mixture of an unknown number of unknown substances, which as a whole presents peculiarities of absorption, to determine whether the whole of these peculiarities are due to the same substance, and if not, what portion are due to one substance, and what portion to another. Little can be done towards the solution of this problem by the mere observation of absorption; we can only say, that some modes of grouping of bands of absorption are common in solutions of pure substances, while others are uncommon, and give rise to the suspicion of a mixture. The phenomena of fluorescence give in some cases material assistance; but in general it is only by combining spectral analysis with processes of chemical separation, especially fractional separation, that a satisfactory conclusion can be arrived at. When a mixture is thus tested in various ways, a conviction, at last approaching certainty, is gradually arrived at, that those bands of absorption which are always found accompanying one another belong to one and the same substance.

For convenience and rapidity of manipulation, especially in the examination of very minute quantities, there is no method of separation equal to that of partition between solvents which separate after agitation. Ether combined with water, either pure or rendered acid or alkaline, is the most generally useful, and the separation, if not otherwise fractional, may be rendered so by introducing the acid or alkali by minute quantities at a time; but other solvents are useful in particular cases. Bisulphide of carbon in conjunction with alcohol enabled the lecturer to disentangle the coloured substances which are mixed together in the green colouring matter of leaves. Solutions of various metallic oxides which are naturally precipitable by an alkali or alkaline carbonate, but are retained in solution by means of a tartrate, are very useful in the examination of the true colouring matters, not merely for producing changes of colour and spectrum without precipitation, but even, in conjunction with ether, for effecting chemical separations; and fractional separation may be effected by making the solution deviate very slightly from perfect neutrality. By combining with ether such a solution of alumina, it was found possible to separate and detect the purpurin, alizarin, and rubiacin present in a portion of powder not exceeding in bulk a fraction of the head of a pin.

II. The phenomenon of fluorescence consists in this, that certain substances in solution, or even in the solid state, when exposed to rays of one kind, emit for the time being rays of another kind, as if they were self-luminous. The following law appears to be universal:—*The emitted rays are of lower refrangibility than the exciting rays.* The light emitted is heterogeneous, even though the active rays be homogeneous, and it is remarkable that it shows no traces of its parentage.

When a solution of a pure fluorescent substance is examined in a pure spectrum formed in the usual way by means of sunlight, it is found that, in passing in the direction from the red to the violet, the fluorescence begins with more or less abruptness at a certain part of the spectrum varying from one substance to another, and continues from thence onwards, though often with considerable fluctuations of intensity. The fluctuations are found to correspond with fluctuations in transparency, so that, in a region of the spectrum where a band of absorption occurs, the fluorescence is unusually strong, while the rays exciting fluorescence are more quickly spent, and accordingly the fluorescence does not extend so far into the solution. Experience shows that with solutions of *a pure substance,* the tint of the fluorescent light is almost perfectly constant throughout the spectrum.

In consequence of this law, the tint of the fluorescent light emitted by a solution becomes a character of importance. This tint, it must be remembered, is that of the light *as emitted,* not as *subsequently modified* by absorption on the part of the solution, in case the solution be sensibly coloured, and some precautions are required in order to observe it correctly*. The fluorescence observed in solutions from the barks of the horse-chestnut, ash, &c., was formerly attributed indiscriminately to the presence of æsculin, whereas a purified solution from the bark of the horse-chestnut exhibits a fluorescence very sensibly different from that of æsculin, which observation alone would suffice to show that the bark must contain some other fluorescent substance besides æsculin.

As in the case of absorption, the nature of the solvent must be attended to. The colour of the fluorescent light is liable to change, not merely in passing from an alkaline to a neutral or acid solution, but even occasionally in passing from one neutral solvent to another. The lecturer has received from Dr Müller a specimen of a substance which in water exhibits a *green,* but in ether a *blue* fluorescence.

The composition of fluorescent light, as revealed by the prism, occasionally presents peculiarities, but in such cases they are found to be connected with peculiarities in the mode of absorption, so that the two are not to be regarded as *independent* characters of a substance; and as the peculiarities in the absorption are, as a general rule, the more easily observed, it is only rarely that the analysis of the fluorescent light is of much use.

The distribution of fluorescence in the spectrum often affords valuable information, but its observation is not of that perfectly simple character, requiring hardly any apparatus, that constitutes one great advantage, for chemical application, of the observation of absorption or of the tint of fluorescent light. The observation is restricted to times when the sun is shining pretty steadily (unless the observer has recourse to electric light, or at least lime-

* See *Quarterly Journal of the Chemical Society,* Vol. XI, p. 19. [*Ante,* p. 114.]

light); it is requisite to reflect the sun's light horizontally, without which the observation would be most troublesome; and unless the reflexion be made by the mirror of a heliostat, the continual change in the direction of the reflected light is most inconvenient. It is requisite to use at least one good prism, better two or three, which must be of tolerable size, in order to have light of sufficient intensity, and the prisms must be combined with a lens, which need not however be achromatic. Hence these observations are not, like the former, adapted to the daily use of every chemist.

It has already been stated, as the result of experience, that the colour of the fluorescent light of a single substance is constant throughout the spectrum, or very nearly so. If, therefore, on examining a solution in a pure spectrum thus formed by projection, we find the fluorescence taking a fresh start *with a different colour*, we may be almost certain that we have to do with a mixture of two different fluorescent substances, the presence of which is thus revealed without any chemical process. If, however, the fluorescence of two fluorescent substances, which may be mixed together, begins at nearly the same point in the spectrum (as commonly happens when there is merely a slight difference of tint in the colour of the fluorescent light of the two substances), the coexistence of the two substances may escape detection when the mixed solution is merely examined in a pure spectrum; and in such cases a combination of processes of fractional separation with the easy observation of the tint of the fluorescent light is more searching. This is the case, for instance, with the mixture of æsculin and fraxin contained in a solution from the bark of the horse-chestnut.

Experience has also indicated a most intimate connexion between the spectral distribution of fluorescence and that of opacity in the case of solutions of pure substances. There are, indeed, theoretical reasons for regarding it as not improbable that instances may yet be found in which absorption, unaccompanied by fluorescence, takes place, in the case of solutions of fluorescent substances, in that part of the visible spectrum which is less refrangible than the point at which the fluorescence commences, but no such instance has yet been observed. Hence from the distribution of the fluorescence, we may infer the character of the absorption belonging to the fluorescent body. For this purpose it is best to make the solution extremely dilute, when any bands of absorption will have their positions indicated by beams of fluorescent light, while in the intermediate parts of relatively great transparency, the fluorescence, in the case of such weak solutions, is almost insensible.

As the occurrence of a decided difference of colour in the fluorescent light seen at two different parts of the spectrum implies, almost to a certainty, the presence of two different fluorescent substances, so, conversely, the exhibition of the same colour is an argument in favour of the identity of the substance producing the fluorescence at the two parts. We cannot indeed say that there may not be two substances present, the fluorescence of which commences at nearly the same part of the spectrum; but assuredly, two different substances, the fluorescence of which commenced at two widely different parts of the spectrum, would reveal themselves by the difference of colour. For experience shows that the refrangibility of the light emitted, at any part of the incident spectrum, by the solution of a pure substance, extends nearly up to that of the point of the incident spectrum at which the fluorescence commences, but not much beyond; and though, in passing from one pure substance to another, variations do occur in the relative brightness of the rays of less refrangibility which compose the fluorescent light, yet, on the whole, there is so close a connexion between the colour of the fluorescent light and the refrangibility of the rays by which the fluorescence is first produced, that no great variation in the one is compatible with constancy or a mere trifling variation in the other.

For an example of the application of these principles, we may refer to the green colouring matter of leaves. The alcoholic solution of this substance exhibits a lively fluorescence of a blood-red colour, and shows also a certain system of bands of absorption. Different chemists in different ways have obtained from it a yellow substance, and M. Frémy, having obtained a yellow substance by the aid of merely neutral reagents, has proved that such a substance pre-exists. The yellow solution was obtained by him in the attempt to divide the green colouring matter into a yellow and a blue; but, by using neutral reagents only, he did not get further in the direction of blue than a green of a bluer shade than at first, which he supposed to be due to the imperfection of the modes of separation. He conceived, however, that he had attained his object by dissolving chlorophyll in a mechanical mixture of ether and hydrochloric acid, the acid stratum, when the fluids separated after agitation, exhibiting a blue colour.

Now, when an alcoholic solution of chlorophyll is examined in a pure spectrum formed by sunlight, in the part of the spectrum extending from the extreme red to the junction of the green and blue the fluorescence exhibits remarkable fluctuations of intensity, corresponding precisely to the bands of absorption in the transmitted light. The red fluorescence is extremely lively in nearly the whole of the blue and in the violet. This proves that the main absorption of these colours cannot be due to the yellow body; for the yellow substance would be either non-fluorescent, or its fluorescence would be of some shade of green or yellow. On the former supposition, the fluorescence of the chlorophyll solution in the blue and violet would be dull, and on the latter would be of some colour different from blood-red, unless the main absorption of that part of the spectrum were due to the substance producing the red fluorescence. In fact, when the yellow body is nearly isolated, it exhibits two characteristic bands of absorption in the blue, to which no fluorescence corresponds, which demonstrates that the slight red fluorescence which the solution may still exhibit is due to the remaining impurity. Hence, if the yellow body were wholly eliminated from chlorophyll, the residue, containing the substance showing the red fluorescence, would still powerfully absorb the blue and violet, and therefore could not be blue, but only a bluer green; so that in seeking to separate chlorophyll into a blue and a yellow substance, M. Frémy was aiming at an impossibility. His phyllocyanin is, in fact, a product of decomposition, and is not blue at all, but merely dissolves in certain acids with a blue colour. It may be mentioned in passing, that the green fluorescent residue is still a mixture, consisting of two different substances, both green, and both exhibiting a red fluorescence.

III. The instances in which substances appear coloured by reflexion are comparatively rare. It is very common in chemical descriptions to read of a solution appearing of such a colour by transmitted, and such a colour by reflected light. In many cases, that is a positive mistake, and the colour described as due to reflexion is really due to transmission. A chemist views a solution contained in a test-tube by transmission, and then by reflexion; and seeing, perhaps, some perfectly different colour in the latter case, describes it as the colour of the solution by reflexion, whereas it is merely the colour by transmission due to a greater thickness, the light having been reflected at the back or bottom of the test-tube, and so having twice passed through the solution. In other cases the colour described as due to reflexion really arises from fluorescence; and though the statement may be true in the sense intended, it seems objectionable to apply the term *reflexion* to a process so utterly different. It is only in the case of metals, such as gold and copper, and of certain other substances such as murexide, platinocyanide of magnesium, &c., that colour is really seen as the result of reflexion.

When this takes place in the case of non-metallic substances, they are found to be endowed, for the colours so reflected, with an intense opacity, comparable with that of metals; while for other parts of the spectrum they may be comparatively transparent, and these parts they reflect with an energy comparable to that of a vitreous substance only. The variations of absorbing power in passing from one part of the spectrum to another, and consequently the variations in reflecting energy, are frequently much more considerable, and accordingly the colour by reflexion is much richer than in the case of metals.

An excellent example of the intimate connexion between metallic reflexion and intense absorption is afforded by crystals of permanganate of potash. These crystals exhibit a green metallic reflexion, and when crushed yield a powder of an intense purple colour by transmitted light. The colour is too intense for spectral analysis, but the solution has a similar colour, merely less intense as corresponds with its smaller concentration, and the analysis of the light transmitted by the solution presents no difficulty. The green is quickly absorbed, but when the solution is sufficiently dilute, five eminently characteristic bands of absorption are seen in that part of the spectrum. A sixth band comes out with a greater thickness or else strength of solution, but even the fifth is somewhat less strong than the others. When the light reflected from a crystal is analysed, four bright bands are seen standing out on a generally luminous ground of inferior brightness. These bright bands correspond in position with the principal *dark* bands in the light transmitted by the solution, and therefore, it may be presumed, by the crystals themselves. When the angle of incidence has a suitable value, and the reflected light is analysed by a Nicol's prism, with its principal plane in the plane of incidence, and then by a common prism, the spectrum is reduced to these four bright bands. A fifth bright band could perhaps be made out, in the case of a fine crystal with a fresh surface. Under the circumstances described, the Nicol's prism would extinguish the light reflected from a vitreous substance, and transmit much of that reflected from a metal. We see, therefore, that, as regards its relations to light, the crystallized body passes repeatedly from the condition of a vitreous to that of a metallic substance and back again, as the refrangibility of the rays, in relation to which it is considered, is continuously increased by a small amount.

The same relation between intense absorption and metallic reflexion exists generally, though it cannot be always studied by means of a solution. The platinocyanides, for example, yield colourless solutions, so that the intense absorption which most of them exercise for certain parts of the spectrum must be attributed to the mode in which the molecules are built up in forming the crystals; but by attending to the colour of the light transmitted by thin crystals, the law is found to be obeyed. Gold can only be obtained, in solution, as gold, by means of the opaque solvent mercury; but its colour by transmission may be studied in gold leaf, or in a chemically deposited film, and is then found to be conformable to the law mentioned, the less refrangible colours, which are those which are the more copiously reflected, being also those which are the more intensely absorbed.

When a body endowed with the property of coloured reflexion, such as permanganate of potash, is dissolved, in consequence of the necessary dilution the opacity of the medium ceases to be, for any part of the spectrum, of that intense kind which is necessary for quasi-metallic reflexion; and accordingly the light reflected by the solution is colourless. Hence coloured reflexion is not available for following a substance through mixtures containing it. The chemist ought, however, to be acquainted with its laws, in order to understand the changes of colour which a substance possessing the property is capable of exhibiting in the solid condition, according to its state of aggregation.

In order that the colour due to reflexion should appear, it is necessary that the substance should have a certain amount of coherence. Thus indigo in the form of a fine loose powder is blue, even when viewed by reflexion. It would be erroneous, however, to describe the body as blue by reflexion, if we were speaking of the properties of the substance, and not the mere crude results of observation made under given circumstances. For though it is true that the light by which the blue colour is seen has undergone reflexion (without which it would not have reached the eye) it is not *in reflexion* that the chromatic selection is made by virtue of which the powder appears blue, but *during transmission.* In fact it is only a small portion of the light that is reflected at the outer irregular surface of the mass; the greater part penetrates a little way, and is reflected at various depths, and in passing through the particles, in going and returning, suffers absorption on the part of the coloured substance. Were the substance intensely opaque for *all* the colours of the spectrum, the powder would be not blue but black, as we see in the case of platinum-black. By burnishing, the powder is reduced to the state of a somewhat coherent mass, and it now begins to exhibit the copper colour due to reflexion. The internal reflexions are at the same time greatly weakened, so that the part of the light which is reflected from beneath and undergoes absorption is much reduced. A pressed mass is not, however, an optically homogeneous medium, so that the colour by reflexion obtained by burnishing cannot in general be quite pure. In the state of a fine crystalline powder, indigo exhibits a mixture of the copper colour due to reflexion, and the blue colour due to transmission, though observed in the light reflected from the mass as a whole; while if the substance could be obtained in large crystals, the colour by reflection would be seen in perfection, and the colour by transmission would disappear, the crystals being sensibly opaque.

On the Reduction and Oxidation of the Colouring Matter of the Blood.

[From the *Proceedings of the Royal Society, June* 16, 1864.]

1. Some time ago my attention was called to a paper by Professor Hoppe*, in which he has pointed out the remarkable spectrum produced by the absorption of light by a very dilute solution of blood, and applied the observation to elucidate the chemical nature of the colouring matter. I had no sooner looked at the spectrum, than the extreme sharpness and beauty of the absorption-bands of blood excited a lively interest in my mind, and I proceeded to try the effect of various reagents. The observation is perfectly simple, since nothing more is required than to place the solution to be tried, which may be contained in a test-tube, behind a slit, and view it through a prism applied to the eye. In this way it is easy to verify Hoppe's statement, that the colouring matter (as may be presumed at least from the retention of its peculiar spectrum) is unaffected by alkaline carbonates and caustic ammonia, but is almost immediately decomposed by acids, and also, but more slowly, by caustic fixed alkalies, the coloured product of decomposition being the hæmatine of Lecanu, which is easily identified by its peculiar spectra. But it seemed to me to be a point of special interest to inquire whether we could imitate the change of colour of arterial into that of venous blood, on the supposition that it arises from reduction.

2. In my experiments I generally employed the blood of sheep or oxen obtained from a butcher; but Hoppe has shown that the blood of animals in general exhibits just the same bands. To obtain the colouring matter in true solution, and at the same time to get rid of a part of the associated matters, I generally allowed the blood to coagulate, cut the clot small, rinsed it well, and extracted it with water. This, however, is not essential, and

* Virchow's *Archiv*, Vol. xxiii, p. 446 (1862). [Cf. *ante*, p. 240.]

blood merely diluted with a large quantity of water may be used; but in what follows it is to be understood that the watery extract is used unless the contrary be stated.

3. Since the colouring matter is changed by acids, we must employ reducing agents which are compatible with an alkaline solution. If to a solution of protosulphate of iron enough tartaric acid be added to prevent precipitation by alkalies, and a small quantity of the solution, previously rendered alkaline by either ammonia or carbonate of soda, be added to a solution of blood, the colour is almost instantly changed to a much more purple red as seen in small thicknesses, and a much darker red than before as seen in greater thickness. The change of colour, which recalls the difference between arterial and venous blood, is striking enough, but the change in the absorption spectrum is far more decisive. The two highly characteristic dark bands seen before are now replaced by a *single* band, somewhat broader and less sharply defined at its edges than either of the former, and occupying nearly the position of the bright band separating the dark bands of the original solution. The fluid is more transparent for the blue, and less so for the green than it was before. If the thickness be increased till the whole of the spectrum more refrangible than the red be on the point of disappearing, the last part to remain is *green*, a little beyond the fixed line *b*, in the case of the original solution, and *blue*, some way beyond *F*, in the

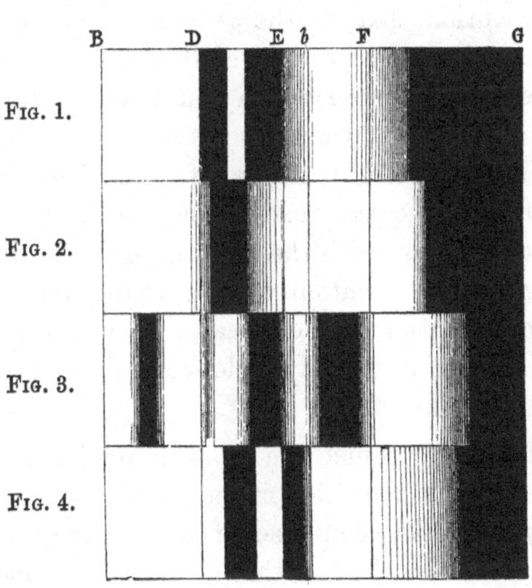

case of the modified fluid. Figs. 1 and 2 in the accompanying woodcut represent the bands seen in these two solutions respectively.

4. If the purple solution be exposed to the air in a shallow vessel, it quickly returns to its original condition, showing the two characteristic bands the same as before; and this change takes place immediately, provided a small quantity only of the reducing agent were employed, when the solution is shaken up with air. If an additional quantity of the reagent be now added, the same effect is produced as at first, and the solution may thus be made to go through its changes any number of times.

5. The change produced by the action of the air (that is, of course, by the absorption of oxygen) may be seen in an instructive form on partly filling a test-tube with a solution of blood suitably diluted, mixing with a little of the reducing agent, and leaving the tube at rest for some time in a vertical position. The upper or oxidized portion of the solution is readily distinguished by its colour; and if the tube be now placed behind a slit and viewed through a prism, a dark band is seen, having the general form of a tuning-fork, like Figs. 1 and 2, regarded now as a single figure, the line of separation being supposed removed.

6. Of course it is necessary to assure oneself that the single band in the green is not due to absorption produced merely by the reagent, as is readily done by direct observation of its spectrum, not to mention that in the region of the previous dark bands, or at least the outer portions of it, the solution is actually more transparent than before, which could not be occasioned by an *additional* absorption. Indeed the absorption due to the reagent itself in its different stages of oxidation, unless it be employed in most unnecessary excess, may almost be regarded as evanescent in comparison with the absorption due to the colouring matter; though if the solution be repeatedly put through its changes, the accumulation of the persalt of iron will presently tell on the colour, making it sensibly yellower than at first for small thicknesses of the solution.

7. That the change which the iron salt produces in the spectrum is due to a simple reduction of the colouring matter, and not to the formation of some compound of the colouring matter with the reagent, is shown by the fact that a variety of reducing agents

of very different nature produce just the same effect. If protochloride of tin be substituted for protosulphate of iron in the experiment above described, the same changes take place as with the iron salt. The tin solution has the advantage of being colourless, and leaving the visible spectrum quite unaffected, both before and after oxidation, and accordingly of not interfering in the slightest degree with the optical examination of the solutions, but permitting them to be seen with exactly their true tints. The action of this reagent, however, takes some little time at ordinary temperatures, though it is very rapid if previously the solution be gently warmed. Hydrosulphate of ammonia again produces the same change, though a small fraction of the colouring matter is liable to undergo some different modification, as is shown by the occurrence of a slender band in the red, variable in its amount of development, which did not previously exist. In this case, as with the tin salt, the action is somewhat slow, requiring a few minutes unless it be assisted by gentle heat. Other reagents might be mentioned, but these will suffice.

8. We may infer from the facts above mentioned that *the colouring matter of blood*, like indigo, *is capable of existing in two states of oxidation, distinguishable by a difference of colour and a fundamental difference in the action on the spectrum. It may be made to pass from the more to the less oxidized state by the action of suitable reducing agents, and recovers its oxygen by absorption from the air.*

As the term *hæmatine* has been appropriated to a product of decomposition, some other name must be given to the original colouring matter. As it has not been named by Hoppe*, I propose to call it *cruorine*, as suggested to me by Dr Sharpey; and in its two states of oxidation it may conveniently be named *scarlet cruorine* and *purple cruorine* respectively, though the former is slightly purplish at a certain small thickness, and the latter is of a very red purple colour, becoming red at a moderate increase of thickness.

9. When the watery extract from blood-clots is left aside in a corked bottle, or even in a tall narrow vessel open at the top, it

[* It was in the year 1864, the date of the present paper, that Hoppe proposed the name *haemoglobin*, now universally employed, Virchow's *Archiv*, XXIX, p. 223.]

presently changes in colour from a bright to a dark red, decidedly purple in small thicknesses. This change is perceived even before the solution has begun to stink in the least perceptible degree. The tint agrees with that of the purple cruorine obtained immediately by reducing agents; and if a little of the solution be sucked up from the bottom into a quill-tube drawn to a capillary point, and the tube be then placed behind a slit, so as to admit of analyzing the transmitted light without exposing the fluid to the air, the spectrum will be found to agree with that of purple cruorine. On shaking the solution with air it immediately becomes bright red, and now presents the optical characters of scarlet cruorine. It thus appears that scarlet cruorine is capable of being reduced by certain substances, derived from the blood, present in the solution, which must themselves be oxidized at its expense.

10. When the alkaline tartaric solution of protoxide of tin is added in moderate quantity to a solution of scarlet cruorine, the latter is presently reduced. If the solution is now shaken with air, the cruorine is almost instantly oxidized, as is shown by the colour of the solution and its spectrum by transmitted light. On standing for a little time, a couple of minutes or so, the cruorine is again reduced, and the solution may be made to go through these changes a great number of times, though not of course indefinitely, as the tin must at last become completely oxidized. It thus appears that purple cruorine absorbs *free* oxygen with much greater avidity than the tin solution, notwithstanding that the oxidized cruorine is itself reduced by the tin salt. I shall return to this experiment presently.

11. When a little acid, suppose acetic or tartaric acid, which does not produce a precipitate, is added to a solution of blood, the colour is quickly changed from red to brownish red, and in place of the original bands (Fig. 1) we have a different system, nearly that of Fig. 3. This system is highly characteristic; but in order to bring it out a larger quantity of substance is requisite than in the case of scarlet cruorine. The figure does not exactly correspond to any one thickness, for the bands in the blue are best seen while the band in the red is still rather narrow and ill-defined at its edges, while the narrow inconspicuous band in the yellow hardly comes out till the whole of the blue and violet, and a good

part of the green, are absorbed. The difference in the spectra Figs. 1 and 3 does not alone prove that the colouring matter is decomposed by the acid (though the fact that the change is not instantaneous favours that supposition), for the one solution is alkaline, though it may be only slightly so, while the other is acid, and the difference of spectra might be due merely to this circumstance. As the direct addition of either ammonia or carbonate of soda to the acid liquid causes a precipitate, it is requisite in the first instance to separate the colouring matter from the substance so precipitated.

This may be easily effected on a small scale by adding to the watery extract from blood-clots about an equal volume of ether, and then some glacial acetic acid, and gently mixing, but not violently shaking for fear of forming an emulsion. When enough acetic acid has been added, the acid ether rises charged with nearly the whole of the colouring matter, while the substance which caused the precipitate remains in the acid watery layer below*. The acid ether solution shows in perfection the characteristic spectrum Fig. 3. When most of the acid is washed out the substance falls, remaining in the ether near the common surface. If after removing the wash-water a solution, even a weak one, of ammonia or carbonate of soda be added, the colouring matter readily dissolves in the alkali. The spectrum of the transmitted light is quite different from that of scarlet cruorine, and by no means so remarkable. It presents a single band of absorption, very obscurely divided into two, the centre of which nearly coincides with the fixed line *D*, so that the band is decidedly less refrangible than the pair of bands of scarlet cruorine. The relative proportion of the two parts of the band is liable to vary. The presence of alcohol, perhaps even of dissolved ether, seems to favour the first part, and an excess of caustic alkali the second, the fluid at the same time becoming more decidedly dichroitic. The blue end of the spectrum is at the same time absorbed. The band of absorption is by no means so definite at its edges as those

* If I may judge from the results obtained with the precipitate given by acetic acid and a neutral salt, a promising mode of separation of the proximate constituents of blood-crystals would be to dissolve the crystals in glacial acetic acid and add ether, which precipitates a white albuminous substance, leaving the hæmatine in solution.

of scarlet cruorine, and a far larger quantity of the substance is required to develope it.

This difference of spectra shows that the colouring matter (hæmatine) obtained by acids is a product of the decomposition, or metamorphosis of some kind, of the original colouring matter.

When hæmatine is dissolved in alcohol containing acid, the spectrum nearly agrees with that represented in Fig. 3.

12. Hæmatine is capable of reduction and oxidation like cruorine. If it be dissolved in a solution of ammonia or of carbonate of soda, and a little of the iron salt already mentioned, or else of hydrosulphate of ammonia, be added, a pair of very intense bands of absorption is immediately developed (Fig. 4). These bands are situated at about the same distance apart as those of scarlet cruorine, and are no less sharp and distinctive. They are a little more refrangible, a clear though narrow interval intervening between the first of them and the line *D*. They differ much from the bands of cruorine in the relative strength of the first and second band. With cruorine the second band appears almost as soon as the first, on increasing the strength or thickness of the solution from zero onwards, and when both bands are well developed, the second band is decidedly broader than the first. With reduced hæmatine, on the other hand, the first band is already black and intense by the time the second begins to appear; then both bands increase, the first retaining its superiority until the two are on the point of merging into one by the absorption of the intervening bright band, when the two appear about equal.

Like cruorine, reduced hæmatine is oxidized by shaking up its solution with air. I have not yet obtained hæmatine in an acid solution in more than one form, that which gives the spectrum Fig. 3, and which I have little doubt contains hæmatine in its oxidized form; for when it is withdrawn from acid ether by an alkali, I have not seen any traces of reduced hæmatine, even on taking some precautions against the absorption of oxygen. As the alkaline solution of ordinary hæmatine passes, with increase of thickness, through yellow, green, and brown to red, while that of reduced hæmatine is red throughout, the two kinds may be conveniently distinguished as *brown hæmatine* and *red hæmatine* respectively*, the former or oxidized substance being the hæmatine of chemists.

13. Although the spectrum of scarlet cruorine is not affected by the addition to the solution of either ammonia or carbonate of soda, yet if after such addition the solution be either heated or alcohol be added, although there is no precipitation decomposition takes place. The coloured product of decomposition is brown hæmatine, as may be inferred from its spectrum. Since, however, the spectrum of an alkaline solution of brown hæmatine is only moderately distinctive, and is somewhat variable according to the nature of the solvent, it is well to add hydrosulphate of ammonia, which immediately developes the remarkable bands of red hæmatine. This is the easiest way to obtain them; but the less refrangible edge of the first band as obtained in this way is liable to be not quite clean, in consequence of the presence of a small quantity of cruorine which escaped decomposition.

Some very curious reactions are produced in a solution of cruorine by gallic acid combined with other reagents, but these require further study.

14. Hoppe proposed to employ the highly characteristic absorption-bands of scarlet cruorine in forensic inquiries. Since, however, cruorine is very easily decomposed, as by hot water, alcohol, weak acids, &c., the method would often be inapplicable. But as in such cases the coloured product of decomposition is hæmatine, which is a very stable substance, the absorption-bands of red hæmatine in alkaline solution, which in sharpness, distinctive character and sensibility rival those of scarlet cruorine itself, may be employed instead of the latter. The absorption-bands of brown hæmatine dissolved in a mixture of ether and acetic acid, or in acetic acid alone, are hardly less characteristic, but are not quite so sensitive, requiring a somewhat larger quantity of the substance.

15. I have purposely abstained from physiological speculations until I should have finished the chemico-optical part of the subject; but as the facts which have been adduced seem calculated to

[* In present-day nomenclature *alkaline hæmatin* and *hæmochromogen* respectively.]

throw considerable light on the function of cruorine in the animal economy, I may perhaps be permitted to make a few remarks on this subject.

It has been a disputed point whether the oxygen introduced into the blood in its passage through the lungs is simply dissolved or is chemically combined with some constituent of the blood. The latter and more natural view seems for a time to have given place to the former in consequence of the experiments of Magnus. But Liebig and others have since adduced arguments to show that the oxygen absorbed is, mainly at least, chemically combined, be it only in such a loose way, like a portion of the carbonic acid in bicarbonate of soda, that it is capable of being expelled by indifferent gases. It is known, too, that it is the red corpuscles in which the faculty of absorbing oxygen mainly resides.

Now it has been shown in this paper that we have in cruorine a substance capable of undergoing reduction and oxidation, more especially oxidation, so that if we may assume the presence of purple cruorine in venous blood, we have all that is necessary to account for the absorption and chemical combination of the inspired oxygen.

16. It is stated by Hoppe that venous as well as arterial blood shows the two bands which are characteristic of what has been called in this paper scarlet cruorine. As the precautions taken to prevent the absorption of oxygen are not mentioned, it seemed desirable to repeat the experiment, which Dr Harley and Dr Sharpey have kindly done. A pipette adapted to a syringe was filled with water which had been boiled and cooled without exposure to the air, and the point having been introduced into the jugular vein of a live dog, a little blood was drawn into the bulb. Without the water the blood would have been too dark for spectral analysis. The colour did not much differ from that of scarlet cruorine; certainly it was much nearer the scarlet than the purple substance. The spectrum showed the bands of scarlet cruorine.

This, however, does not by any means prove the absence of purple cruorine, but only shows that the colouring matter present was chiefly scarlet cruorine. Indeed the relative proportions of the two present in a mixture of them with one another and with colourless substances, can be better judged of by the tint than by the use of the prism. With the prism the extreme sharpness of the bands of scarlet cruorine is apt to mislead, and to induce the observer greatly to exaggerate the relative proportion of that substance.

Seeing then that the change of colour from arterial to venous blood *as far as it goes* is *in the direction* of the change from scarlet to purple cruorine, that scarlet cruorine is capable of reduction even in the cold by substances present in the blood (§ 9), and that the action of reducing agents upon it is greatly assisted by warmth (§ 7), we have every reason to believe that *a portion* of the cruorine present in venous blood exists in the state of purple cruorine, and is reoxidized in passing through the lungs.

17. That it is only a rather small proportion of the cruorine present in venous blood which exists in the state of purple cruorine under normal conditions of life and health, may be inferred, not only from the colour, but directly from the results of the most recent experiments*. Were it otherwise, any extensive hæmorrhage could hardly fail to be fatal, if, as there is reason to believe, cruorine be the substance on which the function of respiration mainly depends; nor could chlorotic persons exhale as much carbonic acid as healthy subjects, as is found to be the case.

But after death there is every reason to think that the process of reduction still goes on, especially in the case of warm-blooded animals, while the body is still warm. Hence the blood found in the veins of an animal some time after death can hardly be taken as a fair specimen as to colour of the venous blood in the living animal. Moreover the blood of an animal which has been subjected to abnormal conditions before death is of course liable to be altered thereby. The terms in which Lehmann has described the colour of the blood of frogs which had been slowly asphyxiated by being made to breathe a mixture of air and carbonic acid seem unmistakeably to point to purple cruorine†.

18. The effect of various indifferent reagents in changing the colour of defibrinated blood has been much studied, but not always with due regard to optical principles. The brightening of

* Funk's *Lehrbuch der Physiologie*, 1863, Vol. I, § 108.

† *Physiological Chemistry*, Vol. II, p. 178.

the colour, as seen by reflexion, produced by the first action of neutral salts, and the darkening caused by the addition of a little water, are, I conceive, easily explained; but I have not seen stated what I feel satisfied is the true explanation. In the former case the corpuscles lose water by exosmose, and become thereby more highly refractive, in consequence of which a more copious reflexion takes place at the common surface of the corpuscles and the surrounding fluid. In the latter case they gain water by endosmose, which makes their refractive power more nearly equal to that of the fluid in which they are contained, and the reflexion is consequently diminished. There is nothing in these cases to indicate any change in the mode in which light is absorbed by the colouring matter, although a change of tint to a certain extent, and not merely a change of intensity, may accompany the change of conditions under which the turbid mixture is seen, as I have elsewhere more fully explained*.

No doubt the form of the corpuscles is changed by the action of the reagents introduced; but to attribute the change of colour to this is, I apprehend, to mistake a concomitant for a cause, and to attribute, moreover, the change of colour to a cause inadequate to produce it.

19. Very different is the effect of carbonic acid. In this case the existence of a fundamental change in the mode of absorption cannot be questioned, especially when the fluid is squeezed thin between two glasses and viewed by transmitted light. I took two portions of defibrinated blood; to one I added a little of the reducing iron solution, and passed carbonic acid into the other, and then compared them. They were as nearly as possible alike. We must not attribute these apparently identical changes to two totally different causes if one will suffice. Now in the case of the iron salt, the change of colour is plainly due to a deoxidation of the cruorine. On the other hand, Magnus removed as much as 10 or 12 per cent. by volume of oxygen from arterialized blood by shaking the blood with carbonic acid. If, as we have reason to believe, this oxygen was for the most part chemically combined, it follows that carbonic acid acts *as if it were* a reducing agent. We are led to regard the change of colour not as a *direct* effect of the *presence of carbonic acid*, but a consequence of the *removal of*

* *Philosophical Transactions*, 1852, p. 527. [*Ante*, Vol. III, p. 359.]

oxygen. There is this difference between carbonic acid and the *real* reducing agents, that the former no longer acts on a dilute and comparatively pure solution of scarlet cruorine, while the latter act just as before.

If even in the case of blood exposed to an atmosphere of carbonic acid we are not to attribute the change of colour to the direct presence of the gas, much less should we attempt to account for the darker colour of venous than arterial blood by the small additional percentage of carbonic acid which the former contains. The ascertained properties of cruorine furnish us with a ready explanation, namely that it is due to a partial reduction of scarlet cruorine in supplying the wants of the system.

20. I am indebted to Dr Akin for calling my attention to a very interesting pamphlet by A. Schmidt on the existence of ozone in the blood*. The author uses throughout the language of the ozone theory. If by ozone be meant the substance, be it allotropic oxygen or teroxide of hydrogen, which is formed by electric discharges in air, there is absolutely nothing to prove its existence in blood; for all attempts to obtain an oxidizing gas from blood failed. But if by ozone be merely meant oxygen in any such state, of combination or otherwise, as to be capable of producing certain oxidizing effects, such as turning guaiacum blue, the experiments of Schmidt have completely established its existence, and have connected it, too, with the colouring matter. Now in cruorine we have a substance admitting of easy oxidation and reduction; and connecting this with Schmidt's results, we may infer that scarlet cruorine is not merely a greedy absorber and a carrier of oxygen, but also an *oxidizing agent,* and that it is by its means that the substances which enter the blood from the food, setting aside those which are either assimilated or excreted by the kidneys, are reduced to the ultimate forms of carbonic acid and water, as if they had been burnt in oxygen.

21. In illustration of the functions of cruorine, I would refer, in conclusion, to the experiment mentioned in § 10. As the purple cruorine in the solution was oxidized almost instantaneously on being presented with free oxygen by shaking with air, while the tin-salt remained in an unoxidized state, so the purple cruorine of

* *Ueber Ozon im Blute,* Dorpat, 1862.

the veins is oxidized during the time, brief though it may be, during which it is exposed to air in the lungs, while the substances derived from the food may have little disposition to combine with free oxygen. As the scarlet cruorine is gradually reduced, oxidizing thereby a portion of the tin-salt, so part of the scarlet cruorine is gradually reduced in the course of the circulation, oxidizing a portion of the substances derived from the food or of the tissues. The purplish colour now assumed by the solution illustrates the tinge of venous blood, and a fresh shake represents a fresh passage through the lungs.

ON A PROPERTY OF CURVES WHICH FULFIL THE CONDITION $\frac{d^2\phi}{dx^2} + \frac{d^2\phi}{dy^2} = 0$. By W. J. MACQUORN RANKINE. (Supplement.)

[From the *Proceedings of the Royal Society*, *May* 9, 1867.]

PROFESSOR STOKES has pointed out to me an extension of the preceding theorem, viz. that at every *multiple* point in a plane curve which fulfils the condition that $\frac{d^2\phi}{dx^2} + \frac{d^2\phi}{dy^2} = 0$, the branches make equal angles with each other, so that, for example, if n branches cut each other at a multiple point, they make with each other $2n$ equal angles of π/n.

On the Internal Distribution of Matter which shall produce a given Potential at the Surface of a Gravitating Mass.

[From the *Proceedings of the Royal Society, May* 16, 1867.]

It is known that if either the potential of the attraction of a mass attracting according to the law of the inverse square of the distance, or the normal component of the attraction, be given all over the surface of the mass, or any surface enclosing it (which latter case may be included in the former by regarding the internal density as null between the assumed enclosing surface and the actual surface), the potential and consequently the attraction at all points external to the surface and at the surface itself is determinate. This proposition leads to results of particular interest when applied to the Earth, as I showed in two papers published in 1849*, where among other things I proved that if the surface be assumed to be, in accordance with observation, of the form of an ellipsoid of revolution, Clairaut's Theorem follows independently of the adoption of the hypothesis of original fluidity, or even of that of an internal arrangement in nearly spherical strata of equal density.

But though the law of the variation of gravity which was originally obtained as a consequence of the hypothesis of primitive fluidity, and was afterwards found by Laplace to hold good, on the condition that the surface be an ellipsoid of revolution as well as a surface of equilibrium, provided only the mass be arranged in nearly spherical strata of equal density, be thus proved to be true whatever be the internal distribution, the question may naturally be asked, Does not the condition that the potential at the surface

* "On Attractions and on Clairaut's Theorem," *Cambridge and Dublin Mathematical Journal*, Vol. IV, p. 194 [*ante*, Vol. II, pp. 104—130]; and "On the Variation of Gravity at the Surface of the Earth," *Cambridge Philosophical Transactions*, Vol. VIII, p. 672 [*ante*, Vol. II, pp. 131—171].

shall have its actual value require that the internal distribution shall be compatible with that of a fluid mass, or at any rate shall be such that the whole mass shall be arranged in nearly spherical strata of equal density? Such a question was in fact asked me by an eminent mathematician at the time to which I have alluded. I replied by referring to the well-known property of a sphere, according to which a central mass may be distributed uniformly over its surface without affecting the external attraction, by applying which proposition to a mass such as the Earth we may evidently, without affecting the external attraction, leave a large excentrically situated cavity absolutely vacuous, the matter previously within it having been distributed outside it. It is known further that the mass of a particle may be distributed over *any surface whatsoever* enclosing the particle without affecting the external attraction, and in this way we see at once that we may leave *any internal space we please*, however excentrically situated, wholly vacuous; nor is it necessary in doing so to introduce an infinite density, by distributing the whole mass previously within that space over its surface, since that mass may be conceived to be divided into an infinite number of infinitely small parts, which are respectively distributed over an infinite number of surfaces surrounding the space in question. These considerations, however, though they readily show that the internal distribution may be widely different from any that is compatible with the hypothesis of primitive fluidity, do not lead to the general expression for the internal density. Circumstances have recently recalled my attention to the subject, and I can now indicate the mode of obtaining the general expression required in the case of any given surface.

Let the mass be referred to the rectangular axes of x, y, z, and let ρ be the density, V be the potential of the attraction. Then for any internal point V satisfies, as is well known, the partial differential equation

$$\frac{d^2V}{dx^2}+\frac{d^2V}{dy^2}+\frac{d^2V}{dz^2}=-4\pi\rho \quad\text{.................}(1),$$

or as it may be written for brevity $\nabla V=0$. This equation may be extended to all space by imagining the body continued infinitely, but having a density which is null outside the limits of the actual body; and by adopting this convention we need not trouble ourselves about those limits. Conversely, if V be a continuously varying function of x, y, z, which vanishes at an infinite distance, and satisfies the partial differential equation (1), V is the potential of the attraction of the mass whose density at the point (x, y, z) is ρ; or, in other words,

$$V = \iiint \frac{\rho'}{r}\, dx'\, dy'\, dz' \quad \text{.....................(2)},$$

where r is the distance between the points (x, y, z) and (x', y', z'), ρ' the density at (x', y', z'), and the limits are $-\infty$ to $+\infty$, is the complete integral of (1) subject to the condition that V shall vanish at an infinite distance.

This may be proved in different ways; most directly perhaps by taking the expression for the potential (U suppose) which forms the right-hand member of (2), substituting for ρ' its equivalent $-\frac{1}{4\pi}\nabla V'$, V' being the same function of x', y', z' that V is of x, y, z, and transforming the integral in the manner done by Green*, when we readily find $U = V$.

Suppose now that we have a given closed surface S containing within it all the attracting matter, and that the potential has a given, in general variable, value V_0 at the surface. For the portion of space external to S, V is to be determined by the general equation $\nabla V = 0$, subject to the conditions $V = V_0$ at the surface, and $V = 0$ at an infinite distance. We know that the problem of determining V under these circumstances admits of one and but one solution, though it is only for a very limited number of forms of the surface S that the solution can actually be effected. Conceive the problem, however, solved, and from the solution let the value of $dV/d\nu$ at the surface be found, ν being measured outwards along the normal. Now complete V for infinite space by assigning to the space within S any arbitrary but continuous† function we

* "Essay on the Application of Mathematical Analysis to the Theories of Electricity and Magnetism," Nottingham, 1828, Art. 3; or the reprint in *Crelle's Journal*, Vol. XLIV, p. 360.

† To avoid prolixity, I include in "continuous" the requirement that the differential coefficients of the function, to any order required, shall vary continuously. What that order may be it is perfectly easy in any case to see. We may of course imagine distributions in which the density becomes infinite at one or more points, lines, or surfaces, but so that a finite volume contains only a finite mass. But such distributions may be regarded as limiting, and therefore particular, cases of a distribution in which the density is finite; and therefore the supposition that ρ is finite does not in effect limit the generality of our results.

please, subject to the two conditions, 1st, that at the surface it is equal to the given function V_0; 2ndly, that it gives for the value of $dV/d\nu$ at the surface that already got from the solution of the problem referred to in this paragraph. This of course may be done in an infinite number of ways, just as we may in an infinite number of ways join two points in a plane by a continuous curve starting from the two points respectively in given directions, which curve may be either expressed by some algebraical or transcendental equation, or conceived as drawn *liberâ manu*, and thought of independently of any idea of algebraical expression. The function V having been thus assigned to the space internal to S, the equation (1) gives, according to what we have seen, the most general expression for the density of the internal matter.

There is, however, no distinction made in this between positive and negative matter, and if we wish to avoid introducing negative matter we must restrict the function V for the space internal to S to satisfy the imparity

$$\frac{d^2V}{dx^2}+\frac{d^2V}{dy^2}+\frac{d^2V}{dz^2} \ngtr 0.$$

It is easy from the general expression to show, what is already known, that the matter may be distributed in an infinitely thin, and consequently infinitely dense stratum over the surface S, and that such a distribution is determinate.

We know that there exists one and but one continuous function applying to the space within S which satisfies the equation $\nabla V=0$, and is equal to V_0 at the surface. Call this function V_1. It is to be remarked that the value of $dV_1/d\nu$ at the surface is not the same as that of $dV/d\nu$, V being the external potential, though V_1 and V are there each equal to V_0. The argument, it is to be observed, does not assume that the two are different; it merely avoids assuming that they are the same; the result will prove that they cannot be the same all over S unless the density, and consequently the potential, be everywhere null, and therefore $V_0=0$. Now attribute to the interior of S a function V which is equal to V_1 except over a narrow stratum adjacent to S, the thickness of which will in the end be supposed to vanish, within which V is made to deviate from V_1 in such a manner as to render the variation of $dV/d\nu$ continuous and rapid instead of abrupt. On applying equation (1), we see that the density is everywhere null except within this stratum, in which it is very great, and in the limit infinite. For the total quantity of matter contained in any portion of the stratum, we have from (1)

$$-\frac{1}{4\pi}\iiint \nabla V\,dx\,dy\,dz,$$

the integration extending over that portion. Let the portion in question be that corresponding to a very small area A of the surface S; we may suppose it bounded laterally by the ultimately cylindrical surface generated by a normal to S which travels round the perimeter of A. Taking now rectangular coordinates λ, μ, ν, of which the last is parallel to the normal at one point of A, since ∇ is not changed in form by referring it to a new set of rectangular axes, we have for the mass required

$$-\frac{1}{4\pi}\iiint\left\{\frac{d^2V}{d\lambda^2}+\frac{d^2V}{d\mu^2}+\frac{d^2V}{d\nu^2}\right\}d\lambda\,d\mu\,d\nu.$$

Of the differential coefficients within brackets, the last alone becomes infinite when the thickness of the stratum, and consequently the range of integration relatively to λ, becomes infinitely small. We have in the limit

$$\int\frac{d^2V}{d\nu^2}d\nu=\frac{dV}{d\nu}-\frac{dV_1}{d\nu},$$

both differential coefficients having their values belonging to the surface. Hence we have ultimately for the mass

$$\frac{A}{4\pi}\left(\frac{dV_1}{d\nu}-\frac{dV}{d\nu}\right).$$

Hence, if ϖ be the superficial density, defined as the limit of the mass corresponding to any small portion of the surface divided by the area of that portion,

$$\varpi=\frac{1}{4\pi}\left(\frac{dV_1}{d\nu}-\frac{dV}{d\nu}\right)\quad\text{.....................(3)},$$

which is the known expression.

In assigning arbitrarily a function V to the interior of S, in order to get the internal density by the application of the formula (1), we may if we please discard the second of the conditions which V had to satisfy at the surface, namely, that $\frac{dV_1}{d\nu}=\frac{dV}{d\nu}$; but in that case to the mass, of finite density, determined by (1) must be added an infinitely dense and infinitely thin stratum extending over the surface, the finite superficial density of this stratum being given by (3).

We have seen that the determination of the most general internal arrangement requires the solution of the problem, To determine the potential for space external to S, supposed free from attracting matter, in terms of the given potential at the surface; and the determination of that particular arrangement in which the matter is wholly distributed over the surface, requires further the solution of the same problem for space internal to S. If, however, instead of having merely the potential given at the surface S we had given a particular arrangement of matter within S, and sought the most general rearrangement which should not alter the potential at S, there would have been no preliminary problem to solve, since V, and therefore its differential coefficients, are known for space generally, and therefore for the surface S, being expressed by triple integrals.

Instead of having the attracting matter contained within a closed surface S, and the attraction considered for space external to S, it might have been the reverse, and the same methods would still have been applicable. The problem in this form is more interesting with reference to electricity than gravitation.

Supplement to a Paper on the Discontinuity of Arbitrary Constants which appear in Divergent Developments.

[From the *Transactions of the Cambridge Philosophical Society*, Vol. XI, Part II. Read *May* 25, 1868.]

In a paper "On the Numerical Calculation of a Class of Definite Integrals and Infinite Series," printed in the IXth volume of the *Transactions* of this Society*, I gave a method by which a definite integral, to which Mr Airy was led in calculating the intensity of light in the neighbourhood of a caustic, may be readily calculated for large values, whether positive or negative, of a certain variable which appears as a constant under the sign of integration. The method consists in forming a differential equation of which the definite integral is a particular solution, obtaining the complete integral of the equation under a form, indicated by the equation itself, involving series according to descending powers of the variable, and determining the arbitrary constants. The equation admits also of integration by means of ascending series multiplied by other arbitrary constants. The ascending series are always convergent, but when the variable is large begin by diverging rapidly: the descending series, on the other hand, are always divergent, but when the variable is large begin by converging rapidly.

The same method was found to apply to several other definite integrals which occur in physical investigations, as well as to differential equations of frequent occurrence. The ascending and descending series are usually both required, the one for application to small, the other to large values of the variable; and it is necessary to connect the arbitrary constants in the descending with those in the ascending series. The analytical determination of the arbitrary constants by which the divergent series are multiplied forms the chief difficulty, a difficulty only partially

[* *Ante*, Vol. II, p. 329.]

overcome in the paper to which I allude. Some years later I resumed the subject, and I then succeeded in overcoming the remaining difficulty. The result is given in the paper* to which the present is a supplement. It opens out some very curious points of analysis, and seems to me to throw new light on the theory of divergent series.

In the latter paper it is shewn that the constants in the descending series are in reality discontinuous, retaining the same value between values of the amplitude of the imaginary variable comprised within certain limits, and then suddenly assuming different values; and a method is given of determining the critical values of the amplitude at which the transition takes place. As shewn in the paper, it is not essential for the application of the method that we should be able to determine analytically the relations between the constants in the ascending and descending series; we may determine the numerical relations by numerical calculation, by calculating from the ascending and descending series separately, and equating the results. With the exception of the comparatively simple integral first discussed, with a view to paving the way to the application of the method to more difficult examples, all the integrals discussed in the paper give rise to differential equations of the second order; and it often happens that it is the differential equations themselves, and not the definite integrals which are particular solutions of them, that present themselves for treatment. As there are two arbitrary constants regarded as unknown which have to be determined in terms of two regarded as known, we require two equations. These would be obtained by calculating the dependent variable for *two* different values of the independent variable from the ascending and descending series separately, and equating the results. We should then have to solve two simple equations by which the two unknown constants are determined.

The object of the present supplement is to shew that by a simple application of the principles laid down in the paper itself the equation expressing the equivalence of the two different developments (by ascending and descending series) of the dependent variable may be split into two *previously to the deter-*

* *Cambridge Philosophical Transactions*, Vol. x, p. 105. [*Ante*, p. 77; see footnote, p. 80.]

mination of the arbitrary constants; and accordingly, in order to determine the two arbitrary constants which are regarded as unknown, that it is sufficient to calculate the dependent variable for *one* value of the independent, from the ascending and descending series separately, and equate the results. Moreover, the necessity of eliminating between two simple equations will thus be avoided, which involves a saving of numerical calculation (in case we should be unable to determine the unknown constants analytically), which is not to be despised, seeing that the coefficients involved are complex imaginaries.

With the exception of one comparatively simple case, all the differential equations considered in my former paper, whether they present themselves directly, claiming the consideration of their complete integrals, or as equations which certain definite integrals that we have to deal with satisfy, and of which those integrals are particular solutions, are particular cases of the differential equation

$$\frac{d^2y}{dx^2} + \frac{A}{x}\frac{dy}{dx} + \frac{B}{x^2}y + Cx^a y = 0 \quad \text{..............(1)},$$

which in the present supplement I propose to consider in its generality.

The four constants A, B, C, a in this equation are easily reduced to one. By writing $x^{2/(2+a)}$ for x (the case $a = -2$ need not be considered) we may get rid of x in the last term. By then writing cx for x we may convert the last coefficient into any constant we please. Lastly, by writing $x^{\alpha}y$ for y we may determine α so as to convert the coefficient of either the second or the third term into zero or any constant we please. I will take as the canonical form of the equation

$$\frac{d^2y}{dx^2} + \frac{1}{x}\frac{dy}{dx} - \frac{n^2}{x^2}y = y \quad \text{....................(2)}.$$

As n may be supposed to be a general constant, capable of being imaginary, no generality is lost by assuming the above as the canonical form. Nevertheless in all the applications of equation (1) that I know of, the coefficient of the third term when the equation is reduced to the form (2) is either zero or negative, so that n is either zero or a finite real quantity, which may be taken positive. Accordingly in discussing the equation (2) I will mainly consider the case in which n is real. The equation (1) being now done with, the letters A, B, C, a are set free for fresh use.

The complete integral of (2) in ascending series, obtained by the usual methods, is

$$y = Ax^n \left\{1 + \frac{x^2}{2(2+2n)} + \frac{x^4}{2.4(2+2n)(4+2n)} + \frac{x^6}{2.4.6(2+2n)(4+2n)(6+2n)} + \dots\right\}$$

$$+ Bx^{-n} \left\{1 + \frac{x^2}{2(2-2n)} + \frac{x^4}{2.4(2-2n)(4-2n)} + \frac{x^6}{2.4.6(2-2n)(4-2n)(6-2n)} + \dots\right\} \quad \dots\dots(3).$$

When n is zero or any integer this integral takes a particular form, which however, being a limiting and therefore particular case of the integral for general values of n, need not at present be separately considered.

The complete integral of (2) expressed in a form which is convenient for calculation when x is large, that is, has a large modulus, obtained as in my former papers, is

$$y = Cx^{-\frac{1}{2}}e^x \left\{1 + \frac{1^2-4n^2}{1.8x} + \frac{(1^2-4n^2)(3^2-4n^2)}{1.2(8x)^2} + \frac{(1^2-4n^2)(3^2-4n^2)(5^2-4n^2)}{1.2.3(8x)^3} + \dots\right\}$$

$$+ Dx^{-\frac{1}{2}}e^{-x} \left\{1 - \frac{1^2-4n^2}{1.8x} + \frac{(1^2-4n^2)(3^2-4n^2)}{1.2(8x)^2} - \frac{(1^2-4n^2)(3^2-4n^2)(5^2-4n^2)}{1.2.3(8x)^3} + \dots\right\} \dots(4).$$

This expression takes no peculiar form when n is integral. The series terminate whenever $2n$ is an odd integer $2i+1$, in which case (2) is the transformation of the well-known integrable form

$$\frac{d^2y}{dx^2} + \frac{2}{x}\frac{dy}{dx} - \frac{i(i+1)}{x^2}y = y.$$

To render everything definite it will be necessary to specify the values of $x^{\pm n}$ and $x^{-\frac{1}{2}}$ which are supposed to be taken. Let*

$$x = \rho\,(\cos\theta + \iota\sin\theta)\;\ldots\ldots\ldots\ldots\ldots\ldots\ldots(5),$$

and starting from a real positive value of x, for which θ is taken $=0$, let any other value be deemed to be arrived at by continuous variations of ρ and θ without suffering ρ to vanish. The amplitude being thus defined, and accordingly amplitudes such as α and $2\pi+\alpha$ being distinguished, notwithstanding that the numerical value of x is the same in the two cases, I will take x^n to be

$$\rho^n(\cos n\theta + \iota\sin n\theta),$$

and not

$$\rho^n\{\cos(n\theta + 2in\pi) + \iota\sin(n\theta + 2in\pi)\}$$

where i is an integer different from zero; and similarly with respect to x^{-n} and $x^{-\frac{1}{2}}$. And the problem, to effect a slight improvement in the solution of which is the object of the present paper, consists in finding the two relations which connect C, D with A, B.

It was shewn in my former paper that the constants C, D are in general discontinuous, changing their values when the amplitude of x passes, for the first through an odd, for the second through an even multiple of π. Let $C', C''\ldots$ be what C becomes when θ passes through $\pi, 3\pi\ldots$ and $D', D''\ldots$ what D becomes when θ passes through $2\pi, 4\pi\ldots.$ Let θ' be an angle lying between 0 and π, and for $\theta=\theta'$ let U, V, u, v denote the functions multiplied by A, B, C, D in (3), (4). We see from (3), (4) that when θ is increased by π the functions multiplied by A, B are reproduced, except as to the constant multiplier $e^{n\pi\iota}$ or $e^{-n\pi\iota}$ introduced, and the functions multiplied by C, D reproduce each other, except as to the multiplier $e^{-\frac{1}{2}\pi\iota}$ introduced. Hence supposing $\theta=\theta'+i\pi$ we have

$$\left.\begin{array}{lll}\text{for } 0<\theta<\pi, & y=AU+BV & =Cu+Dv,\\ \text{for } \pi<\theta<2\pi, & y=Ae^{n\pi\iota}U+Be^{-n\pi\iota}V & =C'e^{-\frac{1}{2}\pi\iota}v+De^{-\frac{1}{2}\pi\iota}u,\\ \text{for } 2\pi<\theta<3\pi, & y=Ae^{2n\pi\iota}U+Be^{-2n\pi\iota}V & =C'e^{-\pi\iota}u+D'e^{-\pi\iota}v,\end{array}\right\}\;(6),$$

and so on indefinitely, the expressions admitting of being continued backwards for negative values of θ in a perfectly similar manner, the constants C, D being changed alternately.

[* $\sqrt{-1}$ has been replaced by ι throughout.]

On account of the linearity of (2), the constants C, D must be linear functions of A, B, vanishing with them. Let then

$$C = pA + qB, \quad D = rA + sB \quad \text{.................(7).}$$

Now it follows from (6) that $De^{-\frac{1}{2}\pi\iota}$, $C'e^{-\frac{1}{2}\pi\iota}$ are composed of $Ae^{n\pi\iota}$, $Be^{-n\pi\iota}$ as C, D are of A, B; that $C'e^{-\pi\iota}$, $D'e^{-\pi\iota}$ are composed of $Ae^{2n\pi\iota}$, $Be^{-2n\pi\iota}$ as C, D of A, B, and so on. Hence we have

$$\left.\begin{aligned} De^{-\frac{1}{2}\pi\iota} &= pAe^{n\pi\iota} + qBe^{-n\pi\iota}, \\ C'e^{-\frac{1}{2}\pi\iota} &= rAe^{n\pi\iota} + sBe^{-n\pi\iota}, \\ C'e^{-\pi\iota} &= pAe^{2n\pi\iota} + qBe^{-2n\pi\iota}, \\ D'e^{-\pi\iota} &= rAe^{2n\pi\iota} + sBe^{-2n\pi\iota} \end{aligned}\right\} \quad \text{...........(8),}$$

and so on. Comparing these equations with (7), we have

$$D = pe^{(\frac{1}{2}+n)\pi\iota} A + qe^{(\frac{1}{2}-n)\pi\iota} B = rA + sB,$$

which since A and B are independent gives

$$r = e^{(\frac{1}{2}+n)\pi\iota} p, \quad s = e^{(\frac{1}{2}-n)\pi\iota} q \quad \text{.................(9).}$$

The comparison of the second and third of equations (8) leads to the same result, and we have

$$\left.\begin{aligned} C &= pA + qB, \\ D &= pe^{(\frac{1}{2}+n)\pi\iota} A + qe^{(\frac{1}{2}-n)\pi\iota} B, \\ C' &= pe^{(1+2n)\pi\iota} A + qe^{(1-2n)\pi\iota} B, \\ D' &= pe^{(\frac{3}{2}+3n)\pi\iota} A + qe^{(\frac{3}{2}-3n)\pi\iota} B \end{aligned}\right\} \quad \text{...........(10),}$$

and so on, an additional factor $e^{(1+2n)\pi\iota}$ or $e^{(1-2n)\pi\iota}$ being introduced whenever the increase of θ through an odd or even multiple of π changes C or D. It is easily seen that the same law applies when θ increases negatively, the factor $e^{-(1\pm 2n)\pi\iota}$ being in this case introduced. The determination of the series of constants C, C', C'', ... D, D', D'', ... in terms of the arbitrary constants A, B is thus reduced to that of the two constants p, q, which depend on n only. In the integrable case, namely, that in which $2n$ is an odd integer, we may see *a priori* that there can be no discontinuity in the constants C, D, and accordingly in this case the factor $e^{(1\pm 2n)\pi\iota}$ is equal to 1.

It was shewn in my former paper that the discontinuity in the expression of a continuous function by means of divergent series is reconciled with the continuity of the function expressed, by the circumstance that the part of the function which contains a constant coefficient that alters discontinuously, which part may be expressed by a divergent series or in finite terms, is accompanied by another part, expressed by means of a divergent series, which in the neighbourhood of the critical value of the amplitude of the variable becomes subject to a certain amount of vagueness, so that its numerical value can only be obtained subject to a certain amount of uncertainty, comparable in amount with the whole value of the part of the function which contains the constant that alters discontinuously*. Hence if there be no such accompanying function the constant that is liable to change discontinuously as θ passes through its critical value cannot do so. This conclusion is easily verified by equations (10), which give

$$C' - C = 2\iota \cos n\pi . D,$$

so that $C' = C$ if $D = 0$. We may notice that if D be real, the discontinuity in the constant C as θ passes through π affects only its imaginary part.

The first of equations (6) combined with the first two of (10) gives, on account of the independence of A, B,

$$\left.\begin{aligned} U &= p\,(u + e^{(\frac{1}{2}+n)\,\pi\iota}\,v) \\ V &= q\,(u + e^{(\frac{1}{2}-n)\,\pi\iota}\,v) \end{aligned}\right\} \quad\quad (11).$$

Hence *previously to the determination*, whether analytical or numerical, *of the arbitrary constants in the descending series*, we are able to tell what combination of functions u, v represents, except as to a constant multiplier, each of the functions U, V. Moreover p, q, on which the complete determination of the arbitrary constants has now been made to depend, are given *separately* by the two equations (11), and may be determined by calculating the four functions U, V, u, v for *one* value of the variable x.

[* This explanation may perhaps be put more concisely in the form that each of the 'semi-convergent series' or 'asymptotic expansions,' above referred to, ceases to be 'uniformly' semi-convergent at certain loci on the diagram of the complex variable, in crossing which its value may in consequence change discontinuously. The special relation in which asymptotic solutions stand to the regular solutions of linear differential equations is of course involved in this remark. The nature of the general phenomenon of non-uniform convergence, described in the text, had been elucidated by Prof. Stokes, as is now well-known, as early as 1847. Cf. *ante*, Vol. I, pp. 279—286; also Vol. IV, p. 80.]

If X_i be the $(i+1)$th term in the first series within brackets in (3)

$$X_i = \frac{x^{2i}}{1.2.3\ldots 2i} \times \frac{1.3.5\ldots(2i-1)}{(2+2n)(4+2n)\ldots(2i+2n)}$$

$$= \frac{x^{2i}}{1.2\ldots 2i} \frac{\Gamma(n+1)\Gamma(i+\frac{1}{2})}{\Gamma(\frac{1}{2})\Gamma(i+n+1)};$$

and $$\frac{\Gamma(n+\frac{1}{2})\Gamma(i+\frac{1}{2})}{(i+n+1)} = 2\int_0^{\frac{\pi}{2}} (\sin\phi)^{2n}(\cos\phi)^{2i}\,d\phi \quad \ldots(12),$$

which gives for the first part of y

$$A\,\frac{\Gamma(n+1)}{\Gamma(\frac{1}{2})\Gamma(n+\frac{1}{2})}\,x^n\int_0^{\frac{\pi}{2}}(\sin\phi)^{2n}(e^{x\cos\phi}+e^{-x\cos\phi})\,d\phi.$$

The second part may be got by writing $-n$ for n and B for A, provided $n<\frac{1}{2}$, so that equation (12) may hold good. Hence subject to this restriction

$$y = \pi^{-\frac{1}{2}}\int_0^{\frac{\pi}{2}}\left\{A\,\frac{\Gamma(1+n)}{\Gamma(\frac{1}{2}+n)}\,x^n(\sin\phi)^{2n}\right.$$
$$\left.+B\,\frac{\Gamma(1-n)}{\Gamma(\frac{1}{2}-n)}\,x^{-n}(\sin\phi)^{-2n}\right\}(e^{x\cos\phi}+e^{-x\cos\phi})\,d\phi \quad \ldots(13),$$

where $\pi^{\frac{1}{2}}$ has been written for $\Gamma(\frac{1}{2})$. This form of the integral of (2) is already known.

We have now to connect the constants in the ascending and descending series, which may be done by the intervention of the integral (13) expressed in the form of definite integrals. This may be effected precisely as in my former papers by seeking the limit of the right-hand member of (13), when x increases indefinitely; only, it will suffice to do this for one amplitude of x. Let then $\theta=0$, and suppose x to increase indefinitely. Then ultimately the difference between the whole integral from 0 to $\frac{1}{2}\pi$ and the part from 0 to ϕ_1, where ϕ_1 is a positive quantity as small as we please, vanishes compared with either; and writing ϕ for $\sin\phi$, $1-\frac{1}{2}\phi^2$ for $\cos\phi$, we shall have for the limit required that of

$$\pi^{-\frac{1}{2}}e^x\int_0^{\phi_1}\left\{A\,\frac{\Gamma(1+n)}{\Gamma(\frac{1}{2}+n)}\,x^n\phi^{2n}+B\,\frac{\Gamma(1-n)}{\Gamma(\frac{1}{2}-n)}\,x^{-n}\phi^{-2n}\right\}e^{-\frac{1}{2}x\phi^2}\,d\phi.$$

Putting $\frac{1}{2}x\phi^2 = \phi'^2$, we shall ultimately have 0 and ∞ for the limits of the integral, however small ϕ_1 may have been taken, and we shall have for the limit of y

$$y = \pi^{-\frac{1}{2}} e^x \int_0^\infty \left\{ A \frac{\Gamma(1+n)}{\Gamma(\frac{1}{2}+n)} x^{-\frac{1}{2}} 2^{n+\frac{1}{2}} \phi'^{2n} \right.$$

$$\left. + B \frac{\Gamma(1-n)}{\Gamma(\frac{1}{2}-n)} x^{-\frac{1}{2}} 2^{-n+\frac{1}{2}} \phi'^{-2n} \right\} e^{-\phi'^2} d\phi',$$

or
$$y = \frac{1}{\sqrt{2\pi x}} e^x \{A 2^n \Gamma(1+n) + B 2^{-n} \Gamma(1-n)\}.$$

Comparing with (4), in which ultimately

$$y = C x^{-\frac{1}{2}} e^x,$$

we have

$$C = \frac{1}{\sqrt{2\pi}} \{2^n \Gamma(1+n) A + 2^{-n} \Gamma(1-n) B\} \ldots\ldots(14);$$

whence by comparison with the first of equations (10),

$$p = \frac{1}{\sqrt{2\pi}} 2^n \Gamma(1+n), \qquad q = \frac{1}{\sqrt{2\pi}} 2^{-n} \Gamma(1-n) \ldots(15),$$

and then from the second of equations (10),

$$D = \frac{1}{\sqrt{2\pi}} e^{-\frac{1}{2}\pi\iota} \{e^{n\pi\iota} 2^n \Gamma(1+n) A + e^{-n\pi\iota} 2^{-n} \Gamma(1-n) B\} \ldots(16).$$

C and D being thus determined from $\theta = 0$ to $\theta = \pi$, the rule already given determines the constants in the descending series for all amplitudes. It must be remembered however that in the establishment of the formulæ (14), (16) n has been restricted to be less than $\frac{1}{2}$. The series (3), (4) are of course subject to no such restriction.

With the view of determining the arbitrary constants in the divergent series when $n^2 > \frac{1}{4}$, let us seek to make the integration of (2) depend on that of a similar equation with a different value of n. Assume

$$y = x^\alpha \int x^\beta z \, dx,$$

where α, β are disposable constants. Substituting in (2) we have

$$x^{\alpha+\beta} \frac{dz}{dx} + (2\alpha + \beta + 1) x^{\alpha+\beta-1} z + (\alpha^2 - n^2) x^{\alpha-2} \int x^\beta z \, dx = x^\alpha \int x^\beta z \, dx.$$

Assume $$\alpha^2 - n^2 = 0 \quad\text{.............................(17)},$$
divide by x^α, differentiate, and then divide by x^β, and we have

$$\frac{d^2z}{dx^2} + (2\alpha + 2\beta + 1)\frac{1}{x}\frac{dz}{dx} + (\beta - 1)(2\alpha + \beta + 1)\frac{z}{x^2} = z;$$

and in order that this may be of the form (2) we must have

$$2\alpha + 2\beta + 1 = 1, \quad \text{or} \quad \beta = -\alpha,$$

when the last equation becomes

$$\frac{d^2z}{dx^2} + \frac{1}{x}\frac{dz}{dx} - \frac{(1+\alpha)^2}{x^2}z = z \quad \text{..............(18)},$$

and the relation between y and z is

$$y = x^\alpha \int x^{-\alpha} z\, d\alpha, \quad \text{or} \quad z = x^\alpha \frac{d}{dx} x^{-\alpha} y \text{..............(19)}.$$

Since from (17) $\alpha = \pm n$, (18) differs from (2) in having n increased or diminished by 1; and by (19) the integral of (2) is deduced from that of (18) by integration, or the integral of (18) from that of (2) by differentiation. Suppose we choose differentiation, and wish n to be lowered in the deduced equation. Then we must take (18) for the deduced equation, and put $\alpha = -n$. Suppose n to lie between $i - \frac{1}{2}$ and $i + \frac{1}{2}$, where i is a positive integer, and repeat the process i times. Then we shall have

$$z = x^{-n+i-1}\frac{d}{dx}x^{n-i+1} . x^{-n+i-2}\frac{d}{dx}x^{n-i+2} \ldots x^{-n}\frac{d}{dx}x^n y,$$

or
$$z = x^{-n+i}\left(\frac{1}{x}\frac{d}{dx}\right)^i x^n y \quad \text{...................(20)},$$

z being the complete integral of

$$\frac{d^2z}{dx^2} + \frac{1}{x}\frac{dz}{dx} - \frac{(n-i)^2}{x^2}z = z \text{.................(21)}.$$

The *form* of the integral of the deduced equation (21) may be got at once from that of (2) by writing $n - i$ for n, and in deducing the integral by the formula (20), with a view to connect the arbitrary constants A, B, C, D with those in the integral of a similar equation in which n falls within the previously prescribed limits, the two relations among which have been already found, we need attend only to the leading terms.

The formula (20) applied to the leading term of the first series in (3) gives

$$2n(2n-2)(2n-4)\ldots(2n-2i+2)Ax^{n-i},$$

or $$2^i \frac{\Gamma(1+n)}{\Gamma(1+n-i)} A x^{n-i}.$$

On applying the formula to the second series, the initial terms successively disappear on differentiation. The first which remains is that which contains x^{-n+2i}, which produces

$$\frac{Bx^{-(n-i)}}{(2-2n)(4-2n)\dots(2i-2n)}, \quad \text{or} \quad 2^{-i}\frac{\Gamma(1-n)}{\Gamma(1-n+i)} Bx^{-(n-i)},$$

provided we take $\Gamma(m)$ for m negative to be defined by the equation

$$\Gamma(m+1) = m\Gamma(m),$$

which being repeated a sufficient number of times reduces the Γ of a negative quantity to the previously defined Γ of a positive quantity.

On applying (19) to (4), we shall evidently get the leading terms by differentiating the exponentials only. We thus get

$$Cx^{-\frac{1}{2}}e^{x} \text{ and } D(-1)^i x^{-\frac{1}{2}} e^{-x};$$

and therefore we have for the complete integral of (21)

$$z = 2^i \frac{\Gamma(1+n)}{\Gamma(1+n-i)} A x^{n-i} \{1+\dots\} + 2^{-i}\frac{\Gamma(1-n)}{\Gamma(1-n+i)} Bx^{-n+i}\{1+\dots\}$$
$$= Cx^{-\frac{1}{2}} e^{x}\{1+\dots\} + D(-1)^i x^{-\frac{1}{2}} e^{-x}\{1-\dots\}.$$

Since $n-i$ lies between the limits $-\frac{1}{2}$ and $+\frac{1}{2}$, the formulæ already investigated will give the two relations between the constants. We have merely to write $2^i \frac{\Gamma(1+n)}{\Gamma(1+n-i)} A$, etc. in place of A, etc., and $n-i$ in place of n. We get thus from (14)

$$C = \frac{1}{\sqrt{2\pi}}\left\{2^{n-i}\Gamma(1+n-i).2^i \frac{\Gamma(1+n)}{\Gamma(1+n-i)} A \right.$$
$$\left. + 2^{-n+i}\Gamma(1-n+i).2^{-i}\frac{\Gamma(1-n)}{\Gamma(1-n+i)} B\right\}$$
$$= \frac{1}{\sqrt{2\pi}}\{2^n\Gamma(1+n)A + 2^{-n}\Gamma(1-n)B\},$$

so that (14) and the equations (15), (16) which follow from it, which were at first proved subject to the restriction $1 > 2n > -1$, are now set free from that restriction, and shown to hold good for all real values of n however great.

When n is imaginary, when for example the coefficient of the third term in (2) is real but positive, we see that the attempt to connect analytically C, D with A, B would lead to the introduction of Γ functions of imaginary variables. As the function Γ has not been investigated and tabulated for imaginary values of the variable, we see that nothing would be gained by the attempt, and we should be obliged to have recourse to numerical calculation. In such case the formulæ (10), (11) would render important service. However in all actual investigations, so far as I am aware, the coefficient is either zero or a negative quantity, so that n is real*.

The solution of the problem proposed may now be deemed complete, but as equations of the form (2) with n integral are of frequent occurrence, it may be well actually to work out the special forms assumed by several of our equations in this case.

Take first the case of $n=0$. Represent for shortness equation (3) by

$$y = Af(n) + Bf(-n),$$

and expand $f(n)$ and $f(-n)$ according to powers of n. Then

$$y = A\{f(0) + f'(0)\,n + \ldots\} + B\{f(0) - f'(0)\,n + \ldots\}.$$

Put $$A + B = A_1, \quad (A - B)\,n = B_1 \ldots\ldots\ldots\ldots\ldots(22),$$

so that

$$A = \frac{1}{2}\left(A_1 + \frac{B_1}{n}\right), \quad B = \frac{1}{2}\left(A_1 - \frac{B_1}{n}\right) \ldots\ldots\ldots(23),$$

and then make n vanish. We get $y = A_1 f(0) + B_1 f'(0)$, or putting for $f(0)$ and $f'(0)$ their values,

$$y = (A_1 + B_1 \log x)\left(1 + \frac{x^2}{2^2} + \frac{x^4}{2^2.4^2} + \frac{x^6}{2^2.4^2.6^2} + \ldots\right)$$
$$- B_1\left(\frac{x^2}{2^2} S_1 + \frac{x^4}{2^2.4^2} S_2 + \frac{x^6}{2^2.4^2.6^2} S_3 + \ldots\right) \ldots\ldots(24),$$

where

$$S_i = \frac{1}{1} + \frac{1}{2} + \frac{1}{3} \ldots + \frac{1}{i}.$$

Substituting in (14) from (23), and then making n vanish, we get

$$C = \frac{1}{\sqrt{2\pi}}\{A_1 + (\log 2 - \gamma)\,B_1\} \ldots\ldots\ldots\ldots\ldots(25),$$

[* Functions of complex order occurred in the problem of a cylindrical pendulum, *ante*, Vol. III, p. 38.]

where γ is Euler's constant ·57721566... the limit of $S_x - \log x$ for $x = \infty$, to which as we know $-\Gamma'(1)$ is equal. The values of D, C', D', etc., may be obtained in a similar manner as limits, or deduced at once from C in the manner before explained. The latter course appears to be the shorter. If U, V be the functions multiplied by A_1, B_1 in (24) for $0 < \theta < \pi$, we see that when θ is increased by π, U recurs, and V becomes $V + \pi\iota U$. Hence from (6), of which the right-hand members remain unchanged, we see that D is composed of $\iota(A_1 + \pi\iota B_1)$, ιB_1 as C of A_1, B_1. Hence

$$D = \frac{1}{\sqrt{2\pi}}\{\iota A_1 + [\iota(\log 2 - \gamma) - \pi] B_1\}\dots\dots\dots(26).$$

Equations (25), (26) agree with the equations (41) of my former paper, the constants C, D having been interchanged in the present supplement to make the law of progress of equations (10) more evident. The agreement is proved by the known relation

$$\pi^{-\frac{1}{2}}\Gamma'(\tfrac{1}{2}) + \log 4 + \gamma = 0.$$

There would be no difficulty after what precedes in forming the expressions for C', C'' ... D', D''..., but I forbear to do so, as it is a matter of curiosity rather than utility.

Take now the case in which n is any positive integer i, and first suppose $n = i + \delta$, where δ is a small quantity which in the end will be made to vanish. Let (3) be denoted by

$$y = Af(n) + \frac{B}{\delta}F(n),$$

and putting $n = i + \delta$ expand $f(n)$, $F(n)$ according to powers of δ. Then observing that

$$F(i) = \frac{(-1)^i}{2 \,.\, 2^2 \,.\, 4^2 \dots (2i-2)^2 \,.\, 2i} f(i),$$

we have

$$y = A\{f(i) + f'(i)\,\delta + \dots\}$$
$$+ \frac{B}{\delta}\left\{\frac{(-1)^i}{2 \,.\, 2^2 \,.\, 4^2 \dots (2i-2)^2 \,.\, 2i} f(i) + F'(i)\,\delta + \dots\right\}.$$

Let $$A + \frac{B}{\delta}\,\frac{(-1)^i}{2 \,.\, 2^2 \,.\, 4^2 \dots (2i-2)^2\, 2i} = A_{\prime\prime} \quad \dots\dots\dots\dots(27),$$

and after eliminating A make δ vanish. Then

$$y = A_{\prime\prime} f(i) + B \left\{ F'(i) + \frac{(-1)^{i+1}}{2 \cdot 2^2 \cdot 4^2 \ldots (2i-2)^2 \, 2i} f'(i) \right\} \ldots (28).$$

We have

$$f(i) = x^i \left\{ 1 + \frac{x^2}{2(2+2i)} + \frac{x^4}{2 \cdot 4 (2+2i)(4+2i)} \cdots \right\} \ldots (29);$$

$$f'(i) = \log x \cdot f(i) - x^i \left\{ \frac{x^2}{2(2+2i)} (S_{i+1} - S_i) + \frac{x^4}{2 \cdot 4 (2+2i)(4+2i)} (S_{i+2} - S_i) \ldots \right\}$$

$$= (\log x + S_i) f(i) - x^i \left\{ S_i + \frac{x^2}{2(2+2i)} S_{i+1} + \frac{x^4}{2 \cdot 4 (2+2i)(4+2i)} S_{i+2} + \ldots \right\};$$

$$F(n) = (n-i) x^{-n} \left\{ 1 - \frac{x^2}{2(2n-2)} + \frac{x^4}{2 \cdot 4 (2n-2)(2n-4)} \cdots + \frac{(-1)^{i-1} x^{2i-2}}{2 \cdot 4 \ldots (2i-2)(2n-2)(2n-4) \ldots (2n-2i+2)} \right\}$$

$$+ \frac{1}{2} \frac{(-1)^i x^{2i-n}}{2 \cdot 4 \ldots 2i (2n-2i+2)(2n-2i+4) \ldots (2n-2)} \left\{ 1 + \frac{x^2}{(2i+2)(2i+2-2n)} + \frac{x^4}{(2i+2)(2i+4)(2i+2-2n)(2i+4-2n)} + \ldots \right\}.$$

Hence differentiating with respect to n, and then putting $n = i$, we have

$$F'(i) = x^{-i} \left\{ 1 - \frac{x^2}{2(2i-2)} + \ldots + \frac{(-1)^{i-1} x^{2i-2}}{2 \cdot 4 \ldots (2i-2)(2i-2)(2i-4) \ldots 2} \right\}$$

$$+ \frac{(-1)^{i+1}}{2 \cdot 2^2 \cdot 4^2 \ldots (2i-2)^2 \, 2i} (\log x + S_{i-1}) f(i)$$

$$+ \frac{(-1)^{i+1} x^i}{2 \cdot 2^2 \ldots (2i-2)^2 \, 2i} \left\{ \frac{x^2}{2(2+2i)} S_1 + \frac{x^4}{2 \cdot 4 (2+2i)(4+2i)} S_2 + \ldots \right\}.$$

We have therefore finally

$$y = A_{\prime\prime} f(i) + Bx^{-i}\left\{1 - \frac{x^2}{2(2i-2)} + \frac{x^4}{2.4(2i-2)(2i-4)}\cdots\right.$$
$$\left. + \frac{(-1)^{i-1}x^{2i-2}}{2.4\ldots(2i-2)(2i-2)(2i-4)\ldots2}\right\}$$
$$+ B\frac{(-1)^{i+1}}{2.2^2.4^2\ldots(2i-2)^2\,2i}\left\{(2\log x + S_{i-1} + S_i)f(i)\right.$$
$$- x^i\left[S_i - S_0 + \frac{x^2}{2(2+2i)}(S_{i+1} - S_1)\right.$$
$$\left.\left. + \frac{x^4}{2.4(2+2i)(4+2i)}(S_{i+2} - S_2) + \ldots\right]\right\} \quad \ldots(30),$$

where $S_0 (=0)$ has been inserted merely for the sake of regularity. Such is the special form assumed by (3) in this case, $f(i)$ being defined by (29). It should be observed that when $i=1$ the denominator $2.2^2\ldots(2i-2)^2\,2i$ is simply 2×2 or 4.

To connect C with $A_{\prime\prime}$, B we must eliminate A from (14) by means of (27), and then pass to the limit by making $n = i$. We have

$$C = \frac{1}{\sqrt{2\pi}}\left\{2^n\Gamma(1+n)A_{\prime\prime} + \left[2^{-n}\Gamma(1-n) - \frac{(-1)^i\,2^n\Gamma(1+n)}{\delta.2.2^2\ldots(2i-2)^2\,2i}\right]B\right\}.$$

Now

$$\Gamma(1-n) = \frac{\Gamma(1+i-n)}{(1-n)(2-n)\ldots(i-n)}$$
$$= \frac{(-1)^i\Gamma(1+i-n)}{\delta(n+1-i)(n+2-i)\ldots(n-1)}$$
$$= \frac{(-1)^i\Gamma(1+i-n)\Gamma(1+n-i)}{\delta\Gamma(n)},$$

which gives for the coefficient of B within the parentheses

$$\frac{(-1)^i}{\delta}\left\{\frac{2^{-n}\Gamma(1+i-n)\Gamma(1+n-i)}{\Gamma(n)} - \frac{2^{-(2i-n)}\Gamma(1+n)}{\Gamma(i)\Gamma(i+1)}\right\},$$

or ultimately

$$\frac{(-1)^i\,2^{-i}}{\{\Gamma(i)\}^2\Gamma(i+1)}\,\text{limit}\,\frac{1}{\delta}\{\Gamma(i)\Gamma(i+1)\,2^{-\delta}\Gamma(1-\delta)\Gamma(1+\delta)$$
$$- 2^{\delta}\Gamma(i+\delta)\Gamma(i+1+\delta)\}$$
$$= \frac{(-1)^{i+1}}{2^i\Gamma(i)}\{\log 4 + \Lambda'(i) + \Lambda'(i+1)\}$$
$$= \frac{(-1)^{i+1}}{2^i\Gamma(i)}\{\log 4 + S_{i-1} + S_i - 2\gamma\},$$

$\Lambda'(i)$ being the derivative of $\Lambda(i)$ or $\log\Gamma(i)$, the value of which is $S_{i-1}-\gamma$. We have then in the limit

$$C=\frac{1}{\sqrt{2\pi}}\left\{2^i\Gamma(1+i)\,A_{\prime\prime}+\frac{(-1)^{i+1}}{2^i\Gamma(i)}(\log 4+S_{i-1}+S_i-2\gamma)\,B\right\}\ (31).$$

To get D let θ be first comprised between 0 and π, and be then increased by π. We see from (29), (30) that setting aside the augmentation of $\log x$ by $\pi\iota$ the functions multiplied by $A_{\prime\prime}$, B will be reproduced, with or without change of sign, according as i is odd or even. And the augmentation of $\log x$ will have the same effect as increasing $A_{\prime\prime}$ by

$$\frac{(-1)^{i+1}\,2^{-(2i-1)}\,\pi\iota}{\Gamma(i)\,\Gamma(i+1)}\,B.$$

Hence $-\iota D$ will be composed of

$$(-1)^i\left\{A_{\prime\prime}+\frac{(-1)^{i+1}\,2^{-(2i-1)}\,\pi\iota}{\Gamma(i)\,\Gamma(i+1)}\,B\right\},\quad (-1)^i B,$$

as C is of $A_{\prime\prime}$, B. Hence

$$D=\frac{1}{\sqrt{2\pi}}\left\{(-2)^i\,\Gamma(1+i)\,\iota A_{\prime\prime}\right.$$
$$\left.+\frac{2^{-i}}{\Gamma(i)}\left[2\pi-(\log 4+S_{i-1}+S_i-2\gamma)\,\iota\right]B\right\}\ \ldots(32).$$

Equations (30), (31), (32) may be simplified a little by including

$$\frac{(-1)^{i+1}\,2^{-2i}}{\Gamma(i)\,\Gamma(i+1)}(S_{i-1}+S_i)\,B$$

in $A_{\prime\prime}$.

Postscript.—After the above paper was read, my attention was called by Professor Cayley to a paper by Professor Kummer, published in *Crelle's Journal**, containing remarkable general transformations of which those of the present paper, considered apart from the question of the discontinuity of the constants involved, are merely particular cases. The discontinuity of the constants, however, the investigation of the nature and consequences of which forms the main object of the present paper and of that to which it is a supplement, has not been treated of, perhaps not even perceived, by the eminent German mathematician.

* Vol. xv, (1836), pp. 39 and 127. [This analysis of Kummer was utilized by Kirchhoff, in his memoir on induced magnetism in cylinders, *Crelle's Journal für Math.*, xlviii, 1853, *Ges. Abhandl.* p. 196, two years subsequent to Prof. Stokes' memoir on pendulums (*ante*, Vol. iii, p. 1), to connect the various expansions of Bessel functions of real argument. See also *supra*, p. 80, footnote.]

On the Communication of Vibration from a Vibrating Body to a surrounding Gas.

[From the *Philosophical Transactions* for 1868. Received and read *June* 18.]

[Abstract, from *Proceedings of the Royal Society*, XVI, pp. 470—471.]

In the first volume of the *Transactions of the Cambridge Philosophical Society* will be found a paper by the late Professor John Leslie, describing some curious experiments which show the singular incapacity of hydrogen either pure or mixed with air, for receiving and conveying vibrations from a bell rung in the gas. The facts elicited by these experiments seem not hitherto to have received a satisfactory explanation.

It occurred to the author of the present paper that they admitted of a ready explanation as a consequence of the high velocity of propagation of sound in hydrogen gas operating in a peculiar way. When a body is slowly moved to and fro in any gas, the gas behaves almost exactly like an incompressible fluid, and there is merely a local reciprocating motion of the gas from the anterior to the posterior region, and back again in the opposite phase of the body's motion, in which the region that had been anterior becomes posterior. If the rate of alternation of the body's motion be taken greater and greater, or, in other words, the periodic time less and less, the condensation and rarefaction of the gas, which in the first instance was utterly insensible, presently becomes sensible, and sound-waves (or waves of the same nature in case the periodic time be beyond the limits of audibility) are produced, and exist along with the local reciprocating flow. As the periodic time is diminished, more and more of the encroachment of the vibrating body on the gas goes to produce a true sound-wave, less and less a mere local reciprocating flow. For a given periodic time, and given size, form, and mode of vibration of the vibrating body, the gas behaves so much the more nearly like an incompressible fluid as the velocity of propagation of sound in it is greater; and on this account the intensity of the sonorous vibrations excited in air as compared with hydrogen may be vastly greater than corresponds merely with the difference of density of the two gases.

It is only for a few simple geometrical forms of the vibrating body that the solution of the problem of determining the motion produced in the gas can actually be effected. The author has given the solution in the two cases of a vibrating sphere and of an infinite cylinder, the motion in the latter case being supposed to take place in two dimensions. The former is taken as the representative of a bell; the latter is applied to the case of a vibrating string or wire. In the case of the sphere, the numerical results amply establish the adequacy of the cause here considered to account for the results obtained by Leslie. In the case of the cylinder they give an exalted idea of the necessity of sounding-boards in stringed instruments; and the theory is further applied to the explanation of one or two interesting phenomena.

In the first volume of the *Transactions of the Cambridge Philosophical Society* is a short paper by Professor John Leslie, "On Sounds excited in Hydrogen Gas," in which the author mentions some remarkable experiments indicating the singular incapacity of hydrogen for becoming the vehicle of the transmission of sound when a bell is struck in that gas, either pure or mixed with air. With reference to the most striking of his experiments the author observes (p. 267), "The most remarkable fact is, that the admixture of hydrogen gas with atmospheric air has a predominant influence in blunting or stifling sound. If one half of the volume of atmospheric air be extracted [from the receiver of the air-pump], and hydrogen gas be admitted to fill the vacant space, the sound will now become scarcely audible."

No definite explanation of the results is given, but with reference to the feebleness of sound in hydrogen the author observes, "These facts, I think, depend partly on the tenuity of hydrogen gas, and partly on the rapidity with which the pulsations of sound are conveyed through this very elastic medium"; and he states that, according to his view, he "should expect the intensity of sound to be diminished 100 times, or in the compound ratio of its tenuity and of the square of the velocity with which it conveys the vibratory impressions." With reference to the effect of the admixture of hydrogen with air he says, "When hydrogen gas is mixed with common air, it probably does not intimately combine, but dissipates the pulsatory impressions before the sound is vigorously formed."

In referring to Leslie's experiment in which a half-exhausted receiver is filled up with hydrogen, Sir John Herschel suggests a possible explanation founded on Dalton's hypothesis that every gas acts as a vacuum towards every other*. According to this

* *Encyclopaedia Metropolitana*, Vol. IV, Art. Sound, § 108.

view there is a constant tendency for sound-waves to be propagated with different velocities in the air and hydrogen of which the mixture consists, but this tendency is constantly checked by the resistance which one gas opposes to the passage of another, calling into play something analogous to internal friction, whereby the sound-vibration though at first produced is rapidly stifled. Air itself indeed is a mixture; but the velocities of propagation of sound in nitrogen and oxygen are so nearly equal that the effect is supposed not to be sensible in this case.

This explanation never satisfied me, believing, as I always have done, for reasons which it would take too long here to explain, that for purely hydrodynamical phenomena (such as those of sound) an intimate mixture of gases was equivalent to a single homogeneous medium. I had some idea of repeating the experiment, thinking that possibly Leslie might not have allowed sufficient time for the gases to be perfectly mixed, though that did not appear likely, when another explanation occurred to me, which immediately struck me as being in all probability the true one.

In reading some years ago an investigation of Mr Earnshaw's, in which a certain result relating to the propagation of sound in a straight tube was expressed in terms among other things of the velocity of propagation, the idea occurred to me that the high velocity of propagation of sound in hydrogen would account for the result of Leslie's experiment, though in a manner altogether different from anything relating to the propagation of sound in one dimension only.

Suppose a person to move his hand to and fro through a small space. The motion which is occasioned in the air is almost exactly the same as it would have been if the air had been an incompressible fluid. There is a mere local reciprocating motion, in which the air immediately in front is pushed forwards, and that immediately behind impelled after the moving body, while in the anterior space generally the air recedes from the encroachment of the moving body, and in the posterior space generally flows in from all sides, to supply the vacuum which tends to be created; so that in lateral directions the motion of the fluid is backwards, a portion of the excess of fluid in the front going to supply the deficiency behind. Now conceive the periodic time of the motion to be continually diminished. Gradually the alternation of movement becomes too rapid to permit of the full establishment of the merely local reciprocating flow; the air is sensibly compressed and rarefied, and a sensible sound-wave (or wave of the same nature, in case the periodic time be beyond the limits suitable to hearing) is propagated to a distance. The same takes place in any gas; and the more rapid be the propagation of condensations and rarefactions in the gas, the more nearly will it approach, in relation to the motions we have under consideration, to the condition of an incompressible fluid; the more nearly will the conditions of the displacement of the gas at the surface of the solid be satisfied by a merely local reciprocating flow.

This explanation when once it suggested itself seemed so simple and obvious that I could not doubt that it afforded the true mode of accounting for the phenomenon. It remained only to test the accuracy of the assigned cause by actual numerical calculation in some case or cases sufficiently simple to permit of precise analytical determination. The result of calculations of the kind applied to a sphere proved that the assigned cause was abundantly sufficient to account for the observed result. I have not hitherto published these results; but as the phenomenon has not to my knowledge been satisfactorily explained by others, I venture to hope that the explanation I have to offer, simple as it is in principle, may not be unworthy of the notice of the Royal Society.

For the purpose of exact analytical investigation I have taken the two cases of a vibrating sphere and a long vibrating cylinder, the motion of the fluid in the latter case being supposed to be in two dimensions. The sphere is chosen as the best representative of a bell, among the few geometrical forms of body for which the problem can be solved. The cylinder is chosen as the representative of a vibrating string. In the case of the sphere the problem is identical with that solved by Poisson in his memoir "Sur les mouvements simultanés d'un pendule et de l'air environnant,"* but the solution is discussed with a totally different object in view, and is obtained from the beginning, to avoid the needless complexity introduced by taking account of the initial circumstances, instead of supposing the motion already going on.

* *Mémoires de l'Académie des Sciences*, t. XI, p. 521.

A. *Solution of the Problem in the case of a Vibrating Sphere.*

Suppose an elastic solid, spherical externally in its undisturbed position, to vibrate in the manner of a bell, the amplitude of vibration being very small. Suppose it surrounded by a homogeneous gas, which is at rest except in so far as it is set in motion by the sphere; and let it be required to determine the motion of the gas in terms of that of the sphere supposed given. We may evidently for the purposes of the present problem suppose the gas not to be subject to the action of external forces.

Let the gas be referred to the rectangular axes of x, y, z, and let u, v, w be the components of the velocity. Since the gas is at rest except as to the disturbance communicated to it from the sphere, u, v, w are by a well-known theorem the partial differential coefficients with respect to x, y, z of a function ϕ of the coordinates; and if a^2 be the constant expressing the ratio of the small variations of pressure to the corresponding small variations of density, we must have

$$\frac{d^2\phi}{dt^2} = a^2\left(\frac{d^2\phi}{dx^2} + \frac{d^2\phi}{dy^2} + \frac{d^2\phi}{dz^2}\right) \quad\text{..................(1)};$$

and if s be the small condensation,

$$s = -\frac{1}{a^2}\frac{d\phi}{dt}.$$

As we have to deal with a sphere, it will be convenient to refer the gas to polar coordinates r, θ, ω, the origin being in the centre. In terms of these coordinates, equation (1) becomes

$$\frac{d^2\phi}{dt^2} = a^2\left\{\frac{d^2\phi}{dr^2} + \frac{2}{r}\frac{d\phi}{dr} + \frac{1}{r^2\sin\theta}\frac{d}{d\theta}\left(\sin\theta\frac{d\phi}{d\theta}\right) + \frac{1}{r^2\sin^2\theta}\frac{d^2\phi}{d\omega^2}\right\} \quad\text{.........(2)};$$

and if u', v', w' be the components of the velocity along the radius vector and in two directions perpendicular to the radius vector, the first in and the second perpendicular to the plane in which θ is measured,

$$u' = \frac{d\phi}{dr}, \quad v' = \frac{1}{r}\frac{d\phi}{d\theta}, \quad w' = \frac{1}{r\sin\theta}\frac{d\phi}{d\omega} \quad\text{..........(3)}.$$

Let c be the radius of the sphere, and V the velocity of any point of its surface resolved in a direction normal to the surface, V being a given function of t, θ, and ω; then we must have

$$\frac{d\phi}{dr} = V, \text{ when } r = c \quad\text{.....................(4).}$$

Another condition, arising from what takes place at a great distance from the sphere, will be considered presently.

The sphere vibrating under the action of its elastic forces, its motion will be periodic, expressed so far as the time is concerned partly by the sine and partly by the cosine of an angle proportional to the time, suppose mat. Actually the vibrations will slowly die away, in consequence partly of the imperfect elasticity of the sphere, partly of communication of motion to the gas, but for our present purpose this need not be taken into account. Moreover there will in general be a series of periodic disturbances co-existing, corresponding to different periodic times, but these may be considered separately. We might therefore assume

$$V = U \sin mat + U' \cos mat,$$

but it will materially shorten the investigation to employ an imaginary exponential instead of circular functions. If we take

$$V = Ue^{imat} \quad\text{.............................(5),}$$

where i is an abbreviation for $\sqrt{-1}$, and determine ϕ by the conditions of the problem, the real and imaginary parts of ϕ and V must satisfy all those conditions separately; and therefore we may take the real parts alone, or the coefficients of i or $\sqrt{-1}$ in the imaginary parts, or any linear combination of these even after having changed the arbitrary constants which enter into the expression of the motion of the sphere, as the solution of the problem, according to the way in which we conceive the given quantity V expressed.

The function ϕ will be periodic in a similar manner to V, so that we may take

$$\phi = \psi e^{imat} \quad\text{..........................(6).}$$

As regards the periodicity merely, we might have had a term involving e^{-imat} as well as that written above; but it will be readily seen that in order to satisfy the equation of condition (4) the sign of the index of the exponential in ϕ must be the same as in V.

On substituting in (2) the expression for ϕ given by (6), the factor involving t will divide out, and we shall get for the determination of ψ a partial differential equation free from t. Now ψ may be expanded in a series of Laplace's Functions so that

$$\psi = \psi_0 + \psi_1 + \psi_2 + \dots \quad\dots\dots\dots\dots\dots\dots(7).$$

Substituting in the differential equation just mentioned, taking account of the fundamental equation

$$\frac{1}{\sin\theta}\frac{d}{d\theta}\left(\sin\theta\frac{d\psi_n}{d\theta}\right) + \frac{1}{\sin^2\theta}\frac{d^2\psi_n}{d\omega^2} = -n(n+1)\psi_n,$$

and equating to zero the sum of the Laplace's Functions of the same order, we find

$$\frac{d^2\psi_n}{dr^2} + \frac{2}{r}\frac{d\psi_n}{dr} - \frac{n(n+1)}{r^2}\psi_n + m^2\psi_n = 0.$$

This equation belongs to a known integrable form. The integral is

$$r\psi_n = u_n e^{-imr}\left\{1 + \frac{n(n+1)}{2\,.\,imr} + \frac{(n-1)n(n+1)(n+2)}{2\,.\,4(imr)^2} + \dots\right\}$$
$$+ u_n' e^{imr}\left\{1 - \frac{n(n+1)}{2\,.\,imr} + \frac{(n-1)n(n+1)(n+2)}{2\,.\,4(imr)^2} - \dots\right\},$$

u_n and u_n' being evidently Laplace's Functions of the order n, since that is the case with ψ_n.

It will be convenient to take next the condition which has to be satisfied at a great distance from the sphere. When r is very large the series within braces may be reduced to their first terms 1, and we shall have

$$r\phi = e^{im(at-r)}\,\Sigma u_n + e^{im(at+r)}\,\Sigma u_n'.$$

The first of these terms denotes a disturbance travelling outwards from the centre, the second a disturbance travelling towards the centre, the amplitude of vibration in both cases, for a given phase, varying inversely as the distance from the centre. In the problem before us there is no disturbance travelling towards the centre, and therefore $\Sigma u_n' = 0$, which requires that each function u_n' should separately be equal to zero. We have therefore simply

$$r\psi_n = u_n e^{-imr}\left\{1 + \frac{n(n+1)}{2 \,.\, imr} + \frac{(n-1)\dots(n+2)}{2 \,.\, 4\,(imr)^2} + \dots + \frac{1 \,.\, 2 \,.\, 3 \dots 2n}{2 \,.\, 4 \,.\, 6 \dots 2n\,(imr)^n}\right\} \quad \dots\dots\dots(8),$$

or, if we choose to reverse the series,

$$r\psi_n = u_n e^{-imr}\frac{1 \,.\, 3 \,.\, 5 \dots (2n-1)}{(imr)^n}\left\{1 + \frac{2n}{1 \,.\, 2n}\, imr + \frac{(2n-2)\,2n}{1 \,.\, 2\,(2n-1)\,2n}(imr)^2 + \dots + \frac{2 \,.\, 4 \,.\, 6 \dots 2n}{1 \,.\, 2 \,.\, 3 \dots 2n}(imr)^n\right\} \dots(9).$$

Putting for shortness $f_n(r)$ for the multiplier of $u_n e^{-imr}$ in the right-hand member of (8) or (9), we shall have

$$\phi = \frac{1}{r}\, e^{im\,(at-r)}\, \Sigma u_n f_n(r).$$

It remains to satisfy the equation of condition (4). Put for shortness

$$\frac{d}{dr}\left\{\frac{1}{r}\, e^{-imr} f_n(r)\right\} = -\frac{1}{r^2}\, e^{-imr}\, F_n(r),$$

so that

$$F_n(r) = (1 + imr) f_n(r) - r f_n'(r) \quad \dots\dots\dots\dots(10),$$

and suppose U expanded in a series of Laplace's Functions,

$$U_0 + U_1 + U_2 + \dots;$$

then substituting and equating the functions of the same order on the two sides of the equation, we have

$$U_n = -\frac{1}{c^2}\, e^{-imc}\, F_n(c)\, u_n,$$

and therefore

$$\phi = -\frac{c^2}{r}\, e^{im\,(at-r+c)}\, \Sigma \frac{U_n}{F_n(c)} f_n(r) \quad \dots\dots\dots\dots(11).$$

This expression contains the solution of the problem, and it remains only to discuss it.

At a great distance from the sphere the function $f_n(r)$ becomes ultimately equal to 1, and we have

$$\phi = -\frac{c^2}{r}\, e^{im\,(at-r+c)}\, \Sigma \frac{U_n}{F_n(c)} \quad \dots\dots\dots\dots\dots\dots(12).$$

It appears from (3) that the component of the velocity along the radius vector is of the order r^{-1}, and that in any direction perpendicular to the radius vector of the order r^{-2}, so that the lateral motion may be disregarded except in the neighbourhood of the sphere.

In order to examine the influence of the lateral motion in the neighbourhood of the sphere, let us compare the actual disturbance at a great distance with what it would have been if all lateral motion had been prevented, suppose by infinitely thin conical partitions dividing the fluid into elementary canals, each bounded by a conical surface having its vertex at the centre.

On this supposition the motion in any canal would evidently be the same as it would be in all directions if the sphere vibrated by contraction and expansion of the surface, the same all round, and such that the normal velocity of the surface was the same as it is at the particular point at which the canal in question abuts on the surface. Now if U were constant the expansion of U would be reduced to its first term U_0, and seeing that $f_0(r) = 1$ we should have from (11)

$$\phi = -\frac{c^2}{r} e^{im(at-r+c)} \frac{U_0}{F_0(c)}.$$

This expression will apply to any particular canal if we take U_0 to denote the normal velocity at the sphere's surface for that particular canal; and therefore to obtain an expression applicable at once to all the canals we have merely to write U for U_0. To facilitate a comparison with (11) and (12) I shall, however, write ΣU_n for U. We have then

$$\phi = -\frac{c^2}{r} e^{im(at-r+c)} \frac{\Sigma U_n}{F_0(c)} \quad \text{.................(13).}$$

It must be remembered that this is merely an expression applicable at once to all the canals, the motion in each of which takes place wholly along the radius vector, and accordingly the expression is not to be differentiated with respect to θ or ω with the view of applying the formulæ (3).

On comparing (13) with the expression for the function ϕ in the actual motion at a great distance from the sphere (12), we see that the two are identical with the exception that U_n is divided by two different constants, namely $F_0(c)$ in the former case and $F_n(c)$ in the latter. The same will be true of the leading terms (or those of the order r^{-1}) in the expressions for the condensation and velocity*. Hence if the mode of vibration of the sphere is such that the normal velocity of its surface is expressed by a Laplace's Function of any one order, the disturbance at a great distance from the sphere will vary from one direction to another according to the same law as if lateral motion had been prevented, the amplitude of excursion at a given distance from the centre varying in both cases as the amplitude of excursion, in a normal direction, of the surface of the sphere itself. The only difference is that expressed by the symbolic ratio $F_n(c) : F_0(c)$. If we suppose $F_n(c)$ reduced to the form $\mu_n(\cos\alpha_n + \sqrt{-1}\sin\alpha_n)$, the amplitude of vibration in the actual case will be to that in the supposed case as μ_0 to μ_n, and the phase in the two cases will differ by $\alpha_0 - \alpha_n$.

If the normal velocity of the surface of the sphere be not expressible by a single Laplace's Function, but only by a series, finite or infinite, of such functions, the disturbance at a given great distance from the centre will no longer vary from one direction to another according to the same law as the normal velocity of the surface of the sphere, since the modulus μ_n and likewise the amplitude α_n of the imaginary quantity $F_n(c)$ vary with the order of the function.

Let us now suppose the disturbance expressed by a Laplace's Function of some one order, and seek the numerical value of the alteration of intensity at a distance, produced by the lateral motion which actually exists.

The intensity will be measured by the *vis viva* produced in a given time, and consequently will vary as the density multiplied by the velocity of propagation multiplied by the square of the amplitude of vibration. It is the last factor alone that is different from what it would have been if there had been no lateral

* Of course it would be true if the *complete* differential coefficients with respect to r of the right-hand members of (12) and (13) were taken, but then the former does not give the velocity u' except as to its leading term, since (12) has been deduced from the exact expression (11) by reducing $f_n(r)$ to its first term 1; nor again is it true, except as to terms of the order r^{-1}, of the actual motion of the unimpeded fluid that the whole velocity is in the direction of the radius vector.

motion. The amplitude is altered in the proportion of μ_0 to μ_n, so that if

$$\mu_n^2/\mu_0^2 = I_n,$$

I_n is the quantity by which the intensity which would have existed if the fluid had been hindered from lateral motion has to be divided.

For the first five orders the values of the function $F_n(c)$ are as follows:—

$$F_0(c) = imc + 1,$$

$$F_1(c) = imc + 2 + \frac{2}{imc},$$

$$F_2(c) = imc + 4 + \frac{9}{imc} + \frac{9}{(imc)^2},$$

$$F_3(c) = imc + 7 + \frac{27}{imc} + \frac{60}{(imc)^2} + \frac{60}{(imc)^3},$$

$$F_4(c) = imc + 11 + \frac{65}{imc} + \frac{240}{(imc)^2} + \frac{525}{(imc)^3} + \frac{525}{(imc)^4}.$$

If λ be the length of the sound-wave corresponding to the period of the vibration, $m = 2\pi/\lambda$, so that mc is the ratio of the circumference of the sphere to the length of a wave. If we suppose the gas to be air and λ to be 2 feet, which would correspond to about 550 vibrations in a second, and the circumference $2\pi c$ to be 1 foot (a size and pitch which would correspond with the case of a common house bell), we shall have $mc = \frac{1}{2}$. The following Table gives the values of the square of the modulus and of the ratio I_n for the functions $F_n(c)$ of the first five orders, for each of the values 4, 2, 1, $\frac{1}{2}$, and $\frac{1}{4}$ of mc. It will presently appear why the Table has been extended further in the direction of values greater than $\frac{1}{2}$ than it has in the opposite direction. Five significant figures at least are retained.

When $mc = \infty$ we get from the analytical expressions $I_n = 1$. We see from the Table that when mc is somewhat large I_n is liable to be *a little* less than 1, and consequently the sound to be *a little* more intense than if lateral motion had been prevented. The possibility of this is explained by considering that the waves of condensation spreading from those compartments of the sphere which at a given moment are vibrating positively, *i.e.* outwards, after the lapse of a half period may have spread over the neighbouring compartments, which are now in their turn vibrating

mc	$n=0$	$n=1$	$n=2$	$n=3$	$n=4$	
4	17	16·25	14·879	13·848	20·177	Values of μ_n^2
2	5	5	9·3125	80	1495·8	
1	2	5	89	3965	300137	
0·5	1·25	16·25	1330·2	236191	72086371	
0·25	1·0625	64·062	20878	14837899	18160×10^6	
4	1	0·95588	0·87523	0·87459	1·1869	Values of I_n
2	1	1	1·8625	16	299·16	
1	1	2·5	44·5	1982·5	150068	
0·5	1	13	1064·2	188953	57669097	
0·25	1	60·294	19650	13965×10^3	17092×10^6	

positively, so that these latter compartments in their outward motion work against a somewhat greater pressure than if each compartment had opposite to it only the vibration of the gas which it had itself occasioned; and the same explanation applies *mutatis mutandis* to the waves of rarefaction. However, the increase of sound thus occasioned by the existence of lateral motion is but small in any case, whereas when mc is somewhat small I_n increases enormously, and the sound becomes a mere nothing compared with what it would have been had lateral motion been prevented.

The higher be the order of the function, the greater will be the number of compartments, alternately positive and negative as to their mode of vibration at a given moment, into which the surface of the sphere will be divided. We see from the Table that for a given periodic time as well as radius the value of I_n becomes considerable when n is somewhat high. However practically vibrations of this kind are produced when the elastic sphere executes, not its principal, but one of its subordinate vibrations, the pitch corresponding to which rises with the order of the vibration, so that m increases with that order. It was for this reason that the Table was extended from $mc = 0{\cdot}5$ further in the direction of high pitch than low pitch, namely, to three octaves higher and only one octave lower.

When the sphere vibrates symmetrically about the centre, *i.e.* so that any two opposite points of the surface are at a given moment moving with equal velocities in opposite directions, or more generally when the mode of vibration is such that there is no change of position of the centre of gravity of the volume, there is no term of the order 1. For a sphere vibrating in the manner of a bell the principal vibration is that expressed by a term of the order 2, to which I shall now more particularly attend.

Putting, for shortness, $m^2c^2 = q$, we have

$$\mu_0^2 = q + 1,\ \mu_2^2 = (q^{\frac{1}{2}} - 9q^{-\frac{1}{2}})^2 + \left(4 - \frac{9}{q}\right)^2 = q - 2 + \frac{9}{q} + \frac{81}{q^2},$$

$$I_2 = \frac{q^3 - 2q^2 + 9q + 81}{q^2(q+1)}.$$

The minimum value of I_2 is determined by

$$q^3 - 6q^2 - 84q - 54 = 0,$$

giving approximately

$$q = 12{\cdot}859,\ mc = 3{\cdot}586,\ \mu_0^2 = 13{\cdot}859,\ \mu_2^2 = 12{\cdot}049,\ I_2 = {\cdot}86941;$$

so that the utmost increase of sound produced by lateral motion amounts to about 15 per cent.

I come now more particularly to Leslie's experiments. Nothing is stated as to the form, size, or pitch of his bell; and even if these had been accurately described, there would have been a good deal of guesswork in fixing on the size of the sphere which should be considered the best representative of the bell. Hence all we can do is to choose such values for m and c as are comparable with the probable conditions of the experiment.

I possess a bell, belonging to an old bell-in-air apparatus, which may probably be somewhat similar to that used by Leslie. It is nearly hemispherical, the diameter is 1·96 inch, and the pitch an octave above the middle C of a piano. Taking the number of vibrations 1056 per second, and the velocity of sound in air 1100 feet per second, we have $\lambda = 12{\cdot}5$ inches. To represent the bell by a sphere of the same radius would be very greatly to underrate the influence of local circulation, since near the mouth the gas has but a little way to get round from the outside to the inside, or the reverse. To represent it by a sphere of half the radius would still apparently be to underrate the effect. Nevertheless for the sake of rather underestimating than exaggerating the influence of the cause here investigated, I will make these two suppositions successively, giving respectively $c = {\cdot}98$ and $c = {\cdot}49$, $mc = {\cdot}4926$, and $mc = {\cdot}2463$ for air.

If it were not for lateral motion the intensity would vary from gas to gas in the proportion of the density into the velocity of propagation, and therefore as the pressure into the square root of the density under a standard pressure, if we take the factor depending on the development of heat as sensibly the same for the gases and gaseous mixtures with which we have to deal. In the following Table the first column gives the gas, the second the pressure p, in atmospheres, the third the density D under the pressure p, referred to the density of air at the atmospheric pressure as unity, the fourth, Q_r, what would have been the intensity had the motion been wholly radial, referred to the intensity in air at atmospheric pressure as unity, or, in other words, a quantity varying as $p \times (\text{the density at pressure } 1)^{\frac{1}{2}}$. Then follow the values of q, I_2, and Q, the last being the actual intensity referring to air as before.

An inspection of the numbers contained in the columns headed Q will show that the cause here investigated is amply sufficient to account for the facts mentioned by Leslie.

It may be noticed that while q is 4 times smaller, and I_2 is 16 or 18 times larger, for $c = {\cdot}49$ than for $c = {\cdot}98$, there is no great difference in the values of Q in the two cases for hydrogen and mixtures of hydrogen and air in given proportions. This arises from the circumstance that q is sufficiently small to make the last terms in μ_0^2 and μ_2^2, namely, 1 and $81q^{-2}$, the most important, so that I_n does not greatly differ from $81q^{-2}$. If this result had been exact instead of approximate, the intensity in different gases, supposed for simplicity to be at a common pressure, would have varied as $D^{\frac{3}{2}}$; and it will be found that for the cases in which $p = 1$ the values of Q in the Table, especially those in the last column, do not greatly deviate from this proportion. But

Gas	p	D	Q_r	$c = {\cdot}98$			$c = {\cdot}49$		
				q	I_2	Q	q	I_2	Q
Air....................................	1	1	1	·2427	1136	1	·06067	20890	1
Hydrogen...........................	1	·0690	·2627	·01674	284700	·001048	·004186	4604000	·001191
Air, rarefied.................. ..	·01	·01	·01	·2427	1136	·01	·06067	20890	·01
The same filled with H	1	·0783	·2798	·01900	220600	·001440	·004751	3572000	·001637
Air of same density ...	·0783	·0783	·0783	·2427	1136	·0783	·06067	20890	·0783
Air rarefied ½	·5	·5	·5	·2427	1136	·5	·06067	20890	·5
The same filled with H	1	·5345	·7311	·1297	4322	·1921	·0324	74890	·2039

the simplicity of this result depends on two things. First, the vibration must be expressed by a Laplace's function of the order 2; for a different order the power of D would have been different; and this is just one of the points respecting which we cannot infer what would be true of a bell of the ordinary shape from what we have proved for a sphere. Secondly, the radius must be sufficiently small, or the pitch sufficiently low, to make q small; at the other extremity of the scale, in which c is supposed to be very large, or λ very small, Q varies nearly as $D^{\frac{1}{2}}$ instead of $D^{\frac{5}{2}}$, whatever be the order of the Laplace's function. Hence no simple relation can be expected between the numbers furnished by experiment and the numerical constants of the gas in such experiments as those of M. Perolle*, in which the same bell was rung in succession in different gases.

B. *Solution of the Problem in the case of a Vibrating Cylinder.*

I will here suppose the motion to be in two dimensions only. In the case of a vibrating string, which I have mainly in view, it is true that the amplitude of excursion of the string varies sensibly on proceeding even a moderate distance along it, and that the propagation of the sound-wave produced by no means takes place in two dimensions only. But the question how far a sound-wave is produced at all, and how far the displacement of the gas by the cylinder merely produces a local motion to and fro, is decided by what takes place in the immediate neighbourhood of the string, such as within a distance of a few diameters; and though the sound-wave, when once produced, in its subsequent progress diverges in three dimensions, the same takes place with the hypothetical sound-wave which would be produced if lateral motion were prevented, so that the comparison which it is the object of the investigation to institute is not affected thereby.

Assuming, then, the motion to be in two dimensions, and referring the fluid to polar coordinates, r, θ, r being measured from the axis of the undisturbed cylinder, we shall have for the fundamental equation derived from (1)

$$\frac{d^2\phi}{dt^2} = a^2\left\{\frac{d^2\phi}{dr^2} + \frac{1}{r}\frac{d\phi}{dr} + \frac{1}{r^2}\frac{d^2\phi}{d\theta^2}\right\} \text{ ....(14)};$$

* *Mémoires de l'Académie des Sciences de Turin*, t. III, (1786–7); *Mém. des Correspondans*, p. 1.

and if u', v' be the components of the velocity along and perpendicular to the radius vector,

$$u' = \frac{d\phi}{dr}, \qquad v' = \frac{1}{r}\frac{d\phi}{d\theta}.$$

If c be the radius of the cylinder, and V the normal component of the velocity of the surface of the cylinder, we must have

$$\frac{d\phi}{dr} = V, \text{ when } r = c.$$

As before, I will suppose the motion of the cylinder, and consequently of the fluid, to be regularly periodic, but instead of using circular functions directly I will employ the imaginary exponential e^{imat}, i denoting as before $\sqrt{-1}$, and will put accordingly $V = e^{imat}\,U$, U being a given function of θ, and $\phi = \psi e^{imat}$. For a given value of r, ψ may by a known theorem be developed in a series of sines and cosines of θ and its multiples, and therefore for general values of r can be so developed, the coefficients being functions of r. If ψ_n be the coefficient of $\cos n\theta$ or $\sin n\theta$, we find

$$\frac{d^2\psi_n}{dr^2} + \frac{1}{r}\frac{d\psi_n}{dr} - \frac{n^2}{r^2}\psi_n + m^2\psi_n = 0 \;\;\ldots\ldots\ldots\ldots(15).$$

If we suppose the normal velocity of the surface of the cylinder to vary in a given manner from one generating line to another, so that U is a given function of θ, we may expand U in a series of the form

$$U = U_0 + U_1 \cos\theta + U_2 \cos 2\theta + \ldots$$
$$+ U_1' \sin\theta + U_2' \sin 2\theta + \ldots$$

On applying now the equation of condition which has to be satisfied at the surface of the cylinder, we see that a term $U_n \cos n\theta$ or $U_n' \sin n\theta$ of the nth order in the expression for U will introduce a function ψ_n of the same order in the general expression for ϕ. Now the only case of interest relating to an infinite cylinder is that of a vibrating string, in which the cylinder moves as a whole. The vibration may be regarded as compounded of the vibrations in any two rectangular planes passing through the axis, the phases of the component vibrations, it may be, being different. These component vibrations may be treated separately, and thus it will suffice to suppose the vibration confined to one plane, which we may take to be that from which θ is measured. We shall accordingly have

$$U = U_1 \cos\theta,$$

U_1 being a given constant, and the only function ψ_n which will appear in the general expression for ϕ will be that of the order 1. Besides this we shall have to investigate, for the sake of comparison, an ideal vibration in which the cylinder alternately contracts and expands in all directions alike, and for which accordingly U is a constant U_0. Hence the equation (15) need only be considered for the values 0 and 1 of n.

For general values of n the equation (15) is easily integrated in the form of infinite series according to ascending powers of r. The result is

$$\left.\begin{aligned}\psi_n = Ar^n &\left\{1 - \frac{m^2r^2}{2(2+2n)} + \frac{m^4r^4}{2.4(2+2n)(4+2n)} - \ldots\right\}\\ + Br^{-n} &\left\{1 - \frac{m^2r^2}{2(2-2n)} + \frac{m^4r^4}{2.4(2-2n)(4-2n)} - \ldots\right\}\end{aligned}\right\} \ldots(16).$$

When n is any integer the integral as it stands becomes illusory; but the complete integral, which in this case assumes a special form, is readily obtained as a limiting case of the complete integral for general values of n.

The series in (16) are convergent for any value of r however great, but they give us no information of what becomes of the functions for very large values of r.

When r is very large, the equation (15) becomes approximately

$$\frac{d^2\psi_n}{dr^2} + m^2\psi_n = 0,$$

the integral of which is $\psi_n = Re^{-imr} + R'e^{imr}$, where R and R' are constant. This suggests putting the complete integral of (15) under the same form, R and R' being now functions of r, which, when r is large, vary but slowly, that is, remain nearly constant when r is altered by only a small multiple of λ. Assuming for R and R' series of the form $Ar^\alpha + Br^\beta + Cr^\gamma \ldots$, where α, β, $\gamma \ldots$ are in decreasing order algebraically, and determining the indices and coefficients so as to satisfy (15), we get for another form of the complete integral

$$\left.\begin{aligned}\psi_n = C(imr)^{-\frac{1}{2}} e^{-imr} &\left\{1 - \frac{1^2 - 4n^2}{1.8imr} + \frac{(1^2-4n^2)(3^2-4n^2)}{1.2(8imr)^2} \right. \\ &\left. - \frac{(1^2-4n^2)(3^2-4n^2)(5^2-4n^2)}{1.2.3(8imr)^3} + \dots\right\} \\ + D(imr)^{-\frac{1}{2}} e^{imr} &\left\{1 + \frac{(1^2 - 4n^2)}{1.8imr} + \frac{(1^2-4n^2)(3^2-4n^2)}{1.2(8imr)^2} \right. \\ &\left. + \frac{(1^2-4n^2)(3^2-4n^2)(5^2-4n^2)}{1.2.3(8imr)^3} + \dots\right\}\end{aligned}\right\} \dots(17).$$

These series, though ultimately divergent, begin by converging rapidly when r is large, and may be employed with great advantage when r is large, if we confine ourselves to the converging part. Moreover we have at once $D=0$ as the condition to be satisfied at a great distance from the cylinder. If mc were large we might employ the second form of integral in satisfying the condition at the surface of the cylinder, and the problem would present no further difficulty. But practically in the case of vibrating strings mc is a very small fraction; the series (16) are rapidly convergent, and the series (17) cannot be employed. To complete the solution of the problem therefore it is essential to express the constants A and B in terms of C and D, or at any rate to find the relation between A and B imposed by the condition $D=0$.

This may be effected by means of the complete integral of (15) expressed in the form of a definite integral. For $n=0$ we know that

$$\psi_0 = \int_0^{\frac{\pi}{2}} \{E + F\log(r\sin^2\zeta)\}\cos(mr\cos\zeta)\,d\zeta \;\dots\dots(18)$$

is a third form of the integral of (15). It is not difficult to deduce from this the integral of (15) in a similar form for any integral value of n. Assuming

$$\psi_n = r^\alpha \int r^\beta \chi_n dr,$$

and substituting in (15), we have

$$r^{\alpha+\beta}\frac{d\chi_n}{dr} + (2\alpha+\beta+1)\,r^{\alpha+\beta-1}\chi_n + (\alpha^2 - n^2)\,r^{\alpha-2}\int r^\beta\chi_n dr + m^2 r^\alpha \int r^\beta \chi_n dr = 0.$$

Assume

$$\alpha^2 - n^2 = 0 \quad \text{..........................(19),}$$

divide the equation by r^α, differentiate with respect to r, and then divide by r^β. The result is

$$\frac{d^2\chi_n}{dr^2} + \frac{2\alpha + 2\beta + 1}{r}\frac{d\chi_n}{dr} + (2\alpha + \beta + 1)(\beta - 1)\frac{\chi_n}{r^2} + m^2\chi_n = 0.$$

This equation will be of the same form as (15) provided

$$\alpha + \beta = 0,$$

which reduces the coefficient of the last term but one to $-(\alpha + 1)^2$. In order that this coefficient may be increased we must choose the positive root of (19), namely n, which I will suppose positive. Hence

$$\psi_n = r^n \int r^{-n}\chi_n dr \quad \text{...................... (20)}$$

gives

$$\frac{d^2\chi_n}{dr^2} + \frac{1}{r}\frac{d\chi_n}{dr} - \frac{(n+1)^2}{r^2}\chi_n + m^2\chi_n = 0,$$

the same equation as that for the determination of ψ_{n+1}. Hence expressing χ_n in terms of ψ_n from (20), writing $n - 1$ for n, and replacing χ_{n-1} by ψ_n, we have

$$\psi_n = r^{n-1}\frac{d}{dr} r^{-(n-1)} \psi_{n-1},$$

a formula of reduction which when n is integral serves to express ψ_n in terms of ψ_0. We have

$$\psi_n = r^n \left(\frac{1}{r}\frac{d}{dr}\right)^n \psi_0 \quad \text{......................(21),}$$

an equation which when applied to (18) gives the complete integral of (15) for integral values of n in the form of a definite integral.

Let us attend now more particularly to the case of $n = 0$. The equation (16) is of the form $\psi_n = Af(n) + Bf(-n)$, $f(n)$ containing r as well as n. Expanding by Maclaurin's Theorem, we have

$$\psi_n = (A + B)f(0) + (A - B)f'(0)n + (A + B)f''(0)\frac{n^2}{1.2} + \ldots.$$

Writing A for $A+B$, $n^{-1}B$ for $A-B$, and then making n vanish, we have

$$\psi_0 = Af(0) + Bf'(0),$$

or

$$\left.\begin{aligned}\psi_0 = (A + B\log r)\left(1 - \frac{m^2r^2}{2^2} + \frac{m^4r^4}{2^2.4^2} - \frac{m^6r^6}{2^2.4^2.6^2} + \ldots\right)\\ + B\left(\frac{m^2r^2}{2^2}S_1 - \frac{m^4r^4}{2^2.4^2}S_2 + \frac{m^6r^6}{2^2.4^2.6^2}S_3 - \ldots\right)\end{aligned}\right\} \ldots(22),$$

where

$$S_n = 1^{-1} + 2^{-1} + 3^{-1} + \ldots + n^{-1}.$$

The integral in the form (17) assumes no peculiar shape when n is integral, and we have at once

$$\left.\begin{aligned}\psi_0 = C(imr)^{-\frac{1}{2}}e^{-imr}\left\{1 - \frac{1^2}{1.8imr} + \frac{1^2.3^2}{1.2(8imr)^2} - \frac{1^2.3^2.5^2}{1.2.3(8imr)^3} + \ldots\right\}\\ + D(imr)^{-\frac{1}{2}}e^{imr}\left\{1 + \frac{1^2}{1.8imr} + \frac{1^2.3^2}{1.2(8imr)^2} - \frac{1^2.3^2.5^2}{1.2.3(8imr)^3} + \ldots\right\}\end{aligned}\right\} \ldots\ldots\ldots(23).$$

I have explained at length the mode of dealing with such functions, and especially of connecting the arbitrary constants in the ascending and descending series, in two papers published in the *Transactions of the Cambridge Philosophical Society**, in the second of which the connexion of the constants is worked out in this very example. To these I will refer, merely observing that while it is perfectly easy to connect A, B with E, F, the connexion of C, D with E, F involves some extremely curious points of analysis. The result of eliminating E, F between the two equations connecting A, B with E, F and the two connecting C, D with E, F is given, except as to notation, in the two equations (41) of my second paper. To render the [present] notation identical with that of the former paper, it will be sufficient to write

$$A - B\log(im) + B\log(imr) \text{ for } A + B\log r,$$

and x for imr. The equations referred to may be simplified by

* "On the Numerical Calculation of a Class of Definite Integrals and Infinite Series," Vol. IX, p. 166 [*ante*, Vol. II, p. 329], and "On the Discontinuity of Arbitrary Constants which appear in Divergent Developments," Vol. X, p. 105 [*supra*, p. 77: see pp. 100—103]. A supplement to the latter paper has recently been read before the Cambridge Philosophical Society [*supra*, p. 283: see pp. 294—5 and 298].

the introduction of Euler's constant γ, the value of which is ·57721566 etc., since it is known that

$$\pi^{-\frac{1}{2}}\Gamma'(\tfrac{1}{2}) + \log 4 + \gamma = 0,$$

$\Gamma'(x)$ denoting the derivative of the function $\Gamma(x)$. Putting

$$A - B \log im = A',$$

we have [here] by equations (41) of the second paper referred to

$$C = (2\pi)^{-\frac{1}{2}} \{iA' + [(\log 2 - \gamma) i - \pi] B\} \text{.........(24)},$$

$$D = (2\pi)^{-\frac{1}{2}} \{A' + (\log 2 - \gamma) B\} \text{ ....(25)},$$

i being written for $\sqrt{-1}$. It is shown in that paper that these values of C, D hold good when the amplitude of the imaginary variable x lies between the limits 0 and π, or that of r (supposed to be imaginary) between the limits $-\frac{1}{2}\pi$ and $\frac{1}{2}\pi$, but in crossing either of these limits one or other of the constants C, D is changed. In the investigation of the present paper r is of course real.

We have now

$$A' = A - B \log im = (\gamma - \log 2) B$$

for the relation between A and B arising from the condition that the motion is propagated outwards from the cylinder; and substituting in (22), we have for the value of ψ_0 subject to this condition

$$\left.\begin{aligned} \psi_0 = B\left(\gamma + \log \frac{imr}{2}\right)\left(1 - \frac{m^2r^2}{2^2} + \frac{m^4r^4}{2^2.4^2} - \dots\right) \\ + B\left(\frac{m^2r^2}{2^2} S_1 - \frac{m^4r^4}{2^2.4^2} S_2 + \dots\right) \end{aligned}\right\} \text{...(26)};$$

or expressed by means of the descending series,

$$\psi_0 = -B\left(\frac{\pi}{2imr}\right)^{\frac{1}{2}} e^{-imr} \left\{1 - \frac{1^2}{1.8imr} + \frac{1^2.3^2}{1.2(8imr)^2} - \frac{1^2.3^2.5^2}{1.2.3(8imr)^3} + \dots\right\} \text{...(27)}.$$

We have from (21)

$$\psi_1 = \frac{d\psi_0}{dr},$$

from which the complete integral of (15) for $n=1$ may be got from that for $n=0$. In the form (17) of the integral the parts arising from differentiation of the parts containing e^{-imr} and e^{imr} respectively will contain those same exponentials, and therefore the complete integral of (15) for $n=1$, subject to the condition that the part containing e^{imr} shall disappear, will be got by differentiating the complete integral for $n=0$ subject to that same condition. The *form* of the integral in the shape of a descending series is given at once by (17). Hence we get by differentiating (26) and (27), and at the same time changing the arbitrary constant by writing $B_1 m^{-1}$ for B,

$$\left.\begin{aligned}\psi_1 = \frac{B_1}{mr}\left\{1 - \frac{m^2r^2}{2^2} + \frac{m^4r^4}{2^2.4^2} - \dots\right\}\\ - B_1\left(\gamma + \log\frac{imr}{2}\right)\left(\frac{mr}{2} - \frac{m^3r^3}{2^2.4} + \frac{m^5r^5}{2^2.4^2.6} - \dots\right)\\ + B_1\left(\frac{mr}{2}S_1 - \frac{m^3r^3}{2^2.4}S_2 + \frac{m^5r^5}{2^2.4^2.6}S_3 - \dots\right)\end{aligned}\right\}\dots(28),$$

$$\psi_1 = B_1\left(\frac{\pi i}{2mr}\right)^{\frac{1}{2}} e^{-imr}\left\{1 - \frac{-1.3}{1.8mr} + \frac{-1.1.3.5}{1.2\,(8mr)^2} - \frac{-1.1.3^2.5.7}{1.2.3\,(8mr)^3} + \dots\right\}\dots(29).$$

To determine the arbitrary constants B_1 and B, the first belonging to the actual motion, the second to the motion which would take place if the fluid were confined by an infinite number of planes passing through the axis, we must have, as before, for $r=c$,

$$\frac{d\psi_0}{dr} = U_1, \qquad \frac{d\psi_1}{dr} = U_1,$$

whence

$$\left.\begin{aligned}\frac{cU_1}{B} = 1 - \frac{m^2c^2}{2^2} + \frac{m^4c^4}{2^2.4^2} - \dots - \left(\gamma + \log\frac{mc}{2} + i\frac{\pi}{2}\right)\left(\frac{m^2c^2}{2} - \frac{m^4c^4}{2^2.4} + \dots\right)\\ + \frac{m^2c^2}{2}S_1 - \frac{m^4c^4}{2^2.4}S_2 + \frac{m^6c^6}{2^2.4^2.6}S_3 - \dots\\ = F_0(mc) + i\frac{\pi}{2}f_0(mc),\ \text{suppose}\end{aligned}\right\}\dots\dots\dots(30);$$

$$\left.\begin{aligned}\frac{cU_1}{B_1} &= -\frac{1}{mc} - \frac{3mc}{2^2} + \frac{7m^3c^3}{2^2.4^2} - \frac{11m^5c^5}{2^2.4^2.6^2} + \dots \\ &\quad - \left(\gamma + \log\frac{mc}{2} + i\frac{\pi}{2}\right)\left(\frac{mc}{2} - \frac{3m^3c^3}{2^2.4} + \dots\right) \\ &\quad - \frac{mc}{2}S_1 - \frac{3m^3c^3}{2^2.4}S_2 + \frac{5m^5c^5}{2^2.4^2.6}S_3 - \dots \\ &= \frac{1}{mc}\left\{F_1(mc) + i\frac{\pi}{2}f_1(mc)\right\}, \text{ suppose}\end{aligned}\right\} \quad \dots\dots\dots(31).$$

If I be the ratio of the intensities at a distance in the supposed and in the actual case, we see from (30) and (31) that I will be equal to the ratio of the squares of the moduli of B and B_1, and we shall therefore have

$$I = \frac{\{4F_1(mc)\}^2 + \pi^2\{f_1(mc)\}^2}{m^2c^2[4\{F_0(mc)\}^2 + \pi^2\{f_0(mc)\}^2]} \quad \dots\dots\dots(32).$$

For a piano string corresponding to the middle C, c may be about ·02 inch, and λ is about 25 inches. This gives $mc = {}$·005027. For such small values of mc, I does not sensibly differ from $(mc)^{-2}$, which in the present case is 39571, so that the sound is nearly 40000 times weaker than it would have been if the motion of the particles of air had taken place in planes passing through the axis of the string. This shows the vital importance of sounding-boards in stringed instruments. Although the amplitude of vibration of the particles of the sounding-board is extremely small compared with that of the particles of the string, yet as it presents a broad surface to the air it is able to excite loud sonorous vibrations, whereas were the string supported in an absolutely rigid manner, the vibrations which it could excite directly in the air would be so small as to be almost or altogether inaudible.

I may here mention a phenomenon which fell under my notice, and which is readily explained by the principles laid down in this paper. As I was walking one windy day on a road near Cambridge, on the other side of which ran a line of telegraph, my attention was attracted by a peculiar sound of extremely high pitch, which seemed to come from the opposite side of the road. On going over to ascertain the cause, I found that it came directly through air from the telegraph wires. On standing near a telegraph post, the ordinary comparatively bass sound with which we are so familiar was heard, appearing to emanate from the post. On receding from the post the bass sound became feebler, and midway between two posts was quite inaudible. Nothing was then heard but the peculiar high-pitched sound, which appeared to emanate from the wires overhead. It had a peculiar metallic ring about it which the ear distinguished from the whistling of the wind in the twigs of a bush. Although the telegraph ran for miles, it was only at one spot that the peculiar sound was noticed, and even there only in certain states of the wind. The wires seemed to be less curved than usual at the place in question, from which it may be inferred that they were there subject to an unusually great tension.

The explanation of the phenomenon is easy after what precedes. The wires were thrown into vibration by the wind, and a number of different vibrations, having different periodic times, coexisted. As regards the vibrations of comparatively long period, the air around the wires behaved nearly like an incompressible fluid, and no sonorous vibrations of sensible amount were produced. These vibrations of the wires were, however, communicated to the posts, which being broad acted as sounding-boards, and thus sonorous vibrations of corresponding period were indirectly excited in the air. But as regards the vibrations of extremely short periodic time, the wires in spite of their narrowness were able by acting directly on the air to produce condensations and rarefactions of sensible amount.

The diameter of the telegraph wire was about ·166 inch; and if we take the C below the middle C of a piano for the representative of the pitch of the lower note, and a note five octaves higher for that of the higher, we have in the first case $\lambda = 50$ inches nearly, and in the second $\lambda = 50 \times 2^{-5}$, giving in the former case $mc = {\cdot}01043$, and in the latter $mc = {\cdot}3338$. The former of these values is so small that we may take $I = (mc)^{-2}$; in the latter case the formula (32) gives for I a value a little less than $(mc)^{-2}$. We find in the two cases $I = 9192$ and $I = 7{\cdot}202$ respectively, so that in the former case the sound is more than 9000 times feebler than that corresponding to the amplitude of vibration of the wire on the supposition of the absence of lateral motion, whereas in the latter case the actual intensity is nearly one-seventh of the full intensity corresponding to the amplitude.

The increase of sound produced by the stoppage of lateral motion may be prettily exhibited by a very simple experiment. Take a tuning-fork, and holding it in the fingers after it has been made to vibrate, place a sheet of paper or the blade of a broad knife with its edge parallel to the axis of the fork, and as near to the fork as conveniently may be without touching. If the plane of the obstacle coincide with either of the planes of symmetry of the fork, as represented in section at A or B, no effect is produced; but if it be placed in an intermediate position, such as C, the sound becomes much stronger.

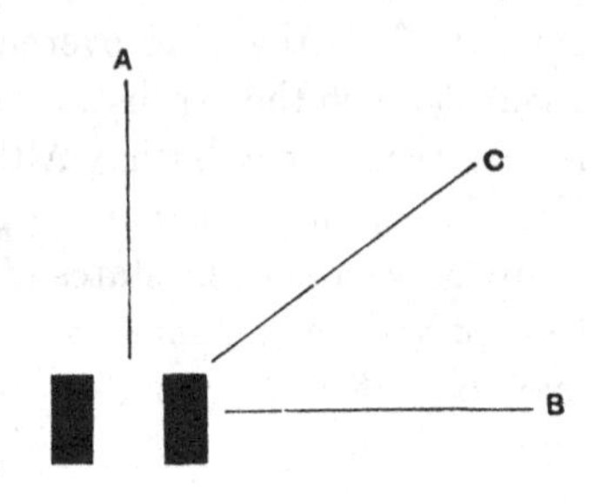

ACCOUNT OF OBSERVATIONS OF THE TOTAL ECLIPSE OF THE SUN....By J. POPE HENNESSY. Note added by Prof. STOKES.

[From the *Proceedings of the Royal Society*, XVII, 1868, pp. 88-89.]

The phenomenon of the sun's crescent reflected on to the disk of the moon would seem to have been something accidental, perhaps (if seen by the writer only) a mere ghost, depending on a double reflection between the glasses of his instrument. The figure represents the "reflected" image as in the same position as the crescent itself, not reversed, indicating either a refraction or a double reflection.

The slender beams of light or shade shooting out from the horns of the crescent would seem to admit of easy explanation, supposing them to have been of the nature of sunbeams, depending upon the illumination of the atmosphere of the earth by the sun's rays. The perfect shadow, or *umbra*, would be a cone circumscribing both sun and moon, and having its vertex far below the observer's horizon. Within this cone there would be no illumination of the atmosphere, but outside it a portion of the sun's rays would be scattered in their progress through the air, giving rise to a faint illumination. When the total phase drew near, the nearer surface of the shadow would be at no great distance from the observer; the further surface would be remote. Attend in the first instance to some one plane passing through the eye and cutting the shadow transversely, and in this plane draw a straight line through the eye, touching the section of the cone which bounds the shadow; and then imagine other lines drawn through the eye a little inside and outside this. In the former case the greater part of the line, while it lay within the lower regions of the atmosphere, would be in shadow, the only part in sunshine being that reaching from the eye to the nearer surface of the shadow; but in the latter case the line would be in sunshine all along. In the direction of the former line, therefore, there would be but little illumination arising from scattered light, while in the direction of the latter the illumination would, comparatively speaking, be considerable. In crossing the tangent there would be a rapid change of illumination. Now pass on to three dimensions. Instead of a tangent line we shall have a tangent plane, and there will of course be two such planes, touching the two sides of the cone respectively. Each of these will be projected on the visual sphere into a great circle, a common tangent to the two small circles, which are the projections of the sun and moon. In crossing either of these there will be a rapid change of illumination (feeble though it be at best) which will be noticed. According as the observer mentally regards darkness as the rule and illumination as the feature, or illumination as the rule and darkness as the feature, he will describe what he sees as a *beam* or a *shadow*. The direction of these beams or shadows given by theory, as just explained, agrees very well with the drawing sent by Governor Hennessy, which does not represent the left-hand beam so distinctly divided as it appears in the woodcut.

ON A CERTAIN REACTION OF QUININE.

[From the *Journal of the Chemical Society*, *May*, 1869: read *Mar.* 18.]

IN the course of two papers on optical subjects, published in the *Philosophical Transactions*, I have mentioned a peculiar reaction of quinine having relation to its fluorescence*. About that time I followed out the subject further, and obtained results which were interesting to myself, especially in relation to a classification of acids which they seemed to afford. Not being a chemist, I did not venture to lay the results before the chemical world. I have, however, recently been encouraged by a chemical friend to think that a further statement of the results might prove of some interest to chemists.

The reaction is best observed by diffused daylight entering a darkened room† through a hole in the shutter, which may be four or five inches square, and which is covered with a deep violet glass, coloured by manganese‡. In front of the hole is

* *Phil. Trans.* for 1852, p. 541, and for 1853, p. 394. [*Ante*, Vol. III, p. 267, Vol. IV, p. 1.]

† In default of a darkened room, a common box, such as an old packing-case, may be readily altered so as to answer very well. The box is sawn obliquely across, the cutting plane being parallel to one edge, as indicated by the figure, which denotes a vertical section. The aperture thus made is covered by a board nailed on, containing a hole, *H*, destined to be covered by the glass plate, which is kept from slipping down by a small ledge, *L*. A portion of the upper covering of the box at *E* is removed to allow the observer to see and manipulate. In observing, the box is placed near a window, with its slant side turned towards the light; the hole is covered with its glass; the object is placed at *O*; and the observer looks in through *E*, covering his head with a dark cloth, to exclude stray light.

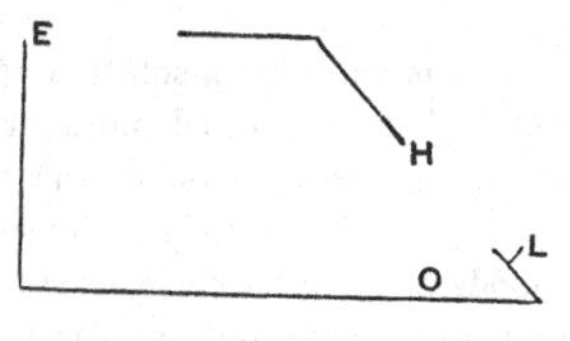

‡ Flint glasses answer best, the colour given by manganese to crown glass being generally somewhat brownish. I have, however, seen one specimen of crown glass coloured by manganese, the colour of which was as fine a purple as that of the flint glasses.

placed a white porcelain tablet, or else one of the porcelain slabs with shallow depressions used for colour tests. A solution of quinine in very weak alcohol is strong enough for the observations, or else very minute fragments may be used. In some cases, as for example with valerianic or benzoic acid, the presence of alcohol interferes with the reaction.

It will conduce to brevity and clearness to describe in the first instance, in a little detail, the phenomena exhibited by two particular acids, say sulphuric and hydrochloric.

Let a series of drops of the quinine solution be deposited on the porcelain. If one of these be touched by a rod dipped in dilute sulphuric acid, the beautiful fluorescence of the quinine is instantly developed. If another drop be similarly touched by a rod moistened with dilute hydrochloric acid, no apparent effect is produced*. Nor is this all. If a little hydrochloric acid be introduced by a moistened rod into the fluorescing drop, the fluorescence is immediately destroyed. If a little of the sulphuric acid be introduced into the drop containing only hydrochloric acid, no effect is produced†.

If a series of drops of a solution of quinine in dilute sulphuric acid be deposited, and a little solution of chloride of potassium, sodium, or ammonium be added, the fluorescence is immediately destroyed. The action of sulphate of potassium, etc., on a solution of quinine in water acidulated, whether with hydrochloric or sulphuric acid, is in each case merely negative‡.

Now, on trying a variety of acids, I found that with hardly an exception, unless when the acid character of the acid was only

* It is true that a solution of quinine in dilute hydrochloric acid is fluorescent, and with concentrated sunlight, or with sunlight uncondensed, but analysed by absorption or dispersion, the fluorescence comes out strongly. It is, however, notably inferior to that produced by sulphuric acid; and for our immediate object a mode of observation in which it hardly, if at all, appears, is even better than one adapted to bring out comparatively feeble degrees of fluorescence. When I speak in the text of fluorescence being *destroyed*, the expression must be understood in this qualified sense.

† It must be understood that I am not here dealing with concentrated acids, nor with any very great preponderance of one kind over another. I suppose all the solutions to be dilute, and the quantity of acid employed, of whatever kind, to be many times that merely required to combine with the quinine.

‡ See the preceding note.

indistinct, the acids ranged themselves with perfect definiteness into two classes, which I will call class A and class B. Those of class A developed fluorescence in a solution of quinine in water just like sulphuric acid, the amount of fluorescence being comparable with that produced with sulphuric acid, and the tint the same. Those of class B not only did not produce it, but destroyed it when produced by acids of the class A. This destruction is produced by the alkaline salts of the acids, as well as by the free acids themselves, and thus we are able to classify acids without having specimens of the free acids at hand.

In the following lists, those acids which were tried only indirectly, by means of one or more of their alkaline salts, are distinguished by an asterisk:—

Class A.	Class B.
Acetic	Hydriodic*
Arsenic	Hydrobromic*
Benzoic	Hydrochloric
Chloric	Hydroferrocyanic*
Citric	Hydropalladiocyanic*
Formic	Hydroplatinocyanic*
Hyposulphuric*	Hydrosulphocyanic*
Iodic	Hyposulphurous*
Malic	
Nitric	
Oxalic	
Perchloric	
Phosphoric	
Silico-fluoric	
Succinic	
Sulphuric	
Tartaric	
Valerianic	

Unless a quinine solution be sufficiently dilute, when alcohol is used, iodic acid produces a precipitate. In what precedes, it must be understood that I refer in all cases to solutions. The character of the fluorescence of the salts of quinine in the solid state varies from salt to salt. The solid iodate obtained by precipitation is strongly fluorescent, with a blue considerably deeper than that of the solutions.

It is not in all cases possible to try all the reactions stated to belong to the acids above mentioned. Thus, in the case of iodic acid, the solution cannot be tested by ferrocyanide of potassium, which instantly decomposes the iodic acid. But the strong fluorescence of the iodic solution, and the immediate destruction of the fluorescence by chloride of sodium, etc., alone suffice to show definitely to which class iodic acid belongs.

The absorption of the fluorogenic* rays by the yellow ferrocyanide of potassium, would itself alone account for the apparent destruction of the fluorescence *if the salt were present in sufficient quantity.* Actually, however, the quantity which suffices to destroy the fluorescence is much less than what would be required to prevent its exhibition by the absorption either of the fluorogenic or of the fluorescent rays, or of both. When the experiment is properly tried, there cannot be a moment's hesitation that the removal of the fluorescence is a true chemical reaction, and not a mere optical effect. This may be further proved by spreading a comparatively large quantity of the ferrocyanide solution in the form of a broad drop on glass, and holding it immediately over the gleaming drop of the quinine solution, when, though the fluorogenic rays entering, and the fluorescent rays leaving the drop, have both to pass through the whole thickness of the absorbing solution, the fluorescence observed is only somewhat reduced. The absorption of the fluorescent rays in this case goes indeed for little; it is the absorption of the fluorogenic rays that we have to consider. That there is a real reaction, and not a mere optical effect, I have further proved by experiments in a pure spectrum, which it would take too long to describe.

Hyposulphurous acid is, it is true, rather easily decomposed, but the destruction of fluorescence by hyposulphite of soda is quite independent of this circumstance. It takes place, for

* By this term I merely mean rays considered in their capacity of producing fluorescence: the introduction of such a term prevents circumlocution. It is convenient also to have a name for the rays emitted by a fluorescent body. If these be called, as they are a little further on, *fluorescent*, no confusion can practically result, though the term has of course a totally different signification as applied to the rays or to the body emitting them.

instance, at once on introducing a very dilute solution of hyposulphite of soda into a drop in which the fluorescence had been excited by very dilute citric or acetic acid.

After these remarks, which were necessary to prevent possible misconception, we may return to our lists. A glance at these lists shows that the classification made by the quinine reaction agrees almost exactly with the old distinction of ox-acids and hydracids. There is, however, one acid, the hyposulphurous, which in the quinine reaction ranges itself with perfect definiteness in class B, but which is not, I believe, usually ranked by chemists with the hydracids, except in so far as acids in general have been regarded from this point of view. This led me to seek whether there might not be other analogies between hyposulphurous acid and the hydracids. I have noticed the two following:—

1. It is known that a solution of chloride of mercury reddens litmus, but the blue colour is restored by an alkaline chloride, though itself neutral to colour tests. Now the very same effect is produced by hyposulphite of soda. This salt and chloride of mercury very readily decompose each other; but if very dilute solutions be used, the solution of chloride of mercury having been coloured by a little litmus, it is easy to observe that the *first* effect of the introduction of the hyposulphite, an effect which takes place immediately, prior to the formation of any precipitate, is to restore the blue colour.

2. It is known that cyanide of mercury is hardly decomposed by ox-acids, so as simply to displace the hydrocyanic acid, but readily by hydracids. Now, if a solution of hyposulphite of soda be added to one of cyanide of mercury, the smell of hydrocyanic acid is immediately perceived.

These circumstances bear out, as to hyposulphurous acid, the classification afforded by the quinine reaction. If there be a real difference of constitution between say sulphuric and hydrochloric acids, expressed in symbols by writing the former (according to the old equivalents) $SO_3 . HO$, and not $SO_4 . H$, and in words by calling it an ox-acid, the mere fact that the radical of hyposulphurous acid, regarded as a hydracid, contains oxygen, does not, of course, oblige us to regard it as an ox-acid. It is that difference of constitution, whatever it may be, which must decide, and if we may trust the quinine reaction, the isolation of S_2O_3, the analogue of chlorine, would seem to be less improbable than that of S_2O_2, the analogue of sulphuric anhydride.

With hydrocyanic and hydrofluoric acids the reaction seemed doubtful. These acids seemed to belong to class B as regards the feeble amount of fluorescence which they developed, but not to prevent the development of strong fluorescence by acetic, sulphuric, etc., acids. Ferridcyanide of potassium had certainly no such action as chloride of sodium, or even ferrocyanide of potassium, in destroying the fluorescence; but the deep colour of the salt prevented a satisfactory decision whether the acid really belonged to class A, or resembled hydrocyanic acid in its action on quinine.

The destruction of the fluorescence of a solution of quinine in a dilute ox-acid on the introduction of a hydracid or its salt, would seem to indicate that the quinine combined by preference with the hydracid. It seemed to me that it would be interesting to restore, if possible, the fluorescence, without precipitation, by the introduction of a substance which should have a preferential affinity for the hydracid. In the case of hydrochloric acid, this may be effected by a salt such as the sulphate or nitrate of the red oxide of mercury. In trying the experiment it is convenient to use only a little hydrochloric acid, or chloride of sodium, etc., otherwise the sparingly soluble chloride of mercury and quinine is precipitated, which, however, redissolves on the addition of more of the mercuric salt. That the restoration of fluorescence is not a mere effect of the acid introduced with the mercuric salt, may be proved by varying the experiment. Let quinine be dissolved with more nitric or sulphuric acid than would otherwise have been necessary; add a little hydrochloric acid so as barely to destroy the fluorescence, and then introduce a little precipitate of oxide of mercury, stirring it up. As the oxide dissolves the fluorescence returns.

Chloride of mercury does not impair the fluorescence of a solution of quinine in an ox-acid; if anything it sometimes seems slightly to increase it. Chloride of barium, strontium, calcium, magnesium, manganese, or zinc, acts like chloride of sodium.

The fluorescence destroyed by a chloride is in some measure restored by nitrate of cadmium.

When the fluorescence is destroyed by bromide or iodide of potassium, it may be restored by oxide of mercury, just as in the case of an alkaline chloride, and under the same conditions.

The precipitate of chloride of mercury and quinine, which is liable to be produced in trying the above reactions, is strongly fluorescent, with a blue which seems to be a trifle greener than that of the solutions. The corresponding precipitate which may be obtained with bromide of potassium, doubtless a bromide of mercury and quinine, is white, and shows a pretty strong orange fluorescence, a very unusual colour for the fluorescence of a white substance. The iodide is pale yellow, and not sensibly fluorescent, at least as examined by daylight with coloured glasses.

Explanation of a Dynamical Paradox.

[From the *Messenger of Mathematics*, New Series, I, 1872, pp. 1—3.]

The answer to the following question, proposed in the Smith's Prize Examination, 1871, is sent in compliance with a request from one of the Editors:

"*In a compound pendulum consisting of masses m, m' attached to strings of length l, l', in which of course the most general small motion in one plane consists of two harmonic vibrations superposed, if the upper mass m be very large compared with the under mass m', it is clear that one of the two periodic times (that corresponding to the mode of vibration in which m is nearly at rest) must be very nearly the same as in a simple pendulum of length l', and the other very nearly the same as in a simple pendulum of length l. By a continuous variation of l', the former may be made to pass continuously from less to greater than the latter, and therefore for some value of l' nearly equal to l the two must be equal. But when a system is in stable equilibrium (as is clearly the case here) the equation the roots of which give the times of vibration cannot have equal roots, for that would imply the transitional condition between stable and unstable.*

"*Point out precisely the fallacy which leads to the above contradiction.*"

The fallacy lies in the tacit assumption that it is *the same* root of the quadratic which determines the times of vibration that correspond throughout to the same approximate physical state, *i.e.* the state in which (not considering the special case in which l, l' are nearly equal) the upper mass is nearly at rest, or the two masses move through comparable spaces, as the case may be. Let T, T' be the two times of vibration (or half periods), τ, τ' the times of vibration of simple pendulums of lengths l, l'; and suppose m, m', l, l' to change continuously, yet so that l always remains distinctly greater than or distinctly less than l'; *i.e.* so that the ratio of $l \sim l'$ to l or l', though absolutely, it may be, small, remains finite while $m' : m$ may be taken as small as we please. Of the two T, T', let T be that which for one set of values of the constants m, m', l, l' is nearly equal to τ; then must the same root T remain throughout that which is nearly equal to τ; for it is obliged to be nearly equal to one of the two τ, τ' which do not become nearly equal to each other. But if we suppose l, l' to change continuously, so that l, from having been distinctly less, becomes distinctly greater than l', and if T be that root which for $l < l'$ is nearly equal to τ, since τ, τ' pass through equality, and T is merely known to be nearly equal to one of them, there is nothing to shew which of the two τ, τ' it is that T is nearly equal to when $l > l'$. By the general principle referred to in the second part of the question, we see that it must be τ'.

The same thing may of course be shewn by the direct solution of the problem. Putting n for π/T, we find by the usual methods

$$mll'n^4 - (m+m')(l+l')gn^2 + (m+m')g^2 = 0,$$

g being gravity, and the roots of this quadratic in n^2 are

$$n^2 = \frac{m+m'}{2m} g \left[\frac{1}{l} + \frac{1}{l'} \pm \sqrt{\left\{\left(\frac{1}{l} + \frac{1}{l'}\right)^2 - \frac{m}{m+m'}\frac{4}{ll'}\right\}}\right].$$

If m' be very small, and l, l' not very nearly equal, the radical becomes very nearly $1/l \sim 1/l'$; and denoting by n^2, n'^2 the roots corresponding to the signs $+$, $-$, respectively, we have very nearly

$$n^2 = \frac{g}{2}\left\{\frac{1}{l} + \frac{1}{l'} + \left(\frac{1}{l} \sim \frac{1}{l'}\right)\right\}, \quad n'^2 = \frac{g}{2}\left\{\frac{1}{l} + \frac{1}{l'} - \left(\frac{1}{l} \sim \frac{1}{l'}\right)\right\}.$$

If $l < l'$, we have

$$n^2 = g/l, \quad n'^2 = g/l';$$

but if $l > l'$,

$$n^2 = g/l', \quad n'^2 = g/l.$$

When l, l' are nearly equal, we can no longer distinguish the two harmonic vibrations by the character, that from one of them m is nearly at rest, while the small mass m' moves considerably, since the motion of m as compared with m' is comparable in the two. In fact, the harmonic vibrations, which, when l, l' are distinctly different, are characterized by the properties above referred to, have their properties interchanged when $l : l'$ passes through 1*.

[* The two roots found above obviously cannot be equal. The limitations to equality of roots being an indication of instability have been determined by Weierstrass and by Routh; cf. Thomson and Tait, *Nat. Phil.* ed. II, § 343.]

ON THE LAW OF EXTRAORDINARY REFRACTION IN ICELAND SPAR.

[From *Proceedings of Royal Society*, XX, pp. 443–4. Received *June* 20, 1872.]

IT is now some years since I carried out, in the case of Iceland spar, the method of examination of the law of refraction which I described in my report on Double Refraction, published in the *Report of the British Association* for the year 1862, p. 272*. A prism, approximately right-angled isosceles, was cut in such a direction as to admit of scrutiny, across the two acute angles, in directions of the wave-normal within the crystal comprising respectively inclinations of 90° and 45° to the axis. The directions of the cut faces were referred by reflection to the cleavage-planes, and thereby to the axis. The light observed was the bright D of a soda-flame.

The result obtained was, that Huygens's construction gives the true law of double refraction within the limits of errors of observation. The error, if any, could hardly exceed a unit in the *fourth* place of decimals of the index or reciprocal of the wave-velocity, the velocity in air being taken as unity. This result is sufficient *absolutely to disprove* the law resulting from the theory which makes double refraction depend on a difference of inertia in different directions†.

I intend to present to the Royal Society a detailed account of the observations; but in the mean time the publication of this preliminary notice of the result obtained may possibly be useful to those engaged in the theory of double refraction‡.

[* *Supra*, p. 187.] [† Cf. *supra*, p. 182.]

[‡ This work was followed up by an investigation by Glazebrook, *Phil. Trans.* CLXXI, 1880, pp. 421—449, with Prof. Stokes' apparatus, which verified the same degree of accuracy over a wide range of measurements, a limit being imposed by imperfections in the lenses employed, and by some measures by Abria and by Kohlrausch also of the same degree of accuracy. Finally C. S. Hastings, *American Journal of Science*, CXXXV, Jan. 1888, pp. 60—73, pushed the verification within two units in the sixth place of decimals in two measurements on Iceland spar. In the work of Abria, referred to in the next paper, the uncertainty was about one per cent.]

SUR L'EMPLOI DU PRISME DANS LA VÉRIFICATION DE LA LOI DE LA DOUBLE RÉFRACTION.

[From *Comptes Rendus*, Vol. LXXVII, *Nov.* 17, 1873, pp. 1150–1152.]

LA Communication de M. Abria (*Comptes rendus*, séance du 13 octobre, p. 814 de ce volume) me détermine à appeler l'attention de l'Académie sur une méthode que j'ai proposée pour le même objet dans un travail sur la double réfraction*, et que j'ai appliquée plus tard au spath calcaire†. Cette méthode me paraît plus facile, plus générale et plus exacte que celle de M. Abria.

Quand on veut mesurer l'indice de réfraction d'une substance ordinaire, on emploie le plus souvent la méthode de la déviation *minimum*. Mais il y a une autre méthode, aussi exacte et presque aussi facile, qui consiste à mesurer la déviation pour un azimut arbitraire du prisme, et en outre l'angle d'incidence ou l'angle d'émergence, suivant que le prisme demeure en repos quand on déplace la lunette, ou qu'il l'accompagne dans son mouvement. Cette méthode n'est pas nouvelle : elle a déjà été employée par M. Swan dans sa vérification de la loi de Snellius pour le rayon ordinaire du spath calcaire‡; mais on n'avait pas, à ma connaissance, indiqué le parti qu'on en pourrait tirer pour la recherche de la loi de la réfraction extraordinaire dans les cristaux. Le phénomène que l'on observe dans le cas d'un cristal est le même que dans le cas d'une substance ordinaire, avec cette seule différence que l'on obtient deux images au lieu d'une seule; on peut encore mesurer la déviation de chacune des deux images, et il ne s'agit que d'interpréter les résultats obtenus. Or, en s'appuyant sur la démonstration qu'a donnée Huyghens pour la

* *Report of the British Association* for 1862, Part I, p. 272. [*Supra*, p. 187.]

† *Proceedings of the Royal Society*, Vol. XX, p. 443 (20 juin 1872). [*Supra*, preceding page.]

‡ *Transactions of the Royal Society of Edinburgh*, Vol. XVI, p. 375.

réfraction en général, démonstration qui, fondée sur le seul principe de la coexistence des petits mouvements, n'exige aucune hypothèse sur la loi de variation des vitesses de propagation dans diverses directions, on démontre facilement que les deux quantités qui représentent pour une substance ordinaire, 1° l'angle de réfraction, 2° l'indice de réfraction, et qui se déduisent des données d'observations par un calcul très-facile, expriment pour un cristal, 1° l'inclinaison de l'onde réfractée à la surface d'incidence, onde qui est nécessairement perpendiculaire au plan d'incidence, 2° le rapport de la vitesse de propagation dans l'air à celle de l'onde réfractée. La direction ainsi déterminée par rapport aux deux faces du prisme est rapportée ensuite, par le calcul, à des directions fixes dans le cristal, l'orientation de chaque face artificielle ayant été déterminée au moyen de la réflexion, par rapport à des faces, soit naturelles, soit de clivage. On peut ainsi examiner un cristal dans une série de directions, au moyen d'un seul angle réfringent, et l'on peut faire tailler deux angles au moins sur un même bloc sans détruire les faces dont on a besoin pour la détermination de l'orientation des plans artificiels.

Je n'ai appliqué jusqu'ici cette méthode qu'au spath calcaire, cristal que j'ai choisi à cause de la facilité avec laquelle on peut s'en procurer de bons échantillons, et de l'énergie de sa double réfraction, qui devrait rendre plus sensibles les écarts par rapport à la loi d'Huyghens, s'il en existait. J'ai trouvé que cette loi représente la réfraction extraordinaire aussi exactement que la loi de Snellius représente la réfraction ordinaire.

L'erreur moyenne de quinze observations du rayon extraordinaire, faites dans des directions qui s'étendaient de 30 à 60 degrés environ de l'axe, et rapprochées de la formule déduite de la construction d'Huyghens en y introduisant les indices principaux, obtenus à 90 degrés de l'axe, ne s'élevait qu'à 0,00013 de l'indice, quantité qui est de l'ordre des erreurs accidentelles de mes observations, et qui correspond à $\frac{1}{1300}$ environ de la différence des indices principaux. L'erreur correspondante de déviation dans un prisme de 45 degrés est d'environ 25 secondes.

NOTICE OF THE RESEARCHES OF THE LATE REV. W. VERNON HARCOURT ON THE CONDITIONS OF TRANSPARENCY IN GLASS AND THE CONNEXION BETWEEN THE CHEMICAL CONSTITUTION AND OPTICAL PROPERTIES OF DIFFERENT GLASSES.

[From the *Report of the British Association for the Advancement of Science*, Edinburgh, 1871.]

THE preparation and optical properties of glasses of various compositions formed for nearly forty years a favourite subject of study with the late Mr Harcourt. Having commenced in 1834 some experiments on vitrifaction, with the object stated in the title of this notice, he was encouraged by a recommendation, which is printed in the 4th volume of the *Transactions of the British Association*, to pursue the subject further. A report on a gas-furnace, the construction of which formed a preliminary inquiry, in which was expended the pecuniary grant made by the Association for this research in 1836, is printed in the Report of the Association for 1844, but the results of the actual experiments on glass have never yet been published.

My own connexion with these researches commenced at the Meeting of the British Association at Cambridge in 1862, when Mr Harcourt placed in my hands some prisms formed of the glasses which he had prepared, to enable me to determine their character as to fluorescence, which was of interest from the circumstance that the composition of the glasses was known. I was led indirectly to observe the fixed lines of the spectra formed by means of them; and as I used sunlight, which he had not found it convenient to employ, I was enabled to see further into the red and violet than he had done, which was favourable to a more accurate measure of the dispersive powers. This inquiry, being in furtherance of the original object of the experiments, seemed far more important than that as to fluorescence, and caused Mr Harcourt to resume his experiments with the liveliest interest, an interest which he kept up to the last. Indeed it was only a few days before his death that his last experiment was made. To show the extent of the research, I may mention that as many as 166 masses of glass were formed and cut into prisms, each mass doubtless in many cases involving several preliminary experiments, besides disks and masses for other purposes. Perhaps I may be permitted here to refer to what I said to this Section on a former occasion* as to the advantage of working in concert. I may certainly say for myself, and I think it will not be deemed at all derogatory to the memory of my esteemed friend and fellow-labourer if I say of him, that I do not think that either of us working singly could have obtained the results we arrived at by working together.

It is well known how difficult it is, especially on a small scale, to prepare homogeneous glass. Of the first group of prisms, 28 in number, 10 only were sufficiently good to show a few of the principal dark lines of the solar spectrum; the rest had to be examined by the bright lines in artificial sources of light. These prisms appeared to have been cut at random by the optician from the mass of glass supplied to him. Theory and observation alike showed that striæ interfere comparatively little with an accurate determination of refractive indices when they lie in planes perpendicular to the edge of the prism. Accordingly the prisms used in the rest of the research were formed from the glass mass that came out of the crucible by cutting two planes, passing through the same horizontal line a little below the surface, and inclined $22\frac{1}{2}°$ right and left of the vertical, and by polishing the enclosed wedge of 45°. In the central portion of the mass the striæ have a tendency to arrange themselves in nearly vertical lines, from the operation of currents of convection; and by cutting in the manner described, the most favourable direction of the striæ is secured for a good part of the prism.

This attention to the direction of cutting, combined no doubt with increased experience in the manufacture of glass, was attended with such good results that now it was quite the exception for a prism not to show the more conspicuous dark lines.

On account of the inconvenience of working with silicates, arising from the difficulty of fusion and the pasty character of the fused glasses, Mr Harcourt's experiments were chiefly carried on

* *Report of the British Association* for 1862, Trans. of Sect. p. 1. [Introductory remarks at the Cambridge meeting.]

with phosphates, combined in many cases with fluorides, and sometimes with borates, tungstates, molybdates, or titanates. The glasses formed involved the elements potassium, sodium, lithium, barium, strontium, calcium, glucinum, magnesium, aluminium, manganese, zinc, cadmium, tin, lead, thallium, bismuth, antimony, arsenic, tungsten, molybdenum, titanium, vanadium, nickel, chromium, uranium, phosphorus, fluorine, boron, sulphur.

A very interesting subject of inquiry presented itself collaterally with the original object, namely, to inquire whether glasses could be found which would achromatize each other so as to exhibit no secondary spectrum, or a single glass which would achromatize in that manner a combination of crown and flint.

This inquiry presented considerable difficulties. The dispersion of a medium is small compared with its refraction; and if the dispersive power be regarded as a small quantity of the first order, the irrationality between two media must be regarded as depending on small quantities of the second order. If striæ and imperfections of the kind present an obstacle to a very accurate determination of dispersive power, it will readily be understood that the errors of observation which they occasion go far to swallow up the small quantities on the observation of which the determination of irrationality depends. Accordingly, little success attended the attempts to draw conclusions as to irrationality from the direct observation of refractive indices; but by a particular method of compensation, in which the experimental prism was achromatized by a prism built up of slender prisms of crown and flint, I was enabled to draw trustworthy conclusions as to the character in this respect of those prisms which were sufficiently good to show a few of the principal dark lines of the solar spectrum*.

Theoretically *any* three different kinds of glass may be made to form a combination achromatic as to secondary as well as primary colour, but practically the character of dispersion is usually connected with its amount, in such a manner that the determinant of the system of three simple equations which must be satisfied is very small, and the curvatures of the three lenses required to form an achromatic combination are very great.

[* *Roy. Soc. Proc.* 1878, to be reprinted *infra.*]

For a long time little hope of a practical solution of the problem seemed to present itself, in consequence of the general prevalence of the approximate law referred to above. A prism containing molybdic acid was the first to give fair hopes of success. Mr Harcourt warmly entered into this subject, and prosecuted his experiments with unwearied zeal. The earlier molybdic glasses prepared were many of them rather deeply coloured, and most of them of a perishable nature. At last, after numerous experiments, molybdic glasses were obtained pretty free from colour and permanent. Titanium had not yet been tried, and about this time a glass containing titanic acid was prepared and cut into a prism. Titanic acid proved to be equal or superior to molybdic in its power of extending the blue end of the spectrum more than corresponds to the dispersive power of the glass; while in every other respect (freedom from colour, permanence of the glass, greater abundance of the element) it had a decided advantage; and a great variety of titanic glasses were prepared, cut into prisms, and measured. One of these led to the suspicion that boracic acid had an opposite effect, to test which Mr Harcourt formed some simple borates of lead, with varying proportions of boracic acid. These fully bore out the expectation; the terborate for instance, which in dispersive power nearly agrees with flint glass, agrees on the other hand, in the relative extension of the blue and red ends of the spectrum, with a combination of about one part, by volume, of flint glass with two of crown.

By combining a negative or concave lens of terborate of lead with positive lenses of crown and flint, or else a positive lens of titanic glass with negative lenses of crown and flint, or even with a negative of very low flint and a positive of crown, achromatic triple combinations free from secondary colour may be formed without encountering (at least in the case of the titanic glass) formidable curvatures; and by substituting at the same time a titanic glass for crown, and a borate of lead for flint, the curvatures may be a little further reduced.

There is no advantage in using three different kinds of glass rather than two to form a fully achromatic combination, except that the latter course might require the two kinds of glass to be made expressly, whereas with three we may employ for two the crown and flint of commerce. Enough titanium might, however be introduced into a glass to render it capable of being perfectly achromatized by Chance's "light flint."

In a triple objective the middle lens may be made to fit both the others, and be cemented. Terborate of lead, which is somewhat liable to tarnish, might thus be protected by being placed in the middle. Even if two kinds only of glass are used, it is desirable to divide the convex lens into two, for the sake of diminishing the curvatures. On calculating the curvatures so as to destroy spherical as well as chromatic aberration, and at the same time to make the adjacent surfaces fit, very suitable forms were obtained with the data furnished by Mr Harcourt's glasses.

After encountering great difficulties from striæ, Mr Harcourt at last succeeded in preparing disks of terborate of lead and of a titanic glass which are fairly homogeneous, and with which it is intended to attempt the construction of an actual objective which shall give images free from secondary colour, or nearly so.

This notice has extended to a greater length than I had intended, but it still gives only a meagre account of a research extending over so many years. It is my intention to draw up a full account for presentation to the scientific world in some other form. I have already said that the grant made to Mr Harcourt for these researches in 1836 has long since been expended, as was stated in his Report of 1844; but it was his wish, in recognition of that grant, that the first mention of the results he obtained should be made to the British Association; and I doubt not that the members will receive with satisfaction this mark of consideration, which they will connect with the memory of one to whom the Association as a body is so deeply indebted.

On the Principles of the Chemical Correction of Object-glasses.

[Lecture to the Photographic Society. From the *Photographic Journal*, *Feb.* 15, 1873.]

When I was invited by your Secretary to deliver a lecture before you, I thought of different subjects, and one occurred to me which I deemed might be of some interest to you. It is one about which I have thought a good deal; and I propose to entitle the lecture "On the Principles of the Chemical Correction of Object-glasses." I shall, however, leave for the present the chemical rays out of consideration, and deal only with the visual rays, because it is easier to explain the portion of the subject which relates to the chemical rays by reference to the visual—since you can more readily picture them to yourselves, and I can more readily refer to them by naming the colours. When once the principles of the mode of proceeding are understood with reference to the visual rays, it is perfectly simple to apply them to the chemical rays. As for the elementary principles of the subject, I shall not dwell upon them, as I presume they are already familiar to you. You well know that Newton discovered the compound nature of white light, which, on passing through a prism, is bent round and decomposed into different parts, producing different impressions of colour. It was already known that, on passing light through different substances, it was not bent round in the same, but in different degrees. Thus, if two kinds of glass, flint and crown, are formed into prisms of the same angle, and light be passed through them, the flint will bend the rays more than the crown. It is usually found that those substances which bend round the rays more than others, also separate the different colours one from another in a greater degree. Newton supposed that the separation of the colours was in all cases proportional to the bending round of the rays as a whole. He appears to have formed this idea from the result of an experiment in which a glass prism was compensated as to refraction by a hollow prism containing water, the angle of which being unsuitable, he dissolved in the water a little sugar of lead; the consequence was that these two media behaved very nearly alike; that is, they separated the different colours in apparently the same proportion as they bent round the light as a whole. Were this generally true, as Newton supposed, the construction of an achromatic telescope composed of two kinds of glass would be impossible,—a very curious example of a great man missing a great discovery through over hasty generalization. Mr Hall, of Worcestershire, was the first to discover Newton's mistake, and he applied his discovery to the construction of an achromatic telescope, and is said to have made several; but the subject was disregarded until the achromatic telescope was rediscovered by Dollond. The possibility of forming it depends upon the amount of separation of the colours produced in the bending round of the light as a whole not bearing in all substances the same proportion to the bending, as Newton supposed. If I take a prism of crown glass and a prism of flint glass, and bend the light passing through the crown glass so as to oppose the flint glass, and imagine the angle of the latter altered until the bending round of the light by the flint just compensates the bending round by the crown, a great deal of colour is apparent, and I must use a slenderer prism of flint if I wish to destroy the separation of colours produced by the crown, which will leave an outstanding deviation in the direction of that produced by the crown glass prism by itself. Now a ray of white light passing through a compound lens consisting of a convex lens of crown glass and a weaker concave of flint will be deflected very nearly in the same way as if it passed through two slender prisms, contained by the tangent planes drawn at the points where the ray of mean refrangibility strikes the four surfaces; and we can thus readily understand how such a compound lens may be achromatic.

I have here, on this diagram, the different lengths of spectra which will be produced by slender prisms of the same angle, of water, crown glass, flint glass, bisulphide of carbon, and oil of cassia. The separation of the colours, you will see, differs enormously; and the bending round of the rays is different also. On comparing the oil of cassia with the water, it will be seen that the bending round of the rays by the former is about twice as great as by the latter, while the separation of the colours is about eight times as great. Substances accordingly differ widely in their *dispersive powers.*

If you take a crown and a flint glass, the separation of colours produced by the flint will be about double the separation produced by the crown, supposing the angle to be small and of the same size in each case; whereas the deviation by the flint will be greater than that by the crown in the ratio of only 6 to 5, or thereabouts. But not only do different substances space out the whole spectrum in different proportions as compared with the bending round of the light as a whole; the different parts are not spaced out in the same relative proportion. In the upper part of the figure the various spectra represented are all reduced in the same proportion, so as to occupy the same length from B to H; and the places of the intermediate lines C, D, E, F, G are laid down from the numbers given by Fraunhofer and by the late Professor Powell. The intermediate parts of the spectra by no means correspond in these various media. If we take the two extremities, water and oil of cassia, the blue end of the spectrum is spaced out far more in proportion by the oil of cassia, while the red end is more spaced out by the water, in consequence of the *irrationality* (or want of proportionality) of dispersion. It is not possible, then, by a combination of two media such as crown and flint glass, to construct an object-glass which shall be *strictly* achromatic. For if the proportion of the whole spacing-out of the colours between flint and crown be two to one (which is near enough for explanation), the proportion at the blue end will be more than two to one, and the proportion at the red end less, and consequently the power of the flint glass which will be requisite to compensate the crown glass, if we want to combine two colours at the blue end, will be somewhat less than would be necessary if we wanted to combine two colours at the red end. It is therefore not so simple a matter as it might at first sight appear to construct an object-glass as nearly achromatic as may be. We may measure the refractive indices of both kinds of glass for various points of the spectrum; but the question is, when we have got them, what are we to do with them? for we may take the refractive indices for one pair of fixed lines, and by employing them for the determination of the dispersive ratio, get one result, while with another pair we should get another result.

In this diagram the figures of the spectra produced by crown and flint glass are reduced in proportion, so as to occupy the same length from B to H; and it will be seen that the intermediate lines in the crown lie a little ahead of those in the flint, the green part of the spectrum being more refracted by the former. The consequence is that in an object-glass, as nearly achromatic as may be, the green part of the spectrum is refracted more than the two ends. The focal length of the combination is a minimum for a certain point within the spectrum. The effect of this is seen quite readily in even a small achromatic object-glass. Turn a small telescope to an object, such as trees in the shade seen against the sky, or a piece of perforated zinc, and focus the telescope on it. Then put it out of focus both ways. When you push in the eyepiece, you will observe the dark parts of the field invaded by a purplish light, and when you draw it out by a green. The reason is that the different parts of the spectrum are not brought strictly to the same focus. The focal length being a minimum within the spectrum, it results that when the eyepiece is put out of focus by pushing in, the two ends of the spectrum are very determinately out of focus, and the blue and red rays invade the dark parts of the field; but when it is drawn out, the green rays are out of focus, and the dark parts of the field are then invaded by a green light. In the case of a large object-glass, the sharpness of the image depends materially upon the part of the spectrum where you turn it back. It may not be, perhaps, very easy to follow with the mind's eye the positions of the different foci of the various colours when you have to conceive them all arranged along the same line, nor is it easy to represent them without confusion on a diagram. But I have here got a figure (Fig. 1) which is partly diagrammatic, and which will serve the purpose. It is not the thing which you actually see; but it is intended to illustrate what takes place. Suppose a horizontal line, BB, representing the axis of an object-glass. Draw vertical lines, and measure along one of them, BH, the position of any particular colour in some standard spectrum. Mark off distances representing the positions of the principal fixed lines from B to H. The lines drawn at these distances parallel to

BB represent and correspond to the lines of Fraunhofer. From the actual place of the focus for any colour in the axis of the lens, draw a line till it meets the horizontal line corresponding to that colour. Do the same for all the other points of the spectrum; then the position of the various foci can be represented to the mind's eye by the curve which you get by joining all these points. Having got the curve, if you want to find the position of the focus of any particular colour in the axis of an object-glass, all you have to do is this:—look out for the place of the colour in the vertical line *BH*, from that place draw a horizontal line to meet the curve, from the point of section let fall a perpendicular on the line *BB*; then the foot of the perpendicular will be the focus of the colour in question.

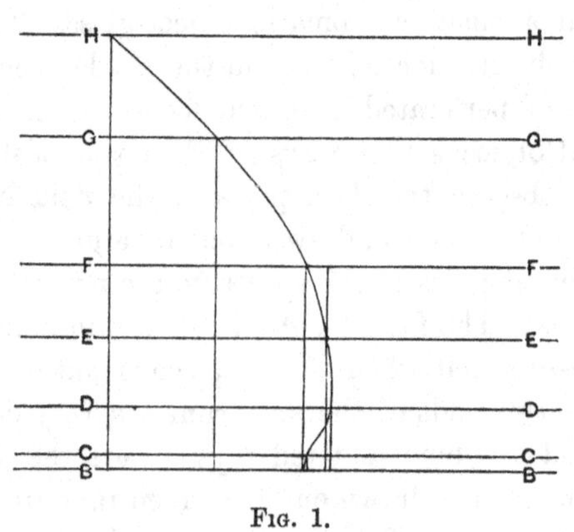

Fig. 1.

This figure, which is carefully drawn to scale, represents the position of the foci of the different colours in an object-glass the data for which are given by Fraunhofer. You see it is very far indeed from being true that the two extremities of the spectrum are collected at the same point in the axis of the lens. What then is the condition that has to be satisfied? According to the data given by Fraunhofer, it appears that the bending back of the spectrum takes place between the lines *D* and *E*, but much nearer to *D*—that is, just at the part of the spectrum where it is brightest. And that appears to be the condition giving the sharpest form possible of the achromatic object-glass, namely to make the focal length of the combination a minimum for the brightest part of the spectrum, which is situated about one-third of the distance from the line *D* towards *E*. The character of the compensation as to colour in an object-glass can be determined by a very simple test. Let the telescope be directed to an object bright on one side and dark on the other, as a chimney seen against the sky, or a piece of white paper on black velvet. Let the edge separating the bright from the dark part be brought into the centre of the field and viewed in focus. Cover half the object-glass by a screen having its edge parallel to the edge of the object, and then the other half, alternately, and you will see the dark edge invaded by the secondary colours—in the one case by the two ends of the spectrum, and in the other case by the part where the spectrum is bent back. When the compensation is such as to render vision sharpest, the tint in the latter case lies about midway between yellow and green. This is an exceedingly delicate test of achromatization. If an object-glass of crown and flint glass be thus constructed, the blue end of the spectrum will be greatly out of focus, and a glass so achromatized will show a very considerable portion of blue. But the subject of the optician is to construct a glass which shall give the sharpest vision, never heeding if there should be some blue outstanding.

In order to unite the neighbouring colours at the blue end, the flint lens must be a little weaker than the above. Imagine a crown glass lens compensated by a flint, the power of which continuously increases. The foci for all the colours will be thrown outwards as the flint lens increases in strength; but the blue end will be thrown out faster than the red. An uncompensated crown lens would have its focus for the blue nearer to the glass than the red; and if you oppose it by a flint concave glass of increasing power, you will throw out both foci, as well as the foci for the intermediate colours, though not exactly at the same rate for all. Reverting now to the diagrammatic construction, you will see that the distribution of the foci is represented at first by a curve not much differing from a straight line. This curve will be drawn outwards; but the combination will begin to be achromatic at the blue end of the spectrum; and the curve having first been nearly straight, will begin to turn at the end representing the blue, and there will be a minimum of focal length at the point of extreme violet. As the flint lens gets more powerful, the focus is bent back near the middle of the spectrum, then near to the red; and by bending it back altogether, you would again get nearly a straight line, the blue foci being now further from the glass than the red. The only option the optician has is to choose what part of the spectrum it shall be where the focal length of the combination is a minimum. If you wish to determine very exactly what the requisite powers of the two lenses shall be, there is a very simple way of effecting it. Suppose you take the crown and flint glasses to be employed, and form little prisms of both for the purpose of measurement. Measure the refractive indices of the two glasses for each of three fixed lines (bright or dark) in the spectrum. It does not matter what lines are taken, so that they divide the spectrum pretty fairly; and bright lines, which are easily produced artificially, will do just as well as dark. By means of these six indices the ratio of the dispersive powers which must be employed to give the best result may be simply determined, as I have shown in the *Report of the British Association* for 1855, Part II, p. 14*. I have also employed another method in determining the ratios of the dispersive powers of different kinds of glass, which I devised in the course of some researches made by the late Mr Vernon Harcourt and myself, and which is no less accurate than that which involves the measurement of fixed lines, and is also easier, as it can be used with common daylight. This, however, I shall not trouble you with, and I shall now pass on to speak of the chemical rays.

What do we mean by chemical rays? We know that light or, possibly, something which accompanies light, is capable of producing chemical changes, such as, for instance, those effected on salts of silver, on which the whole practice of photography is founded. At one time it was supposed by many that certain rays were distinctly chemical in their action, while others accompanying them were simply luminous; but I suppose hardly anyone imagines this now-a-days. When a pure solar spectrum is formed, we see in it the fixed lines called Fraunhofer's lines. If the spectrum, instead of being viewed, is received on a sensitized plate, we get an impression photographed in which far more dark lines are exhibited than seen directly by the eye, the lines in fact extending a long way beyond the extreme violet. The spectrum given by the electric discharge goes enormously beyond even this —to the extent of six or eight times the length of the visible

[* *Supra*, p. 64.]

spectrum. This, however, need not be considered as part of my subject, and I shall simply confine myself to the solar spectrum. Whether you see the spectrum with the eye or obtain an impression of it on a sensitized plate, you get the same series of dark lines in all that part of the spectrum that affects both the eye and the plate; the only difference is that the part of the spectrum towards the red end is more or less inefficient in producing chemical changes (to what extent depends upon the particular nature of the preparation), while at the other end of the spectrum the chemical effects are produced far beyond the extreme violet. There is no reason then for supposing there are distinct rays, capable of producing chemical effects, mixed with the visual rays; but everything leads us to believe that excitement of the sensation of light and the production of chemical changes are merely different effects produced by the same agent. The spectral rays are also capable of producing other effects, as when certain objects placed in part of the spectrum give out light for the time being, as though they were self-luminous, the light so emitted being different altogether in its nature from those rays which are active and which cause the objects to give out that light. The light is in general of a different colour from that of the incident rays; and not only so, but it is a matter of perfect indifference whether the rays which produce it are visible or not. Through the kindness of Mr Spiller I am enabled to show you this property. [*Experiment.*]

When I spoke just now of chemical rays my meaning was, rays contemplated with reference to their chemical and not to their visual effects. I cannot, of course, speak *absolutely* of the amount of such effects, as that would depend upon the nature of the particular substance acted upon. In the case of chlorophyll, for example, Sir J. Herschel found by experiment that it was nearly the extreme red which acted most powerfully in bleaching the substance. On the other hand, salts of silver are influenced chiefly by the rays at the violet or more refrangible end; and even among the silver salts there is a difference of action, to a certain extent, according to the salt. Thus the iodide of silver is affected powerfully by the violet rays from G to H, while the bromide appears to be affected a great deal lower down. I may also add that Dr Draper found that the process which lies at the foundation of all life, the decomposition of the carbonic acid of the air by vegetables under the influence of the solar rays, is produced chiefly by the brightest part of the spectrum. You will understand why I call this process fundamental as regards life, because the growth of vegetables depends upon it—with the exception of the fungi, which may be called vegetables of prey; and animals live, either directly or indirectly, upon vegetables. The chemical action, then, of rays depends upon the substance acted upon; and I will now confine myself to the salts of silver used in photography, which are mainly affected by the violet end of the spectrum, the mean centre of action being perhaps midway between G and H, or perhaps nearer to G. Suppose you wanted to construct an object-glass in which the most efficient chemical rays should be as much as possible concentrated to one focus; you must proceed according to the very same principle as you did with reference to the visual rays; only instead of bending back the spectrum (or making the focal length of the combination a minimum) for the brightest part of the spectrum considered visually, you must take the brightest part of the spectrum considered chemically, bending it back somewhere between G and H; that is to say, the lens must be very determinately under-corrected. If an objective be as nearly as possible chemically achromatic, it will be inconvenient for common use in photography, because the chemical and visual foci will not coincide. By the visual focus of an imperfectly compensated lens I mean that place where the best visual image is produced, which would lose sharpness by either going outwards or inwards. If a lens so achromatized were focused it would give you a sharp chemical image; but the chemical and visual foci would not coincide, the chemical lying *inside* the visual, whereas in a lens corrected so as to give sharpest vision the chemical focus would lie a little *outside* the visual; and of course for some intermediate correction the two foci would coincide.

The distance between the chemical and visual foci, supposing they do not coincide, will not be the same for objects at different distances, but will vary nearly as the square of the distance between the image and the object-glass. The inconvenience thence arising in the ordinary practice of photography, combined with the circumstance that for pleasing effects the last degree of sharpness of image is not required, renders it undesirable, in aiming at an ideal perfection, to surrender the convenience of making the two foci coincide. If your object, however, were celestial photography, the case would be different; for there the object is always practically at a uniform distance, and therefore, if the chemical and visual foci do not coincide, the distance between the two will always be the same, and can easily be allowed for; and in a large instrument our aim should be to obtain the sharpest possible chemical definition. The part of the spectrum where it must be doubled back in order to give the sharpest chemical definition, or else a coincidence of the chemical and visual foci, must be determined by experiments made once for all; and then an object-glass corrected in either of these ways can be constructed by simple rules. I should suppose that for the former the place of minimum focal length ought to lie about midway between G and H, and for the latter about one-third of the distance FG from F towards G.

The effect of irrationality constitutes the chief obstacle to the perfection of a properly constructed achromatic telescope; and it becomes an interesting question whether it is possible to remove this defect. The possibility of doing so was shown long ago by Dr Blair in the case of lenses composed of at least one fluid. He even constructed one in which the aperture was as much as one-third of the focal length, and found there was no perceptible colour; but no solid media possessing similar properties had been discovered. Mr Vernon Harcourt for a long period was engaged on this very subject; and in the course of that research nearly 200 prisms, composed of different experimental glasses, were formed and examined. For a long time there appeared to be very little hope of getting rid of the secondary spectra; there seemed to be a general law, connecting the relative extension of the blue and red ends of the spectrum with the dispersive power of the medium, which would render the destruction practically impossible. Hardly anything can be done in this direction with flint glasses of greater or less dispersive power; you must have recourse to glass of different composition altogether. It was found at length that molybdic and titanic acids have the power of conferring on glass the property of extending the blue, as compared with the red end of the spectrum, much more than corresponds to the dispersive power; while on the other hand terborate of lead, which in dispersive power nearly agrees with flint glass, lies decidedly nearer to crown than to flint in point of irrationality. By the use of such glasses it would be possible to destroy the secondary dispersion in an object-glass. Terborate of lead is slightly yellow, is too soft to be easily worked to a correct form, and is rather liable to tarnish, though the last defect does not much signify when the terborate is used for the middle lens of a triple, of which the adjacent surfaces fit each other and are cemented. The most promising combination, however, appeared to be a titanic glass in lieu of crown, achromatized by a light flint. In consequence of the elevation of dispersive power produced by the introduction of titanic acid, the curvature would be rather severe if the objective were constructed of only two lenses; but by making the objective a triple, the convex lens being replaced by two convex lenses placed first and third, the curvatures to be encountered would be no greater than in an ordinary double.

On the Improvement of the Spectroscope. By Thomas Grubb, F.R.S. Note appended by Prof. Stokes.

[From the *Proceedings of the Royal Society*, xxii, *April* 30, 1874, p. 309.]

If a ray of light be refracted in any manner through any number of prisms arranged as in a spectroscope, undergoing, it may be, any number of intermediate reflections at surfaces parallel to the common direction of the edges of the prisms—or, more generally, if a ray be thus refracted or reflected at the surfaces of any number of media bounded by cylindrical surfaces in the most general sense (including, of course, plane as a particular case), the generating lines of which are parallel, and for brevity's sake will be supposed vertical, and if α be the altitude of the ray in air, $\alpha', \alpha'', \ldots$, its altitudes in the media of which the refractive indices are $\mu', \mu'', \ldots$, then

(1) The successive altitudes will be determined by the equations

$$\sin \alpha = \mu' \sin \alpha' = \mu'' \sin \alpha'' = \ldots,$$

just as if the ray passed through a set of parallel plates.

(2) The course of the horizontal projection of the ray will be the same as would be that of an actual ray passing through a set of media of refractive indices $\frac{\mu' \cos \alpha'}{\cos \alpha}, \frac{\mu'' \cos \alpha''}{\cos \alpha}, \ldots$ instead of $\mu', \mu'', \ldots$. As $\alpha' < \alpha$, the fictitious index is greater than the actual, and therefore the deviation of the projection is increased by obliquity.

These two propositions, belonging to common optics, place the justice of Mr Grubb's conclusions in a clear light*.

[* Namely, that the curvature of the lines, when the slit is straight, is a function of (in fact proportional to) the dispersion. For developments, cf. Lord Rayleigh, *Phil. Mag.* viii, 1879, *Scientific Papers*, i, p. 542, and Larmor, *Proc. Camb. Phil. Soc.* vii, 1890, p. 85.]

ON THE CONSTRUCTION OF A PERFECTLY ACHROMATIC TELESCOPE.

[From the *Report of the British Association for the Advancement of Science*, Belfast, 1874.]

AT the Meeting of the Association at Edinburgh in 1871, it was stated that it was in contemplation actually to construct a telescope, by means of disks of glass prepared by the late Mr Vernon Harcourt, which should be achromatic as to secondary as well as primary dispersion. This intention was subsequently carried out, and the telescope, which was constructed by Mr Howard Grubb, was now exhibited to the Section. The original intention was to construct the objective of a phosphatic glass containing a suitable percentage of titanic acid, achromatized by a glass of terborate of lead*. As the curvature of the convex lens would be rather severe if the whole convex power were thrown into a single lens, it was intended to use two lenses of this glass, one in front and one behind, with the concave terborate of lead placed between them. It was found that, provided not more than about $\frac{1}{3}$ of the convex power were thrown behind, the adjacent surfaces might be made to fit, consistently with the condition of destroying the spherical as well as the chromatic aberration. This would render it possible to cement the glasses, and thereby protect the terborate, which was rather liable to tarnish.

At the time of Mr Harcourt's death, two disks of the titanic glass had been prepared, which it was hoped would be good enough for employment, as also two disks of terborate. These were placed in Mr Grubb's hands. On polishing, one of the titanic disks was found to be too badly striated to be employed; the other was pretty fair. As it would have required a rather

* The percentage of titanic acid was so chosen that there should be no irrationality of dispersion between the titanic glass and the terborate.

severe curvature of the first surface, and an unusual convexity of the last, to throw the whole convex power into the first lens, using a mere shell of crown glass behind to protect the terborate, Prof. Stokes thought it more prudent to throw about $\frac{1}{6}$ of the whole convex power into the third or crown-glass lens, though at the sacrifice of an *absolute* destruction of secondary dispersion, which by this change from the original design might be expected to be just barely perceptible. Of the terborate disks, the less striated happened to be *slightly* muddy, from some accident in the preparation; but as this signified less than the striæ, Mr Grubb deemed it better to employ this disk.

The telescope exhibited to the Meeting was of about $2\frac{1}{2}$ inches aperture and 21 inches focal length, and was provided with an objective of the ordinary kind, by which the other could be replaced, for contrasting the performance. When the telescope was turned on to a chimney seen against the sky or other suitable object, and half the object-glass covered by a screen with its edge parallel to the edges of the object, in the case of the ordinary objective vivid green and purple were seen about the two edges, whereas with the Harcourt objective there was barely any perceptible colour. It was not, of course, to be expected that the performance of the telescope should be good, on account of the difficulty of preparing glass free from striæ, but it was quite sufficient to show the possibility of destroying the secondary colour, which was the object of the construction.

On the Optical Properties of a Titano-Silicic Glass.

[From the *Report of the British Association for the Advancement of Science*, Bristol, 1875.]

At the Meeting of the Association at Edinburgh in 1871, Professor Stokes gave a preliminary account of a long series of researches in which the late Mr Vernon Harcourt had been engaged on the optical properties of glasses of a great variety of composition, and in which, since 1862, Professor Stokes had cooperated with him*. One object of the research was to obtain, if possible, two glasses which should achromatize each other without leaving a secondary spectrum, or a glass which should form with two others a triple combination, an objective composed of which should be free from defects of irrationality, without requiring undue curvature in the individual lenses. Among phosphatic glasses, the series in which Mr Harcourt's experiments were for the most part carried on, the best solution of this problem was offered by glasses in which a portion of the phosphoric was replaced by titanic acid. It was found, in fact, that the substitution of titanic for phosphoric acid, while raising, it is true, the dispersive power, at the same time produces a separation of the colours at the blue as compared with that at the red end of the spectrum, which ordinarily belongs only to glasses of a much higher dispersive power. A telescope made of disks of glass prepared by Mr Harcourt was, after his death, constructed for Mrs Harcourt by Mr Howard Grubb, and was exhibited to the Mathematical Section at the late Meeting in Belfast. This telescope, which is briefly described in the 'Report'†, was found fully to answer the expectations that had been formed of it as to destruction of secondary dispersion.

Several considerations seemed to make it probable that the substitution of titanic acid for a portion of the silica in an ordinary crown glass would have an effect similar to what had

* *Report* for 1871. Transactions of the Sections, p. 38 [*supra*, p. 339].

† *Ibid.* 1874, Trans. Sect. p. 26 [*supra*, p. 357].

been observed in the phosphatic series of glasses. Phosphatic glasses are too soft for convenient employment in optical instruments; but should titano-silicic glasses prove to be to silicic what titano-phosphatic glasses had been found to be to phosphatic, it would be possible, without encountering any extravagant curvatures, to construct perfectly achromatic combinations out of glasses having the hardness and permanence of silicic glasses; in fact the chief obstacle at present existing to the perfection of the achromatic telescope would be removed, though naturally not without some increase to the cost of the instrument. But it would be beyond the resources of the laboratory to work with silicic glasses on such a scale as to obtain them free from striæ, or even sufficiently free to permit of a trustworthy determination of such a delicate matter as the irrationality of dispersion.

When the subject was brought to the notice of Mr Hopkinson, he warmly entered into the investigation; and, thanks to the liberality with which the means of conducting the experiment were placed at his disposal by Messrs Chance Brothers, of Birmingham, the question may perhaps be considered settled. After some preliminary trials, a pot of glass free from striæ was prepared of titanate of potash mixed with the ordinary ingredients of a crown glass. As the object of the experiment was merely to determine, in the first instance, whether titanic acid did or did not confer on the glass the unusual property of separating the colours at the blue end of the spectrum materially more, and at the red end materially less, than corresponds to a similar dispersive power in ordinary glasses, it was not thought necessary to employ pure titanic acid; and rutile fused with carbonate of potash was used as titanate of potash. The glass contained about 7 per cent. of rutile; and as rutile is mainly titanic acid, and none was lost, the percentage of titanic acid cannot have been much less. The glass was naturally greenish, from iron contained in the rutile; but this did not affect the observations, and the quantity of iron would be too minute sensibly to affect the irrationality.

Out of this glass two prisms were cut. One of these was examined as to irrationality by Professor Stokes, by his method of compensating prisms*, the other by Mr Hopkinson, by accurate

[* *Supra*, p. 341

measures of the refractive indices for several definite points in the spectrum. These two perfectly distinct methods led to the same result—namely, that the glass spaces out the more as compared with the less refrangible part of the spectrum no more than an ordinary glass of similar dispersive power. As in the phosphatic series, the titanium reveals its presence by a considerable increase of dispersive power; but, unlike what was observed in that series, it produces no sensible effect on the irrationality. The hopes, therefore, that had been entertained of its utility in silicic glasses prepared for optical purposes appear doomed to disappointment.

P.S.—Mr Augustus Vernon Harcourt has now completed an analytical determination which he kindly undertook of the titanic acid. From 2·171 grammes of the glass he obtained ·13 gramme of pure titanic acid, which is as nearly as possible 6 per cent.

On a Phenomenon of Metallic Reflection.

[From the *Report of the British Association for the Advancement of Science*, Glasgow, 1876.]

THE phenomenon which I am about to describe was observed by me many years ago, and may not improbably have been seen by others; but as I have never seen any notice of it, and it is in some respects very remarkable, I think that a description of it will not be unacceptable.

When Newton's rings are formed between a lens and a plate of metal, and are viewed by light polarized perpendicularly to the plane of incidence, we know that, as the angle of incidence is increased, the rings, which are at first dark-centred, disappear on passing the polarizing angle of the glass, and then reappear white-centred, in which state they remain up to a grazing incidence, when they can no longer be followed. At a high incidence the first dark ring is much the most conspicuous of the series.

To follow the rings beyond the limit of total internal reflection we must employ a prism. When the rings formed between glass and glass are viewed in this way, we know that as the angle of incidence is increased the rings one by one open out, uniting with bands of the same respective orders which are seen beneath the limit of total internal reflection; the limit or boundary between total and partial reflection passes down beneath the point of contact, and the central dark spot is left isolated in a bright field*.

Now when the rings are formed between a prism with a slightly convex base and a plate of silver, and the angle of incidence is increased so as to pass the critical angle, if common light be used, in lieu of a simple spot we have a ring, which becomes more conspicuous at a certain angle of incidence well

[* *Ante*, Vol. II, p. 358.]

beyond the critical angle, after which it rapidly contracts and passes into a spot.

As thus viewed the ring is, however, somewhat confused. To study the phenomenon in its purity we must employ polarized light, or, which is more convenient, analyze the reflected light by means of a Nicol's prism.

When viewed by light polarized in the plane of incidence, the rings show nothing remarkable. They are naturally weaker than with glass, as the interfering streams are so unequal in intensity. They are black-centred throughout, and, as with glass, they open out one after another on approaching the limit of total reflection and disappear, leaving the central spot isolated in the bright field beyond the limit. The spot appears to be notably smaller than with glass under like conditions.

With light polarized perpendicularly to the plane of incidence, the rings pass from dark-centred to bright-centred on passing the polarizing angle of the glass, and open out as they approach the limit of total reflection. The last dark ring to disappear is not, however, the first, but the second. The first, corresponding in order to the first bright ring within the polarizing angle of the glass, remains isolated in the bright field, enclosing a relatively, though not absolutely, bright spot. At the centre of the spot the glass and metal are in optical contact, and the reflection takes place accordingly and is not total. The dark ring, too, is not absolutely black. As the angle of internal incidence increases by a few degrees, the dark ring undergoes a rapid and remarkable change. Its intensity increases till (in the case of silver) the ring becomes sensibly black; then it rapidly contracts, squeezing out, as it were, the bright central spot, and forming itself a dark spot, larger than with glass, isolated in the bright field. When at its best it is distinctly seen to be fringed with colour, blue outside, red inside (especially the former), showing that the scale of the ring depends on the wave-length, being greater for the less refrangible colours. This rapid alteration taking place well beyond the critical angle is very remarkable. Clearly there is a rapid change in the reflective properties of the metal, which takes place, so to speak, in passing through a certain angle determined by a sine greater than unity.

I have described the phenomenon with silver, which shows it best; but speculum-metal, gold, and copper show it very well, while with steel it is far less conspicuous. When the coloured metals gold and copper are examined by the light of a pure spectrum, the ring is seen to be better formed in the less than in the more refrangible colours, being more intense when at its best; while with silver and speculum-metal there is little difference, except as to size, in the different colours. Hæmatite and iron pyrites, which approach the metals in opacity and in the change of phase which they produce by reflection of light polarized parallel relatively to light polarized perpendicularly to the plane of incidence, do not exactly form a ring isolated in a bright field; but the spot seen with light polarized perpendicularly to the plane of incidence is abnormally broad just about the limit of total reflection, and rapidly contracts on increasing the angle of incidence.

It seemed to me that a sequence may be traced from the rapidly contracting rings of diamond seen in passing the polarizing angle of that substance, through the abnormally broad and rapidly contracting spot seen with iron pyrites just about the limit of total reflection, and the somewhat inconspicuous ring of steel seen a little beyond the limit, to the intense rapidly contracting ring of silver seen considerably beyond the limit. If so, the full theory of the ring will not be contained in the usually accepted formulæ for metallic reflection, modified, as in the case of transparent substances, in accordance with the circumstance that the incidence on the first surface of the plate of air is beyond that of total reflection.

MacCullagh was the first to obtain the formulæ for metallic reflection, showing that they were to be deduced from Fresnel's formulæ by making the refractive index a mixed imaginary, though they are usually attributed to Cauchy, who has given formulæ differing from those of MacCullagh merely in algebraic detail. As regards theory, Cauchy made an important advance on what MacCullagh had done in connecting the peculiar optical properties of metals with their intense absorbing power*. Now Fresnel's formulæ do not include the phenomena discovered by

* The apparent difference between MacCullagh and Cauchy as to the values of the refractive indices of metals is merely a question of arbitrary nomenclature.

Sir George Airy, which are seen in passing the polarizing angle of diamond, and which have been more recently extended by M. Jamin to the generality of transparent substances*; and if these pass by regular sequence to those I have described as seen with metals beyond the limit of total internal reflection, it follows that the latter would not be completely embraced in the application of Fresnel's formulæ, modified to suit an intensely absorbing substance and an angle of incidence given by a sine greater than unity†.

[* Traced by Lord Rayleigh for the case of water, mainly, but not entirely, to the presence of a surface film of transition arising from contamination, *Phil. Mag.* 1892, *Scientific Papers*, III, p. 496; reference is there made to similar observations by Drude on recently formed cleavage planes in crystals.]

† It was long ago observed, both by Professor MacCullagh and Dr Lloyd, that when Newton's rings are formed between a glass lens and a metallic plate, the first dark ring surrounding the central spot, which is comparatively bright, remains constantly of the same size at high incidences, although the other rings, like Newton's rings formed between two glass lenses, dilate greatly as the incidence becomes more oblique. See *Proceedings of the Royal Irish Academy*, Vol. I, p. 6.

Preliminary Note on the Compound Nature of the Line-Spectra of Elementary Bodies. By J. N. Lockyer, F.R.S. (Extract.)

[From the *Proceedings of the Royal Society*, XXIV, 1876, p. 352.]

March 3, 1876.

My dear Lockyer,—You might perhaps like that I should put on paper the substance of the remarks I made last night as to the evidence of the dissociation of calcium.

When a solid body such as a platinum wire, traversed by a voltaic current, is heated to incandescence, we know that as the temperature increases, not only does the radiation of each particular refrangibility absolutely increase, but the proportion of the radiations of the different refrangibilities is changed, the proportion of the higher to the lower increasing with the temperature. It would be in accordance with analogy to suppose that as a rule the same would take place in an incandescent surface, though in this case the spectrum would be discontinuous instead of continuous*. Thus if A, B, C, D, E denote conspicuous bright lines, of increasing refrangibility, in the spectrum of the vapour, it might very well be that at a comparatively low temperature A should be the brightest and the most persistent; at a higher temperature, while all were brighter than before, the relative brightness might be changed, and C might be the brightest and the most persistent, and at a still higher temperature E. If, now, the quantity of persistence were in each case reduced till all lines but one disappeared, the outstanding line might be A at the lowest temperature, C at the higher, E at the highest. If so, in case the vapour showed its presence by absorption but not emission, it follows, from the correspondence between absorption and

[* Kayser regards this as still unsettled, *Handbuch der Spectroscopie*, II, 1902, p. 331.]

emission, that at one temperature the dark line which would be the most sensitive indication of the presence of the substance would be A, at another C, at a third E. Hence, while I regard the facts you mention as evidence of the high temperature of the sun, I do not regard them as *conclusive* evidence of the dissociation of the molecule of calcium.

Yours sincerely,

G. G. STOKES.

APPENDIX.

Correspondence of Prof. G. G. Stokes and Prof. W. Thomson on the nature and possibilities of spectrum analysis.

[The following letters from Lord Kelvin came to light in arranging the scientific correspondence of Sir George Stokes. On supplying their dates to Lord Kelvin, he was able to extend the record. The parts relating to spectrum analysis are here printed, with Lord Kelvin's permission, as supplementary to the extracts contained in pp. 127—136 of this volume: cf. p. 136. It may be recalled that Prof. Stokes became Secretary of the Royal Society in 1854.]

---

2 College, Glasgow
Feb. 20, 1854.

My dear Stokes

It is a long long time since I have either seen you or heard from you, and I want you to write to me about yourself and what you have been doing since ever so long. Have you made any more revolutions in Science? or done any of the exp^l research on the friction of air? I saw a notice of your lecture at the R. I. Tell me any new discoveries you have made, &c. However I do not mean to impose upon you by demanding all this, but if there is anything short and good you can tell me I shall be glad to hear it. I want to ask you about artif^l lights and the solar dark lines. Is there any other substance than soda that is related to *D*? Are bright lines corresponding to it to be seen where soda is not present? Have any terrestrial relations to any other of the solar dark lines been discovered (or to the dark lines of any of the stellar spectra)? Are all artificial lights subject to dark lines? I should be greatly obliged by your telling me in a word or two what is known on these questions, which I suppose you will easily do.

....

Yours always truly
William Thomson

PEMBROKE COLL. CAMBRIDGE
Feb. 24*th*, 1854.

MY DEAR THOMSON

....

Now for your questions. I am not aware that there is any pure substance known to produce the bright line D except soda. See end *. It would be extremely difficult to prove, except in the case of gases or substances volatile at a not very high temperature, that the bright line D, if observed in a flame, was not due to soda, such an infinitesimal quantity of soda would be competent to produce it. It is very common in ordinary artificial flames (such as a candle &c.) but I think in such cases it may be attributed with probability to soda. In a spirit lamp I feel satisfied it is derived from the wick, for I find that alcohol burnt in a clean saucer does not give it, except perhaps a flicker now and then. Miller told me (and I have verified the observation) that it is not found in an oil lamp, and I find that when the wick of a candle is cut short, so as to be surrounded by gas, and not to project into the luminous envelope where the combustion goes on, D disappears.

Sir D. Brewster states (*Brit. Ass. Report*, 1842, p. 15 of the 2nd part) that the flame of deflagrating nitre contains bright lines corresponding to the dark lines A and B of Fraunhofer[1], and implies that other of the bright lines of this flame correspond to the dark lines of the solar spectrum. I saw somewhere a statement, I think, by Sir D. B., that the flame, I think of burning potassium, certainly some flame in which potash was concerned, gave 7 bright lines corresponding to the dark lines forming Fraunhofer's group a. These are the only cases I know of in which identification has yet been established, but the subject has barely yet been attacked. A vast deal of measurement has yet to be gone through. I think it likely that very interesting results will come out.

You will find in Moigno's *Répertoire d'optique moderne* (part III p. 1237) much information on the subject of your questions.

You ask "Are all artificial lights subject to dark lines?" No, it is quite the exception. When there are lines of any kind it is usually *bright* lines. The flame of nitrate of strontia (i.e. a flame coloured red by nitrate of strontia) shews such dark lines in the red, but then these same dark lines are found in the spectrum of

[1 See however p. 135 *supra*.]

common light transmitted across the flame, so that they appear to be due, not to the non-production of light of definite or almost definite refrangibility, but to its absorption by a certain gas or gases produced by combustion.

. . . .

* Moigno mentions (p. 1244) that Foucault in experiments with the electric pile, as well as Miller in observations on the flame of a spirit lamp to which various salts had been added, was struck with the constant occurrence of the bright line *D*.

Yours very truly
G. G. STOKES

---

2 COLLEGE, GLASGOW
March 2, 1854.

MY DEAR STOKES

Many thanks for your most satisfactory answers to all my questions. It was by a "lapsus pennae" that I wrote *dark* lines instead of bright lines, when I asked if all artif[l] sources are subject to them.

I think it is really a splendid field of investigation, that of the relations betw. the bright lines of artif[l] light and the dark lines of the solar spectrum. Dont you think anyone who takes it up might find a substance for almost each one of the principal dark lines by examining the effects of all salts on the flame of burning alcohol, or on other artif[l] lights? I think it will lead us to a qualitative analysis of the sun's atmosphere. Do you think the supposed tendency to vibrate *D* of other salts besides common salt mentioned by Moigno sufficiently established to want no confirmation. I am much disposed to doubt it, from what you have told me. I have tried a spirit lamp behind a slit through which sun light is coming and the effect is most decided. If I remember right the magnifying power I used was sufficient to divide *D*; but certainly whatever it was the light of the sp. lamp corresponded as exactly as could be observed with the dark of the solar *D*. It was curious to observe, the dark line not sensibly illuminated by the full light of the sp. lamp coming through it, (the brightness on each side was so great) but a line of light above and below the solar spectrum appeared as an exact continuation of the dark line *D*. Will you not take up the whole subject of spectra of solar and artif[l] lights, since you have already done so much on it? I am quite impatient to get another undoubted substance besides vapour of soda in the sun's atmosphere. What you tell me looks very like as if there is potash too. I think copper &c. would be hopeful. The galvanic arc shows rather broad bands of green for copper than fine lines. Salts of copper sh[d] be tried on flame. Do try iron too. There must be a great deal of that about the sun, seeing we have so many iron meteors falling in, and there must be immensely more such falling in to the sun. I find the heat of combustion of a mass of iron w[d] be only about $\frac{1}{34000}$ of the heat derived from potential energy of gravitation, in approaching the sun. Yet it w[d] take 2000 pounds of meteors per sq. foot of the sun, falling annually to account for his heat by gravitation alone.

....

Yours very truly
W. THOMSON

PEMBROKE COLL. CAMBRIDGE
March 7th, 1854.

MY DEAR THOMSON

There are one or two points which occur to me with reference to your last letter which I wish to mention.

Miller told me of an experiment of his which he performed many years ago for testing the coincidence of the dark double line *D* of the solar spectrum with the bright line in the light of a spirit lamp. The sun's light was introduced by a slit, and refracted by 3 good prisms, and then viewed through a telescope with a pretty high mag. power. The two lines forming the line *D* were "like that" (as he said holding two of his fingers about 3 inches apart) and he counted *six fixed lines between them.* The whole apparatus was left untouched till dusk, and then a spirit lamp was placed behind the slit. This gave two bright lines coinciding, as near as measurement could give, with the two dark lines *D*.

Miller seemed disposed to regard this as an accidental coincidence. It seemed to me that a plausible physical reason might be assigned for it by supposing that a certain vibration capable of existing among the ultimate molecules of certain ponderable bodies, and having a certain periodic time belonging to it, might either be excited when the body was in a state of combustion, and thereby give rise to a bright line, or be excited by luminous vibrations of the same period, and thereby give rise to a dark line by absorption.

But we must not go on too fast. This explanation I have not seen, so far as I remember, in any book, nor do I know a single experiment to justify it. I am not aware that any absorption bands seen in the spectrum of light transmitted across any vapour that has been examined have been identified with *D*.

I gave you Moigno's statement about Foucault's exper[t] but I confess I am sceptical. I want more explicit proof of the absence of soda in some shape.

If I remember right chloride of copper on packthread, put into the flame of a spirit lamp, gave two broadish green luminous bands (besides two more refrangible) which are each resolved, even by the naked eye with a highly dispersive prism of 60°, into a system of lines. Metallic copper in the voltaic arc is I believe somewhat different.

....

Yours very truly

G. G. STOKES

---

2 COLLEGE, GLASGOW
March 9, 1854.

MY DEAR STOKES

It was Miller's experiment (w[h] you told me about a long time ago), which first convinced me there must be a physical connection between agency going on in and near the sun, and in the flame of a sp. lamp with salt on it. I never doubted after I learned Miller's experiment that there *must* be such a connection, nor can I conceive of any one knowing Miller's experiment and doubting. There is I suppose something in Miller's mind inconceivable to me. You told me too your mechanical explanation, which struck me at the time, and for years has been taking a deeper and deeper hold [on] me.

If it could only be made out that the bright line *D* never occurs without soda, I sh[d] consider it as perfectly certain that there is soda or sodium in some state in or about the sun. If bright lines in any other flames can be traced as perfectly as Miller did in his case, to agreement with dark lines in the solar spectrum, the connection w[d] be equally certain, to my mind. I quite expect a qualitative analysis of the sun's atmosphere by experiments like Miller's on other flames.

Could you make anything decided (mathematical investigation) of your mechanical theory? Can you investigate mechanically the undulatory theory of radiant heat, e.g. a hot black ball, in the centre of a hollow black sphere, each of given temperature, what is the wave length, or lengths, of the undulations? The wave lengths, as experiment shows, are less the higher the temperature of the hot body. It is a splendid subject for mathematical investigation.

....

Yours always truly
W. THOMSON

SENATE-HOUSE CAMBRIDGE
March 8*th*, 1854.

MY DEAR THOMSON

....

Miller is now in the Senate-House examining for the Natural Sciences Tripos. I spoke to him about the fixed line *D*. I find he has not published the observation. I find I did him wrong in supposing that he regarded the coincidence as accidental: he supposes that there is some cause for it.

Yours very truly
G. G. STOKES

PEMBROKE COLL. CAMBRIDGE
March 28*th*, 1854.

MY DEAR THOMSON

Since I wrote last I have been somewhat of an invalid. Between the sailing of the Baltic fleet one day, Westminster Abbey the next, and one or two other things of a similar nature, I caught cold which produced a severe inflammation in my left eye. I was confined to my room for a week, and spent some days in darkness, not however making experiments on invisible light. My eye is now well, so that I can read and write as usual.

I certainly think that Foucault's &c. observations as to the general occurrence of the fixed line *D* require confirmation. I am disposed to suspect contamination with sodium, or some compound of sodium. At the same time, although the compounds of sodium do produce this bright line in flames, I do not think that we can necessarily infer the presence of sodium or any compound of sodium from the appearance of this bright line. I will explain my meaning presently.

....

As to the augmentation of the refrangibility of the emitted rays when the temperature of a body is more and more raised, I would refer you to some exper[ts] of Draper's, in case you should not have seen them. They were published a few years ago in the *Phil. Mag.*

As to the cause of this result I can state nothing positive as you can conceive. It falls in very well with certain conjectures which I have made in my paper On the Change of Refrangibility of Light (art. 230 &c.) but these are only conjectures.

I was greatly struck with the enormous length of the spectrum which I obtained with my quartz train with the powerful battery belonging to the Royal Institution. It far surpassed in length the solar spectrum, I mean even taking in that highly refrangible part which a quartz train is required to show. In the case of metallic points this spectrum consisted of isolated bright lines. I cannot help thinking that decompositions of a very high order may be going on in such an arc (the voltaic arc I mean) and that a careful examination of these lines may lead to remarkable inferences respecting the bodies at present regarded as elementary. There is nothing extravagant in this supposition: few chemists I imagine believe that the so-called elements are all really such.

Now it is quite conceivable that chemically pure metals should agree with compounds of sodium in giving the bright line *D*. If this were made out I should say that perhaps these metals were compounds of sodium, but more probably they and sodium were compounds of some substance yet more elementary.

Yours very truly

G. G. STOKES

24—3

69 ALBERT ST. REGENT'S PARK
LONDON *Nov* 26, 1855

MY DEAR THOMSON

I write to refer you to a paper of Foucault's in which he finds that the voltaic arc *produces* by absorption the fixed line D in light in wh. it did not before exist, and that the *bright* line D is of constant occurrence in the arc. L'Institut N° 788. Jan. or Feb. 1849.

Yours very truly
G. G. STOKES

---

69 ALBERT STREET REGENT'S PARK
LONDON *Dec* 6*th* 1855

MY DEAR THOMSON

....

I was speaking to Foucault about the artificial dark line D. It is easily produced by arranging so that the coke pole shall be seen *through* the arc, the pole itself giving an uninterrupted spectrum in which the *dark* line D is seen by absorption while in the less bright spectrum formed by the arc itself is seen the *bright* line D as a continuation of the other.

....

Yours very truly
G. G. STOKES

ADMIRALTY
LONDON *July* 7/71

DEAR STOKES

Many thanks for your letter and printed paper which are most valuable to me.

It was certainly prior to the summer of 1852 that you taught me solar and stellar chemistry. I never was at Cambridge from the summer of 1852 until we came there in 1866, and the conversation was certainly in Cambridge. I had begun teaching it to my class in session 1852–3, (as an old student's note book which I have shows*) and probably earlier although of this I have not evidence. I used always to show a spirit lamp flame with salt on it, behind a slit prolonging the dark line *D* by bright continuation. I always gave your dynamical explanation, always asserted that certainly there was sodium vapour in the sun's atmosphere and in the atmospheres of stars which show presence of the *D*'s, and always pointed out that the way to find other substances besides sodium in the sun and stars was to compare bright lines produced by them in artificial flames with dark lines of the spectra of the lights of the distant bodies.

You very probably learned more from Foucault of what he had done when you met him in 1855, but you certainly told me of a Frenchman (I don't recollect the name you then gave me or even that you mentioned the name though it is most probable you did) having obtained the dark line by absorption.

Yours truly
W. THOMSON

P.S. I think of using almost all you say of Herschel, in my address as from you.

* I speak from memory. I believe it was of session 1852–3. I have it in Glasgow.

CAMBRIDGE 11*th Jany* 1876

MY DEAR THOMSON

You have pushed me far too much forward as to spectrum analysis. Whitmell sent me a syllabus of his lectures, in wh. I was mentioned far too prominently, and this led me to write to him my recollection of my ideas (and it is a distinct one) when we talked about the matter. It was *you* who proposed to extend the method so much. I rather thought you were allowing your horse to run away with you.

Most of the absorbing (optically not mentally) matters that then came across me were things that would not stand the fire, and I don't know that I ever asked myself the question What would take place if a coloured glass were heated? Anyhow Stewart's extension of Prevost's Theory of Exchanges in the Proc. R. S. x 385 form (for I was not *then* acquainted with his Edin. Phil. Trans. papers though I might have been) came before me with all the freshness of a new discovery. I don't know whether you had known it before.

As to the bright line D of a spirit lamp with salted wick. I thought the flame gave such a shake to the Na Cl molecule as to make the Na bell ring, but did not think the more ethereal vibrations of D pitch could do it unless the Na were free, which I did not think of its being in the flame of a spirit lamp, though I thought the tremendous actions on the sun's surface might set it free. When later (1855) I heard from Foucault's lips of his remarkable discovery I still thought of the sodium being set free by the tremendous action of the voltaic arc of a battery of 40 or 50 elements of Grove or Bunsen.

I thought from memory that it was *when* Foucault came to receive the Copley medal that he told me of it, though all I could feel certain of from memory that it was in the evening at Dr Neil Arnott's. On referring to your Glasgow address*, I was going to strike out the date, when I obtained proof positive that I was right. I feel *certain* that I should have mentioned Foucault's remarkable discovery in what I drew up for the President about his researches if I had then known of it (Proc. R. S. VII, 571).

Yours sincerely

G. G. STOKES

[* Edinburgh address, 1871, *loc. cit.* p. 135 *supra*?]

INDEX TO VOLUME IV.

[*The references are to pages.*]

CAMBRIDGE: PRINTED BY J. AND C. F. CLAY, AT THE UNIVERSITY PRESS.

Printed in the United States
By Bookmasters